ESPRIT Basic Research Series

Edited in cooperation with
the European Commission, DG III

Editors: P. Aigrain F. Aldana H. G. Danielmeyer
O. Faugeras H. Gallaire R. A. Kowalski J. M. Lehn
G. Levi G. Metakides B. Oakley J. Rasmussen J. Tribolet
D. Tsichritzis R. Van Overstraeten G. Wrixon

J. L. Crowley   H. I. Christensen (Eds.)

Vision as Process

Basic Research
on Computer Vision Systems

With Editorial Assistance by:
Alain Chehikian   Jan-Olof Eklundh   Gösta Granlund
Erik Granum   Josef Kittler

With 204 Figures and 25 Tables

 Springer

Volume Editors

James L. Crowley
LIFIA-IMAG
I. N. P. Grenoble
46, Ave. Félix Viallet
F-38031 Grenoble Cedex

Henrik I. Christensen
Aalborg University
Laboratory of Image Analysis
Fr. Bajers Vej 7D
DK-9220 Aalborg

Editorial Assistance

Alain Chehikian
Laboratorie de Traitment d'Images et
Reconnaissance de Formes
Institut National Polytechnique de
Grenoble (LTIRF, INPG)
Grenoble, France

Jan-Olof Eklundh
Computer Vision and Active
Perception Laboratory
Kungliga Tekniska Högskolan
(CVAP, KTH)
Stockholm, Sweden

Gösta Granlund
Computer Vision Laboratory
Linköping University (CVL, LiTH)
Linköping, Sweden

Erik Granum
Laboratory of Image Analysis
Aalborg University (LIA, AUC)
Aalborg, Denmark

Josef Kittler
Department of Electrical and Electronic
Engineering
University of Surrey (UOS)
Guildford, England

CR Subject Classification (1991): I.2.10, I.4.8, I.5, 1.2.9, 1.2.11

ISBN 978-3-642-08197-2

CIP data applied for

Publication No. EUR 15917 EN of the European Commission, Dissemination of Scientific and Technical Knowledge Unit, Directorate-General Information Technologies and Industries, and Telecommunications, Luxembourg. Neither the European Commission nor any person acting on behalf of the Commission is responsible for the use which might be made of the following information.

© ECSC – EC – EAEC, Brussels – Luxembourg, 2010
Printed in Germany

Foreword

Human and animal vision systems have been driven by the pressures of evolution to become capable of perceiving and reacting to their environments as close to instantaneously as possible. Casting such a goal of reactive vision into the framework of existing technology necessitates an artificial system capable of operating continuously, selecting and integrating information from an environment within stringent time delays. The VAP (Vision As Process) project embarked upon the study and development of techniques with this aim in mind.

Since its conception in 1989, the project has successfully moved into its second phase, VAP II, using the integrated system developed in its predecessor as a basis. During the first phase of the work the "vision as a process paradigm" was realised through the construction of flexible stereo heads and controllable stereo mounts integrated in a skeleton system (SAVA) demonstrating continuous real-time operation. It is the work of this fundamental period in the VAP story that this book aptly documents.

Through its achievements, the consortium has contributed to building a strong scientific base for the future development of continuously operating machine vision systems, and has always underlined the importance of not just solving problems of purely theoretical interest but of tackling real-world scenarios. Indeed the project members should now be well poised to contribute (and take advantage of) industrial applications such as navigation and process control, and already the commercialisation of controllable heads is underway.

From the point of view of encouraging competitive Information Technology in Europe, the project is an elegant example of return on investment in a basic research activity, and we look forward to seeing the present and future endeavours of the consortium members maturing and coming to bear fruit.

September 1994

Jakub Wejchert
European Commission

Contents

Part IV Active Gaze Control and 3D Scene Description

Part V Active Scene Interpretation

Part VI Lessons Learned 419

Introduction

James L.Crowley * and Henrik I. Christensen†
* LIFIA, INPG
† LIA, AUC

During the last few years, there has been a growing interest in the use of active control of image formation to simplify and accelerate scene understanding. Basic ideas suggested by [Bajcsy 88] and [Aloimonos et al. 87] have been extended by several groups including [Ballard 91], and [Pahlavan-Eklundh 92]. Brown [Brown 90] has demonstrated how multiple simple behaviours may be used for control of saccadic, vergence, vestibulo-ocular reflex, and neck motion.

This trend has grown from several observations. For example, Aloimonos and others observed that vision cannot be performed in isolation. Vision should serve a purpose [Aloimonos et al. 87], and in particular should permit an agent to perceive its environment. This leads to a view of a vision system which operates continuously and which must furnish results within a fixed delay. Rather than obtain a maximum of information from any one image, the camera is an active sensor giving signals which provide only limited information about the scene. Bajcsy [Bajcsy 88] observed that many traditional vision problems, such as stereo matching, could be solved with low complexity algorithms by using controlled sensor motion. Examples of such processes were presented by Krotkov [Krotkov 90]. Ballard and Brown [Ballard 88] [Brown 90] demonstrated this principle for the case of stereo matching by restricting matching to a short range of disparities close to zero, and then varying the camera vergence angles. The development of robotic camera heads has led to the possibility of exploiting controlled sensor motion and control of processing to construct continuously operating real-time vision systems.

At the same time, research in applying artificial intelligence techniques to machine vision led to an emphasis on the use of declarative knowledge to control the perceptual process. Systems such as the Schema System [Draper et al. 89] developed a black-board architecture in which multiple independent knowledge sources attempted to segment and interpret an image. A major problem in such systems is control of perception. Such systems emphasise explicit representation of goals and goal directed processing which direct the focus of attention to accomplish system tasks. It has not been obvious how such a knowledge based approach to control of attention could be married to a real-time continuously operating system.

In July 1989, the European Commission funded a consortium of six laboratories[1] to investigate control of perception in a continuously operating vision system. The consortium partners set out to

---

[1]The Partners in project ESPRIT BR 3038 are Laboratory of Image Analysis, Aalborg University (LIA, AUC, DK), Dept. of Electrical and Electronic Engineering, University of Surrey (UOS, UK), Computer Vision and Active Perception Laboratory, Kungliga Tekniska Högskolan, Stockholm, (CVAP, KTH, S), Computer Vision Laboratory,

build a test-bed vision system for experiments in control and integration. A system was constructed which integrates a 12-axis robotic stereo camera head mounted on a mobile robot, dedicated computer boards for real-time image acquisition and processing, and a distributed system for image description. The distributed system includes independent modules for 2D tracking and description, 3D reconstruction, object recognition, and control. On March 18, 1992, a fully integrated testbed system was demonstrated to the European Commission, operating at speeds of 1 Hz to 10 Hz depending on the task. This book reports on the development of this system and the research which the system makes possible in control of a real-time vision system.

Real-time response requires limiting the amount of information processed at any instant. Applying this idea to stereo vision leads to a system in which reconstruction is limited to the region of a scene around a fixation point. At a macroscopic level, perception is then controlled as a serial process in which fixation is momentarily posed on interesting points in the scene, and the local structure is described. This idea of controlling of fixation as a serial process propagates throughout a vision system, from image level description to scene interpretation. In addition to the problem of control of fixation, the construction of an "active" 3D vision system required solutions to a number of technical problems, some of which are new and some of which are classic problems posed in a new form.

The Project Vision as Process

The starting point for this project was the objective of demonstrating a "high-level" vision system capable of continuous real-time operation. It was quickly realised that such an ambition raises two problems:

1) The technical problem of integrating processes which model the environment in terms of descriptions which are qualitatively different.
2) The problem of controlling the "attention" of a continuously operating system.

Concerning the first problem, different robotic tasks require qualitatively different measurements of the contents of a scene. Such descriptions can include 2D image description, 3D scene descriptions, and symbolic labelling of the components of a scene. Such processes are complementary and mutually supportive. However, they tend to operate cyclically with different cycle times. A framework was required which would permit the integration of multiple vision processes. This can be considered an "engineering" problem.

University of Linköping (CVL, LiTH, S), Laboratorie de Traitment d'Images et, Reconnaissance de Fromes, INPG, Grenoble, (LTIRF, INPG, F) and Laboratorie Informatique Fundamentale Intelligence Artificielle-INPG (LIFIA,INPG, F). The project has been continued under EP-7108 with the addition of University of Genoa (I).

The second problem is both subtle and fundamental. Many of the algorithms used in vision have computational costs which depend on the quantity of data. In the best cases the relation is linear, but in many cases it is quadratic, cubic, or even exponential. Real-time response requires that the processing time for any part of the system be bounded by a hard limit. This requires that the amount of data considered during each processing cycle be bounded. This in turn raises the problem of which subset of the available data the system should attend to at any cycle. This is part of the large problem of controlling perception. A general purpose real-time vision system requires some kind of solution to this problem.

These observations led the consortium to develop a long-term work plan with both an engineering and a scientific goal. These are:

Engineering Goal:
Develop techniques for integrating cyclic real-time processes for description of a scene in terms of 2D images, 3D structure, and labelled objects using active control of a camera head.

Scientific Goal:
Develop methods (and eventually a theory) of control of attention in perceptual processes.

With these twin goals, the consortium has developed a long-term plan leading to the demonstration of methods for the construction of integrated continuously operating vision systems, and the elaboration of a theory for the control of such systems. This book is concerned with the design of such a system.

The Nature of Basic Research in a Consortium

The process of performing basic research as a multi-year project by a consortium distributed over a large geographic area is a somewhat different experience than the more traditional style of individual work. One of the dominant characteristics of a multi-year, multi-partner project is stability. This can be both a benefit and a handicap. Before a consortium can begin constructing a system it must have a well-thought-out work program so that the partners can know what they can obtain from other partners. Designing the workplan forces each scientist to carefully work out the activities for his research and to justify each part of it among his partners. Criticism and suggests from collaborators can be introduced at a very early stage, avoiding unnecessary expenditures of resources. Within the consortium partners criticize each other and help each other to remain objective. The disadvantage is a loss in flexibility. It is difficult to foresee and plan for changes of insight or unexpected results. To counteract this, the consortium must expect to revise their workplan at regular intervals to take into account changes in the viewpoint, as well as new results from both within and without the consortium.

The VAP consortium developed the process of "cooperative competition" in constructing its work plan. In such a process, partners work together to develop a common definition of a research problem, possible approaches to a solution, and evaluation criteria for success. Partners explore individual solutions and the results are then brought together and compared. An objective comparison of results often leads to a synthesis of the possible solutions, bringing together the most successful aspects of each.

The spirit of cooperative competition is evident in the chapters of this book. For a number of the problems that have been addressed, several solutions have often been explored. In Part I, we present two individual experiments in software integration environments, MNT (Chapter 3) and Vipwob (Chapter 4). Techniques from these two systems were integrated into the common testbed system SAVA II. In a similar manner, competing approaches to image description are presented in Parts II and III. The approach in Part II is a more classical approach permitting a quick implementation of a real-time system, while Part III presents a longer-term approach which offers the potential to provide description which is both more robust and more appropriate to the problems of continuously operating vision. Three different physical binocular camera heads and one simulated head were developed by the consortium. Part IV presents the most successful of these heads, as well as the description of a control architecture designed to map onto any of the heads. Parts IV and V present alternative approaches to interpretation and control based on a common definition of the problem. In each of these cases, competition led to a more robust systems design and a more objective understanding of the results.

Contents of Chapters

This book presents the results from the first three years of the ESPRIT project "Vision as Process". The presentation is organised as five parts which reflect the components developed for the resulting experimental testbed system. In some cases, these parts also present advanced techniques which have been developed for inclusion in later versions of this system.

Part I describes the consortium efforts in systems integration to produce a testbed for experiments in control of perception. Chapter 1 discusses the problem of integration and control in an integrated continuously operating system. It presents the architecture of the VAP testbed system in term of the component processes and their communication links. This model serves as a reference for the later chapters. Chapter 2 describes the skeleton system which was produced in order to furnish the consortium with a testbed for experiments in integration and control. The architecture of the SAVA system evolved over the course of the project and furnished a reference for development of components as well as control techniques. The communications protocol in SAVA II drew heavily from the MNT system developed at KTH. MNT is described in Chapter 3. This section discusses problems inherent to multi-process communication, particularly in the context of a distributed software development project. The SAVA II system also exploited techniques for control and communications for distributed processes developed for the VIPWOB system at the University of Aalborg. This system is described in Chapter 4.

Two approaches to image description were pursued by the consortium. A short-term approach was designed to provide real-time image processing based on classical techniques, and a longer range investigation into techniques based on spatio-temporal filtering. Part II presents the technical details developed in the near-term effort. This near-term effort concentrated on the use of the gradient computed from a multiple resolution image pyramid as a low level image description technique.

Chapter 5 gives an overview of the image processing and description module of the SAVA system. This module integrates the hardware and software components provided by the partners. It employs signal processing hardware for pyramid computation and edge extraction, as well as an I860 coprocessor to provide fast computation of procedures provided by the consortium partners for such things as extraction of ellipses, interest point calculation, and image processing to support reactive control of focus, aperture, and vergence. Chapter 6 presents the algorithm and hardware used for computing a 12-level half-octave binomial pyramid of stereo images as they are acquired. All image processing operations were computed from multiple resolution region of interest (a log-Cartesian fovea) extracted from these pyramids. Chapter 7 describes the algorithm for fast extraction of edge segments. The gradient magnitude produced by this hardware was also used to compute edge chains for ellipse fitting, as well as measures for control of aperture and focus. Chapter 8 describes the SAVA module for tracking and grouping of edges. This module illustrates the use of the "standard module architecture" developed in Part I. It also illustrates the role of "demons" as perceptual agents and grouping for associative access of the image description. Chapter 9 describes a particularly powerful grouping technique developed by the University of Surrey for image interpretation.

Part III presents a set of the longer range image description techniques based on spatio-temporal filtering. Chapter 10 describes the background and approach of the technique developed by the University of Linköping for representing images with spatio-temporal filters. Chapter 11 presents the tensor signal representation developed by this group for interpreting the results of filtering. Chapter 12 describes a technique for using the phase of even and odd filters to estimate disparity. This technique led to a simple fast measure for phase which was incorporated into the image processing module (Part 5) to reactively control stereo vergence. Chapter 13 presents the results of a very novel technique to use image measurements to direct the focus of attention of an active vision system. This system is demonstrated with a simulation of a binocular head similar to that used in the demonstration system.

Part IV is concerned with gaze control and 3D scene description. Chapter 14 presents the KTH head-eye system. This binocular stereo head, with 13 degrees of freedom, represents the state of the art in binocular camera heads. Chapter 15 describes the hierarchical camera control architecture which was developed for the SAVA demonstration system. This architecture maintains and controls an estimate of the 3D fixation point. The device level controller has been implemented using the concept of a virtual head so that it can be mapped on to any of the heads produced by the consortium. Chapter 16 is concerned with construction and maintenance of a 3D scene description using an active head. An auto-calibration technique is presented which permits the stereo cameras to be calibrated dynamically to reference frames defined by objects in the scene. Chapter 17 describes a system for representing 3D objects from Geons using both top-down expectations and bottom-up image processing. This system was implemented as a module in the demonstration system.

Part V describes techniques for control of perception in an active scene interpretation system. The problem of control of perception is discussed in Chapter 18. This chapter reviews different control models which were developed by the consortium for exploration. It then presents a rule based system which was developed in the CLIPS language for the integrated skeleton system. This rule base is capable of controlling processes for finding, watching, tracking, and recognising objects. Chapter 19 addresses the control problem from the point of view of the scene interpretation module. It describes a rule base which was demonstrated in the integrated system for finding and watching objects on a table top. This system makes use of perceptual procedures for recognising particular classes of objects. Chapter 20 presents the recognition procedure which was used for polyhedral objects. Chapter 21 describes the recognition procedure for cylindrical objects. These chapters include a number of results of experiments with these techniques. Chapter 22 addresses the problem of intentional control of camera gaze direction and view point for interpretation. This is a system which makes use of context to direct fixation and processing for recognition.

Part VI closes the book with a discussion of lessons learned and directions for future research. This part compares the original plans of the consortium for the system and the techniques resulting from three years of fruitful collaborative research.

References

Aloimonos, J.Y. , I. Weiss and A. Bandopadhay: "Active Vision", *Internat. J. on Computer Vision*, 1(3), pp. 333-356, 1987.

Bajcsy, R.: "Active Perception", *IEEE Proceedings*, 76(8), pp. 996-1006, Aug. 1988.

Ballard, D.H. and Ozcandarli, A.: "Eye Fixation and Early Vision: Kinematic Depth", *IEEE 2nd Internat. Conf. on Comp. Vision*, Tarpon Springs, Fla., pp. 524-531, Dec. 1988.

Brown C., "Prediction and Cooperation in Gaze Control", *Biological Cybernetics* 63, pp. 61-70, 1990.

Draper, B., R.T. Collins, J. Broloi, A.R. Hanson, & E. Riseman: "The Schema System", *Internat. J. on Computer Vision,*, 2(3), p. 209-250, Jan. 1989.

Krotkow, E., Henriksen, K. and Kories, R., "Stereo Ranging from Verging Cameras", *IEEE Trans. on PAMI*, 12(12), pp. 1200-1205, Dec. 1990.

Pahlavan, K. and J.-O. Eklundh, "Head, Eye and Head-Eye System", SPIE Applications of AI, X: Machine Vision and Robotics, Orlando, Fla., pp. 14-25, April 1992.

Part I

Integration and Control of Perception

The original goal of the VAP project was to develop techniques for integration of a continuously operating vision system. It was soon realised that continuous operation required control. Thus the consortium defined a work plan to develop an integrated test-bed system which could be used for experiments in integration and control.

Chapter 1 discusses the problems of integration and control in an integrated active vision system. The architecture of the VAP test bed system is presented. This architecture composes a vision system as a distributed set of Unix processes communicating via message passing and with dedicated high-band width channels. The components of this system and the techniques which were used to implement them are the subject of the different parts of this book.

Chapter 2 describes the skeleton system which was produced in order to furnish the consortium with a test bed for experiments in integration and control. The architecture of the SAVA system evolved over the course of the project and furnished a reference for development of components as well as control techniques. The framework for continuous perception is presented based on a cyclic process composed of the phases predict-match-update. An architecture for a standard module based on this framework is then presented.

The communications machanisms used in the SAVA system were developed in a collaborative effort between LIFIA, AUC and KTH. The communications protocol drew heavily from the MNT system developed at KTH. MNT is described in chapter 3. This section discusses problems inherent to multi-process communication, particularly in the context of a distributed software development project. The SAVA II system also exploited techniques for control and communications for distributed processes developed for the VIPWOB system at the University of Aalborg. This system is described in chapter 4.

1. System Integration and Control

Henrik I. Christensen [†], Erik Granum [†] and James L. Crowley[*]
† LIA, AUC
* LIFIA, INPG

1.1 Introduction

The aim for a continuously operating vision system, raises a number of problems. When the VAP project was conceived (1988), only a small number of vision systems were capable of performing symbolic interpretation, and they were designed for interpretation of single (static) images. The well known examples included VISIONS [Hanson-Riseman, 1978], ACRONYM [Brooks, 1981], 3DPO [Bolles-Horaud, 1984], and ALV [DARPA, 1987].

Most work on analysis of image sequences had been carried out on pre-recorded images and the level of description was almost entirely parametric. i.e., systems could describe regions or features with independent motion in terms of their image- or 3D-velocity. A review of the state of the art is provided by [Huang 1981;1983]. Granum & Christensen [1988] reviewed motion analysis and found that most work had been dedicated to motion detection, while some efforts were concerned with registration of motion trajectories and taking some temporal context into account. E.g. [Sethi & Jain 1987] exploits a concept of path coherence to resolve simultaneous and crossing motion trajectories of objects.

Continuous and real-time observation of a dynamically changing scene were rarely set as a target unless for very dedicated applications. When thinking in more general and/or multi-purpose performance the major problems encountered were:

1. Processing of the enormous amount of information coming from the continuous sampling of the scene.
2. Interpretation and structuring of all the information flowing into the system.

The VAP project was based on a set of hypotheses to address such problems arising from continuous operation.

1.1.1 Fundamental VAP Hypotheses

One principle acknowledged by the VAP consortium was that vision should be studied in the context of its purpose, i.e. its use by a another process. Without dedicating to any specific application this implies that visual processing can be controlled to concentrate on the subset of visual information which is considered relevant to the current goal as defined by a user process.

Another principle is the likely coherence in the dynamic evolution in a scene being observed. In a continuously operating system temporal context will typically be available such that changes in the scene can be predicted and computational resources can be directed to confirm expectations rather than exploring from scratch whenever new images are acquired. This implies that tracking is basic operation within a continuously operating system.

In a continuously operating system, processing is carried out in a temporal context where each image is a sample of a dynamically changing scene. Given the domain of application and continuous operation, the change between images can be assumed limited and localised. In a quasi-structured environment, it may further be assumed that the set of objects present is organised and that contextual information may be used as cueing features. To cope with the immense amount of information present in a sequence of images, it is necessary that processing be limited and simplified through exploitation of contextual information.

Most of the algorithms used in vision involve some form of matching, in particular when used in a temporal context. However, matching has typically a worst-case computational complexity which is NP. To bound the computational complexity and to ensure continuous operation it is necessary to limit the size of the internal models. A fundamental principle is thus that continuous operation requires introduction of a mechanisms for forgetting that allows maintenance of limited sized internal models.

1.1.2 Outline Of The Chapter

In the following sections the problem of combining techniques into a fully integrated system is discussed, including considerations concerning operation in a temporal context. Based on a defined structure for the system and its breakdown into modules the problem of control is discussed as an introduction to some of the later chapters. A more elaborate discussion on control is provided in chapter 18. Throughout the presentation various references will be made to the following chapters in the book.

1.2 System Approaches

In the construction of a fully integrated system it is necessary to define a suitable system architecture that facilitates investigation of the problems related to continuous operation. This issue is addressed in detail below. For formulation of component modules for a system it is argued that the most coherent manner approach is achieved through introduction of a standard module architecture that may be replicated at each of the levels in the system. A standard module architecture, which has been used in the initial system, is introduced in section 1.2.5, and described in more detail in chapters 2 and 19.

1.2.1 The Reconstruction Approach

The most frequently used method of structuring is according to the standard set of representations, suggested by [Marr 1982]. In this model, processing is organised as sequential processes which gradually provides information at still more complete levels of representation of the external environment. The levels suggested by Marr are: Images, Primal Sketch (intrinsic images), the 2.5-D sketch (viewer centred depth map), 3D map and symbolic description. In the Marr model, processing is data-driven, in the sense that recognition and description are based on complete descriptions constructed at the lower levels in the system. In general this model is computationally demanding and it has proven difficult (if not impossible) to provide image descriptors that are sufficiently robust to allow characterisation of all phenomena in a natural environment.

The Marr processing model is termed a *reconstruction approach* as it aims at a full reconstruction of the environment. The processing model is purely data driven, and thus poses a problem in terms of computational resources. The Marr model assumes all processing may be carried out as a sequential process. This implies that a module uses the representation(s) at the level just below as a basis for its processing and the result is stored in the next higher representational level. The interfaces are consequently well defined. A simplified model of the processing is shown in figure 1.1.

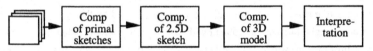

Fig. 1.1 A processing model for the reconstruction approach

In terms of representations, this processing model implies that information needed to perform recognition and interpretation of settings must be available as part of the 3D model. i.e., a diverse set of descriptors must be tagged onto the 3D model representation to facilitate recognition and description. This duplication of information up through the system and the unavailability of pixel level primitives might pose a problem in terms of model size and maintenance over time. Though the processing model might not be suitable for continuous operation the levels of representation may provide a possible structuring of a system.

1.2.2 The Non-Committal Approach

In the VISIONS system [Hanson-Riseman, 1978;1987] the representational levels outlined above are used, but the processing model is not necessarily sequential. In this system data are stored on a blackboard, a common storage. All modules in the system can access information at any of the representational levels. This implies that information does not have to be replicated up through the system to be made available for recognition procedures. A simplified model of the non-committal approach is illustrated in figure 1.2.

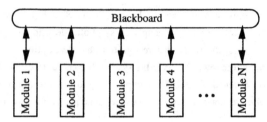

Fig. 1.2. Simplified architecture for the non-committal approach

In this architecture each of the modules in the system has control information that specify the information that must be available before the module may carry out its task. In addition it has control information that specifies the information which may be provided by the module. Through use of a control executive it is consequently possible to perform both goal directed and data driven processing through use of this control information.

This processing model imposes few constraints on the representations used in the systems and it is simple to add new modules to the system. The common blackboard is a a potential problem for a continuously operating system. All information generated and used by the system passes through the blackboard. It might thus pose a problem with respect to information bandwidth. To indicate the amount of information, which may be generated directly from images one may look at the Image Understanding Architecture, described by [Weems et al. 1989]. In that system the storage reserved for intermediate representations is 4 GB and the system is only aimed at analysis of single images. Such an architecture will require extensive use of special purpose hardware, in particular when applied in the temporal context.

1.2.3 The Purposive Approach

Introduction of goal directed operation and use in a limited and well defined domain of application allow synthesis of a vision system which is composed of a set of specifically engineered modules. Such modules may be designed to be computationally well behaved, in the sense that the computational complexity is bounded and often robust representations can be provided. This approach to construction of vision system has been promoted by Bajcsy [Bajcsy-Rosenthal, 1980], [Bajcsy, 1988], [Ballard 1989] and [Aloimonos et al. 1987;1989]. Although the approach exploits specific vision modules, Bajcsy [1988] has tried to enumerate a set of modules that might form a general purpose system. The use of dedicated modules is a way to provide robust information and computationally tractable techniques. Well known examples of dedicated modules used for robot navigation are optical flow modules that can compute the position of the focus of expansion for the optical flow field, and modules which can compute the time to contact from features in an image sequence.

In this approach to system construction, which is termed *the purposive* or *the animate* approach, it is envisaged that the construction and analysis of specific modules gradually will provide insight that will allow definition of modules applicable in general vision systems. The convergence towards a standard set of modules through analysis of diverse application domains might provide valuable insight, but it is not obvious that convergence will be achievable.

In the purposive approach to system construction the information needed for computation of particular descriptors are used, i.e., there is no imposed restrictions on use of information in the system. This implies that exploitation of information available is task driven. The exploitation of information available may consequently be very different from one task to the next. The basic system architecture should thus be flexible and facilitate dynamic change of the information flow.

In a purposive system there are in principle no imposed requirements to the representations used. In practical systems a number of modules may exploit the same representation and once a system has been defined an analysis of the representational requirements may point to the definition of a set of standard representations. Given present state of the art no such general representations are known.

This approach to vision rejects most of the established techniques. A very direct approach to computation of usable information is used. This implies, for now, that modules must be built from scratch whenever a new type of information/representation is required. There is consequently a concern that from this approach little insight will be gained in terms the general vision problem.

1.3 The VAP Architecture

The consortium "Vision as Process" was formed in response to a call for proposals for a "High Level Vision System". Inspired by recent results in active vision, the consortium sought to to develop techniques to integrate and control a continuously operating vision system with the ability to observe an environment with 2D measurements, 3D models and recognition of objects. One of the first tasks of the consortium was to design an architecture which permitted integration, control and continuous operation.

1.3.1 The system architecture

At the outset of the VAP project , the consortium chose to begin with levels of representation inspired by those of Marr. This decision was partly motivated by a desire to avoid "recreating" all of computer vision; the consortium wanted to reuse existing techniques whenever possible, and most techniques were based on the levels defined by the Marr model. In terms of the architectural requirements, neither the Marr model nor the non-committal approach seemed appropriate. While the strict hierarchy in the Marr model seemed to be suitable for pre-attentive (default) processing, the flexibility of the non-committal approach is appealing as it provides a flexibility that allows a less constrained investigation of different control methods. To facilitate such investigation it was decided to use a hybrid system architecture which has facilities for both reconstructive and non-committal processing. An outline of the hybrid system architecture and the set of component modules is shown in figure 3

The architecture in figure 1.3 has a data flow part, which is similar to the Marr processing model, that constitutes the main flow of data up through the system. It should be noted that the data flow is not only bottom-up. The flow may be both bottom-up and top-down (the channels are bi-directional). This implies that top-down expectations (derived from the present set of goals and contextual information) can be used to direct/control processing at lower levels, while detected event at the same time can drive a reconstructive mode of processing.

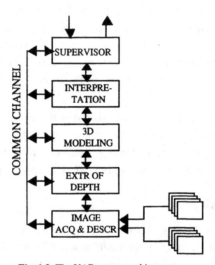

Fig. 1.3. The VAP system architecture

The architecture contains also a common communication channel that allow communication between any two modules in the system. This communication channel may be used both for investigation of a non-committal processing model, and for investigation of purposive systems, as the component processes in the system in figure 1.3 may either general purpose or dedicated. The basic architecture imposes thus few constraints on the design of component modules and it provides flexibility for the investigation of system level control issues. Initially it was envisaged that the main flow of data would exploit the communication links between adjacent modules, while only control information would be communicated through the common channel. During the execution of the project it was realised that a more flexible processing model was needed to make computations both efficient and robust. The system architecture and its implementation is described in further details in chapters 2, 3, 4 and 19.

1.3.2 The Module Architecture

A basic issue to be investigated in control has been the possibility of defining a standard module architecture that may be replicated at all levels of the system. The use of a common architecture at all level would provide insight into general processing models for visual functionalities. The definition of such a module architecture should pay particular attention to temporal integration as it is essential that the methods used has such facilities for continuous operation. i.e., in the development of a standard architecture it must be ensured that it is general so that few constraints are imposed on the techniques used at the different levels of the system.

A review of the literature which describes analysis of sequences of images, in particular for computation of motion, reveals that many systems use a cycle composed of the phases predict-match-update This architecture was introduced for analysis of images in the late eighties [Granum-Christensen 1988;1990], [Crowley et al. 1988], based on techniques used in the control community since the early sixties [Kalman, 1963]. The architecture is shown in figure 1.4.

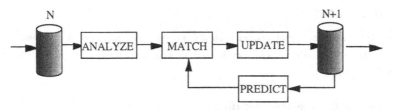

Fig. 1.4. Basic Predict-Match-Update architecture for temporal processing

The *analysis* block, in figure 1.4, is responsible for the frame by frame analysis, which generated a set of primitives, that must be co-ordinated with information in the temporal context. The correspondence with information in the temporal context is performed in the *match* block. To simplify matching the information in the temporal context is used in a prediction of the expected content of the next frame. Once correspondence has been established the information contained in the internal models must be updated to reflect the new information contained in the new frame. Once updating has been carried out the cycle may start all over again. The architecture in figure 1.4 is only used for data driven processing as there are no facilities for goal directed control.

However, as a model at level N+1 is used for prediction of primitives in the next frame, the predictor may also be given other types of input which can be used for guidance of processing. Introduction of goal derived information into the model at level N+1 will consequently allow top-down/attention based control of processing. Likewise may the goal influenced prediction be transformed into a representation that is compatible with the one used the level below, so that it may drive processing at the level below. A module architecture that allow processing using such a scheme is shown in figure 1.5.

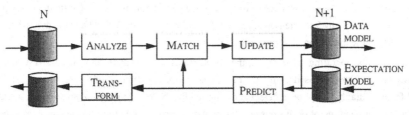

Fig. 1.5. Module architecture that allow goal directed control of processing

The prediction sub-module includes here both top-down and bottom-up (data-driven) information. Through careful control it is possible to dynamically change the balance between these two types of information. The introduction of information into the *expectation model* may be performed using both the data-flow path in the architecture in figure 1.3 or the common channel, which facilitates investigation of different control strategies.

In chapter 2 and 5 it is described in detail how the architecture has been implemented and how it may be used for multi-modal integration, while chapter 18 discusses the module architecture in more depth.

1.4 Control Issues

Construction of an operational system involves not only definition of system and module architectures but also overall issues in control to ensure a) continuous operation and b) satisfaction of user defined goals.

1.4.1 Achieving Continuous Operation

Achieving continuous operation at the different levels of the system may require different actions depending on the level of description. The issue of continuous operation has mainly been addressed at the level of image primitives and in the estimation of scene depth. A study of the approaches adopted in these area can provide cues for possible approaches to achievement of continuous operation at the levels of object and scene phenomena.

Low Level Exploitation of Context

In image level processing a frequently used processing model is the one presented in section 1.2.5. Both [Granum-Christensen 1988;1990], [Crowley et al. 1988], and [Wiklund-Granlund 1987] have demonstrated that the computations may be simplified through use of such a model. The prediction of the content of the next image provides a basis for specification of the regions in the image where relevant features are expected to appear. The use of predictions simplifies processing as the matching becomes less ambiguous. If estimation of time varying parameters is based on recursive statistical estimation techniques, such as Kalman filtering, the associated covariance matrix provides an indication of the estimation uncertainty. This uncertainty indicates the size of the window in which one may expect to locate predicted features. Through selection of a confidence level and evaluation of the associated χ^2 distribution it is possible to compute the appropriate size of attention regions. This allow control of the computational complexity of the image description process, as described by [Christensen et al. 1992]. [Deriche-Faugeras 1991] have formulated similar constraints for the extraction of 3D information.

In continuous operating systems another problem is maintenance of limited sized models. [Crowley et al. 1988] has used a simple recency criteria for pruning of model size through rejection of all features that have not been detected during the last N (N=5) images. This provides a simple mechanism for handling of occlusions and segmentation errors, while limiting the temporal window of interest to N images. When information in the temporal context is used for definition of focus of attention regions, such an approach might be too simple. I.e., use of predifined temporal windows might not be efficient as the phenomena of interest may appear at a range of different temporal scales. The model pruning should consequently take the present set of goals into consideration. Information should be evaluated both with respect to recency and relevance to provide an intelligent and dynamic mechanism for model pruning, as described by [Hayes-Roth 1991]. The use of such a strategy is illustrated in Chapters 7 and 8.

In signal level description of images it has been argued [Granlund, 1978] that adaptation of a hierarchical feature representation and analysis in a spatio-temporal context provides for a more robust description and both data and goal driven specification of information of interest. Multi-scale image description allows initial processing to be carried out at a coarse scale. Processing may be carried out fast due to the limited amount of information. It provides at the same time an expectation that may drive processing at higher resolutions. The multi-scale representation can be used both for spatial and temporal description. When continuous operation is required the time-frame available for processing of single images is bounded, but with a multi-scale description the level of detail can be varied depending on the content and the dynamics of the scene. The multi-scale representation also allows for use of model pruning techniques where high resolution information is invalidated before coarse resolution information. This implies that an overview of the scene may be maintained while detailed information about individual objects is pruned whenever the size of the model must be reduced. This idea correspond to the concept of spatio-temporal scale space as described by [Koenderick 1988], [Christensen-Granum 1988] and [Knutsson-Granlund 1989]. Experiments have shown that adaptation of this idea and use of temporal context for definition of adaptive procedures enables control of the computational process [Knutsson et al., 1991], [Westelius et al., 1991], [Haglund, 1992], [Andersson, 1992]. Chapters 11-13 describe examples of how such signal level procedures may be combined with multi-scale representations to facilitate controllable performance.

Symbolic Context
In terms of high level processing one problem in continuous operation is related to the search for object models and integration of such descriptions into scene models. Little work has been conducted to achieve this as few systems available have considered the problem of continuous operation. Notable exceptions are the ALVEN system [Tsotsos, 1987] and the CVS system [Fischler-Strat, 1989].

To facilitate continuous operation and limit the need for computational resources the VAP consortium have adopted a contextual approach where the domain of application is described in terms of a hierarchically organised context description. In this description the top-level description contains a specification of the major structures in the environment. These objects should be recognised initially in provide a context for recognition of other objects. I.e. a ground plane should be established early in the interpretation process, as it may provide a reference system for other analysis/description tasks.

Identified major structures allow generation of expectations as to where objects of interest are likely to occur, which in turn allow specification of focus of attention regions. Such use of contextual information has for example been reported by [Rimey-Brown 1991]. The use of contextual information in interpretation is described in further detail in Chapters 18 and 19.

Detected events may give rise to information that is not related to any of the current goals. Such information may result in a recognition task involving the well-known indexing problem, but given contextual information, techniques like geometric hashing provide mechanisms for recognition in linear time, as described by [Lamdan-Wolfson 1988].

At the symbolic level it is also necessary to limit the size of internal models to maintain continuous operation. This may be achieved with the method suggested by [Hayes-Roth 1989]. In this method the items in a model are ranked according to the following 4 criteria:

Recency
Relevance to present goals
Importance to present goals
Cost of computation of the information

Based on a linear combination of these evaluation criteria it is possible to rank the information and use a simple overflow mechanism for pruning of the models. I.e., items which are ranked at the bottom are first rejected during the pruning operation. This mechanism can ensure that only the most relevant information which has been used recently will be maintained by the system.

The above set of methods constituted a reference for the VAP effort in the investigation of continuous operation at the symbolic level of the vision system. Possible implementations of such mechanisms are described in Chapters 17, 18, 19, and 21.

1.4.2 Goal Directed Processing

Goals are widely recognised as a fundamental component of intelligent systems. A fundamental hypothesis in VAP is consequently that processing should be goal directed. This implies that the system should not try to provide a full reconstruction of the scene, but it should concentrate on providing information in response to user defined queries. Most of the systems reported in the literature have been aimed at use in a well defined domain. I.e., in VISIONS [Hanson-Riseman, 1978] the domain was analysis of images from a rural area in Massachusetts with houses and roads, in ALVEN [Tsotsos, 1987] the domain was images of the left ventricular, and in CVS [Fischler-Strat, 1991] the domain is navigation in a known room. In VAP the aim is simple adaptability to a variety of domains. This implies that goal directed processing, to the extent possible, should be based on a set of generic goals. For this the consortium initially defined a set of general goal commands [Jensen, 1990], which is:

search(X):   Is X present in the scene?

find(X):   Where is X, given X has been identified earlier?

relate(X,Y):   What is the spatial relation between X and Y?

describe(X,Y):   Determine property Y for object X

watch(X):   Allocate resources for notification of changes for X

track(X):   Maintain a dynamic description of object X.

This set of goals defines the user level interface to the system. Based on the potential goals, the system must be able to allocate its resources for optimal satisfaction of the concurrent goal(s). Allocation of resources in response to goals is a well known problem which may be approaches using a number of different approaches.

In the vision literature a number of approach to goal directed processing have been reported. Most of these efforts include use of a cyclic process, in which data received are matched against expectations. Depending on success or failure in the matching process an updating or event detection process is used to drive the next cycle of processing. An example of such a cyclic process is described by [Tsotsos 1987] for the ALVEN system. This model of processing fits conveniently with the architecture introduced in section 1.3.2.

When the VAP effort was initiated the use of production systems and reasoning under uncertainty appeared the most promising in terms of providing insight into the problem of the cycle of planning, sensing and interpretation. A set of initial considerations are provided in section 1.4.3 and more detailed discussions are presented in Chapters 18 and 19.

1.4.3 Approaches to System Level Control

In system level control externally defined goal commands, and implicit system strategies, must be changed into actions that can be executed by one or more system modules. The internal handling of such actions is an issue that is resolved for each of the modules.

The conversion of external goal commands into system actions may be based on planning. Planning is here formulation of a sequence of actions that may be expected to allow completion of a goal.

Planning has been an area of active research in the Artificial Intelligence community for some time. Early work in planning was based on planning at a single abstraction level, as for example described for the STRIPS system [Fikes-Nielsson, 1971]. In planning using a single abstraction level a problem is that all actions must be included in the plan. If a hierarchical approach is adopted it is possible initially to perform planning at an abstract level, where a limited number of actions are sufficient to allow completion. As actions are carried out they are expanded through invocation of sub-plans. Such an approach is efficient in terms of both speed and memory. Planning as described above is similar to cost driven graph traversal. Actions are here assumed to have a unity utility and cost which implies that the plan with the minimum number of actions have precedence. I.e., when several different plans may result in completion the shortest plan (the one with the smallest number of actions) is chosen.

Actions will typically have different utilities, when executed, and it is consequently desirable to include such utility information in the planning. Utility or confidence may be expressed using different paradigms such as confidence factors [MYCIN, 1982], interval logic (i.e., Dempster-Schafer [Schafer 1986;1987]), and Baysian probability theory [Pearl, 1990], [Rimey-Brown, 1992]. In VAP two different approaches have been investigated, one based on rule based planning using a set of pre-defined plans, and another based on reasoning under uncertainty using belief networks. The first approach is described in details Chapters 18 and 19, while the latter is described in [Jensen et al 1990].

1.5 Summary

Based on a review of approaches to construction of vision systems a flexible architecture was outlined, and it was descripbed how a standard module architecture may be used in the context of the larger system. The architeture outlined has been used in the efforts described in the remainder of this book and it was provided the basis for the integrated system which was demonstrated in March 1992 to representatives for the European Commission and its reviewers.

In addition a strategy for control of a system has been outlined. The method has bene used with success in the VAP project as described in chapters 2 and 19.

A basic problem which remains to be solved to adoptation of more formal methods for integration of visual modules and a common paradigm for description and reasoning about control. Both issues will be addressed in the continuation project EP-7108-VAP II which was initiated July 1992.

References

[André et al 1988] André, E.; Herzog, G. and Rist. T. On the Simultaneous Interpretation of Real World Image Sequences and their Natural Language Description: The System SOCCER. In: *Proc. of the 8th European Conference of Artificial Intelligence,* Munich, West Germany, 1988, pp. 449-454.

[Aloimonos et. al. 1988] Aloimonos, J. Y., I. Weiss, and A. Bandyopadhyay, "Active Vision", International Journal of Computer Vision, Vol. 1, No. 4, Jan. 1988.

[Aloimonos-Schulman 1989] J. Y. Aloimonos & D. Schulman, Integration of Visual Modules, Academic Press, Orlando, Fl., 1989.

[Aloimonos 1990] J.Y. Aloimonos, Purposive and Qualitative Vision, DARPA Image Understanding Workshop 1990, Morgan-Kuffman Publ., Philadelphia, Penn, September 1990.

[Bajcsy 1988] R. Bajcsy, Active Perception, IEEE Proceedings, Vol. 76, No 8, pp. 996-1006, August 1988.

[Bajcsy-Rosenthal 1980] R. Bajcsy & D Rosenthal, Visual and Conceptual Focus of Attention, in Structured Computer Vision, S. Tanimoto & A. Klinger (Eds.), Academic Press, New York, N.Y., 1980.

[Ballard 1989] D.H. Ballard, R.C. Nelson, and B. Yamauchi. Animate Vision. Optics News, 15(5):17-25, 1989.

[Bar-Shalom & Fortmann 1987] Y. Bar-Shalom & B. Fortmann, Tracking and Data Association, Academic Press, Boston, Mass, 1987.

[Bergevin-Levine 1989] R. Bergevin & M. Levine, Generic Object Recognition: Building Coarse 3D Descriptions from Line Drawings, CVGIP, June 1992.

[Biederman 1987] I. Biederman, Matching Image Edges to Object Memory, Internat. Conference on Computer Vision, IEEE CS Press, London, June 1987

[Bolles-Horaud 1984] Bolles, R. C. and Horaud, P., Configuration Understanding in Range Data, Second ISRR, August, 1984.

[Bowyer 1991] K. Bowyer, P. Flynn, A. Kak, State of The Art In Cad Based Vision Systems, In proc. IEEE workshop on CAD Based Vision, IEEE CS Press, Maui, June 1991.

[Brooks 1981] Brooks R.A., Symbolic Reasoning Among 3D Models and 2-D Images, Artificial Intelligence, Vol. 17, pp 285-348, 1981.

[Brown 1990] C. Brown, Prediction and Co-operation in Gaze Control, Biological Cybernetics 63, 1990.

[Brownston et. al. 1985] Brownston, L. , R. Farrell, E. Kant and N. Martin, Programming Expert Systems in OPS-5, Addison Wesley, 1985.

[Buchanan-Shortliffe 1984] Buchanan, B. G. and E. H. Shortliffe, Rule Based Expert Systems, Addison Wesley, 1985

[Burt 1988] P. J. Burt. Smart Sensing in Machine Vision. Academic Press, 1988.

[Califano 1990] A. Califano, R. Kjeldsen, & R.M. Bolle, Data and Model Driven Foviation, 10 ICPR, Atlantic City, June 1990.

[Christensen 1989] H.I. Christensen, Aspects of Real Time Sequence Analysis, Ph.D. Dissertation (2nd Edition), Institute of Electronic Systems, Aalborg University, December 1989.

[Christensen et al. 1992] H.I. Christensen, C.S. Andersen, & E. Granum, Control of Perception in Dynamic Computer Vision, First Danish Conference on Pattern Recognition and Image Analysis, Copenhagen, June 1992.

[Crowley et.al. 1988] Crowley, J. L., P. Stelmaszyk and C. Discours, Measuring Image Flow by Tracking Edge Lines, 2nd ICCV, Tarpon Springs, Fl. Dec. 1988.

[Dickinson 1991] S. Dickinson, A. Pentland & A. Rosenfeld, 3D Object Recognition using Distributed Aspect Graph Matching, University of Maryland Tech Report, May 1991.

[Dickinson 1992] S. Dickinson, A. Pentland & A. Rosenfeld, Qualitative 3D Shape Recovery using Distributed Aspect Matching, IEEE Trans on PAMI, June 1992.

[Dickmanns 1988] E.D. Dickmanns. 4D-Dynamic Scene Analysis with Integral Spatio-Temporal Models. In R. Bolles and B. Roth, editors, Proc. 5th Int. Symposium on Robotics Research, Tokyo (Japan), 1988. MIT Press.

[Pahlavan 1992] K. Pahlavan & J-O Eklundh, A Head-Eye System — Analysis and Design, CVGIP, Vol. 56. No. 1, pp. 41-56 July 1992.

[Fikes-Nilsson 71] Fikes, R. E. and N. J. Nilsson, STRIPS: A New Approach to the Application of Theorem Proving to Problem Solving, Artificial Intelligence, Vol 2, No. 3-4, Winter 1971.

[Fischler-Strat 1989] M.A. Fischler & T.A. Strat. Recognising objects in a Natural Environment; A Contextual Vision System (CVS). DARPA Image Understanding Workshop, Morgan Kauffman, Los Angeles, CA. pp. 774-797, 1989.

[Granlund 1978] G. H. Granlund, In search of a General Picture Processing Operator" Computer Graphics and Image Processing, Vol. 8, No. 2, pp 155-178, October 1978.

[Granum-Christensen 1990] E. Granum & H.I. Christensen, Dynamic Robot Vision, In: Traditional and Non-traditional Robotics Sensors, T. Henderson (Ed.), NATO ASI Series in Computer Science, Vol. 63, Springer-Verlag, 1990.

[Haglund 1989] L. Haglund, Hierarchical Scale Analysis of Images Using Phase Description, Thesis 168, LIU-TEK-LIC-1989:08, Linköping University, Sweden, 1989.

[Hanson-Riseman 1978] Hanson, A.R. & Riseman, E.M., VISIONS: A Computer Vision System for Interpreting Scenes, in Computer Vision Systems, A.R. Hanson & E.M. Riseman, Academic Press, New York, N.Y., pp. 303-334, 1978.

[Hanson-Riseman 1987] A.R. Hanson & E. Riseman, A Knowledge Based Approach to Vision. In: Vision, Brain and Co-operative Computation, (Eds.) M.A. Arbib & A.R. Hanson, MIT Press, Cambridge, Mass, 1987.

[Hayes-Roth 1989] B. Hayes Roth, Architectural Foundations for Real-Time Performance in Intelligent Agents, Tech Report KSL 89-63, Stanford University, Ca. December 1989.

[IPS2-2152] ESPRIT Information Processing System ESPRIT II project, Visual Inspection and Evaluation of Wide-Area Scenes (VIEWS).

[Jensen et al 1990] F.V. Jensen, J. Nielsen, H.I. Christensen, Use of Causal Probabilistic Networks as High Level Models in Computer Vision, Tech Report R-90-39, Aalborg University, Institute of Electronic Systems, November 1990.

[Knutsson 1990] H. Knutsson, L. Haglund and G. H. Granlund, Tensor Field Controlled Image Sequence Enhancement. In SSAB Symposium on Image Analysis, Linköping, Sweden, March 1990.

[Koenderick-Doorn 1986] Koenderick, J.J., van Doorn, A.J., Dynamic Shape, Biological Cybernetics, 53, 1986, 383-396

[Kriegman et. al. 1987] Kriegman, D.J., Triendl, E., Binford, T.O., A Mobile Robot: Sensing, Planning And Locomotion, Proc Intl. Conf Robotics and Automation, IEEE Press, 1987, 402-408

[Lamdan-Wolfson 1988] Y. Lamdan & H.J. Wolfson. Geometric Hashing: A General and Efficient Model-Based Recognition Scheme. In: 2nd Intl. Conf on Computer Vision, Tarpon Springs, Fl., Dec. 1988. pp. 238-246.

[Lindeberg 1991] T. Lindeberg, Discrete Scale-Space Theory and the Scale-Space Primal Sketch , PhD Thesis, Dept of Numerical Analysis and Computing Science, Royal Institute of Technology, Stockholm, Sweden. 1991

[Lindley 1985] Lindley, D.V. *Making Decisions,* John Wiley & Sons, London, 1985 (first edition 1971).

[Lowe 1987] Lowe D.G. Three Dimensional Object Recognition from Single Two Dimensional Images, Artificial Intelligence, Vol. 31, pp 233-395,1987.

[Marr 1982] Marr, D., Vision, W. H. Freeman, San Francisco, 1982.

[Mohnhaupt-Neumann 1990] M. Mohnhaupt & B. Neumann, Understanding Object Motion: Recognition, Learning, and Spatio-Temporal Reasoning, FBI-HH-B-145/90, Hamburg University, March 1990.

[Newell 1973] Newell, A. "Production Systems: Models of Control Structures", in Visual Information Processing, W.G. Chase (Ed), Academic Press, New York, 1973.

[Newell 1981] Newell, A. "The Knowledge Level", Artificial Intelligence, 2(2), 1981.

[Nilsson 1980] Nilsson, N. J. Principles of Artificial Intelligence, Tioga Press, Palo Alto, Ca., 1980.

[Pearl 1988] J. Pearl, Probabilistic Reasoning in Intelligent Systems, Morgan Kauffmann, Los Angeles, CA., 1988.

[Rao-Jain 1988] A.R. Rao & R. Jain, Knowledge Representation and Control in Computer Vision Systems, IEEE Expert, Spring 1988, pp. 64-79.

[Retz-Schmidt 1991] Retz-Schmidt, G. Recognising Intentions, Interactions and Causes for Plan Failures, In *User Modelling and User Adapted Interaction*, Vol. 1, 1991, pp. 173-205.

[Rosenfeld 1984] A. Rosenfeld (Ed) Multi-resolution Image Processing and Analysis. Springer-Verlag, NY/Berlin, 1984.

[Shafer 76] Shafer, G. A Mathematical Theory of Evidence, Princeton, NJ: Princeton University Press.

[Shafer 87] Shafer, G. Probability judgement in artificial intelligence and expert systems. Statistical Science, 3, 3-16.

[Tistarelli & Sandini 1991] M. Tistarelli and G. Sandini, Direct estimation of time-to-impact from optical flow, In IEEE Workshop on Motion, Princeton (N.J.), October 5-7 1991.

[Tsotsos 1987] J.K. Tsotsos, Representational Axes and Temporal Co-operative Processes, In: Vision, Brain and Co-operative Computation, (Eds.) M.A. Arbib & A.R. Hanson, MIT Press, Cambridge, Mass, pp. 361-418, 1987.

[Wilkes-Tsotsos 1992] Wilkes, D. and Tsotsos, J. Active Object Recognition, In Proc. Computer Vision and Pattern Recognition, Urbana, Ill., June 1992.

2. The SAVA Skeleton System

James L. Crowley and Christophe Discours
LIFIA, INPG

To study methods for integration and control of a continuously operating vision system, the project required a test-bed system for experiments. It was decided to construct an empty "skeleton" system, which partners would then "fill in" the functional parts needed for their experiments. A distributed system composed of independent Unix processes was specified in order to accommodate processes written in different programming environments and using both Lisp and C.

A distributed skeleton system was constructed during the first year of the project and made available to the partners. This system was named SAVA, for the french acronym "Squelette d'Application pour la Vision Active". Use of the SAVA skeleton system during the second year revealed a number of problems. To address these problems, a series of design meetings were held, accompanied by extensive exchanges of electronic mail. The result was the specification and realization of a simpler and more robust communications library, and the redesign of standard module components built around a finite state "scheduler". This new system was released under the name SAVA II.

Experience with SAVA II showed the importance of demons for combining purposive and event driven control of the perceptual processes. However, programming control experiments in SAVA II was a somewhat difficult task. Control knowledge was embedded in procedural code and thus hard to understand or change. It was decided to design a control system based on an interpreter for declarative expressions of the control logic. At the same time, a design effort was made to create an interpreter for demons. From these two needs emerged the idea of using the CLIPS 5.1 rule interpreter as the scheduler, the control component and the demon interpreter within each module. CLIPS 5.1 is written in C and is provided with the full source code. As a result, it was extremely easy to integrate the SAVA modules into the CLIPS environment.

A new version of the skeleton system, SAVA III, has been created based on this principle. In SAVA III, almost all of the procedures for processing and communication within a module are explicitly declared to the CLIPS 5.1 rule interpreter. Rules are written using these procedures. The basic processing cycle is built as a sequence of states with transitions managed by rules. The processing performed within a state can be easily changed based on either perceptual events or external commands. Because the control rules are interpreted, the control sequence for a module may be changed dynamically, without recompiling a module. It is even possible for a module to send another module function definitions as ascii messages, using the CLIPS deffunction facility.

The rule based scheduler is particularly useful for the implementation of demons. Demons may be programmed as rules which react to the contents of the model as well as to external messages. The use of a rule based system as the scheduler and control component of a module has brought numerous unexpected dividends.

This chapter presents an overview of the SAVA system. It first gives a brief summary of the modules which make up the demonstration system. It then describes the mathematical foundations for a system which maintains a description of the external environment by cyclic modeling process. In section 2.3 this model is used to design a standard SAVA module. The use of demons for pre-attentive and post-attentive control is described in section 2.4. An example of the use of this module to build a module for image tracking and description is provided in chapter 8.

2.1 An Integrated Continuously Operating System

In order to demonstrate integration of a continuously operating vision system, the VAP consortium has developed a demonstration systems composed of:

  1) Control of the image acquisition process (gaze, vergence, focus, iris)
  2) Image description hardware, operating at multiple resolutions
  3) Processes for tracking a 2-D description of the contents of the image
  4) Processes for maintaining a 3-D description around a fixation point
  5) Processes for maintaining a symbolic interpretation of the scene
  6) System Supervisor for coordinating goals and actions of the system modules

This demonstration system, illustrated in figure 2.1, was constructed in the SAVA skeleton system. Each process dynamically maintains a description of a limited region of the scene at different levels of abstraction. This system includes independent processes for camera control, image acquisition and tracking, 3D modeling, symbolic interpretation, and system supervision. Each module is a continuously running cyclic process, implemented as a separate Unix process.

The six processes which compose the system architecture are described as follows:

Camera Control Unit

The camera control unit provides a standard interface to the device controller for the VAP binocular stereo heads. This module maintains a copy of the current state of the gaze point and the component axes for the binocular head. It receives commands in the form of tasks expressed either in either device or motor coordinates. Commands are communicated to the binocular head, the arm (neck) or the mobile platform using the device control protocol.

The Camera control unit also contains facilities for programming procedural style "perceptual actions". Such perceptual actions are reflex procedures that command the state of the binocular head at either the device level or the motor level based on measurements made from images. Examples of low level perceptual actions include ocular reflexes for servoing aperture, focus and vergence. Other examples include procedures for servoing the gaze point along a 3D contour using the disparity gradient, and procedures for physically tracking a moving object. A hierarchical camera controller for SAVA is described in chapter 15. The KTH head-eye system is described in chapter 16.

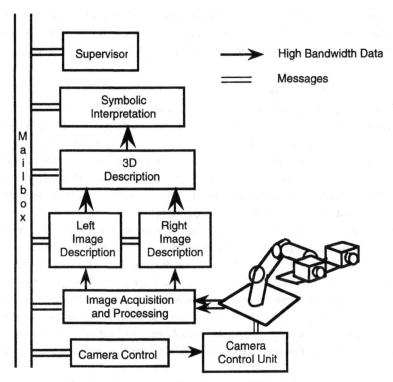

Fig. 2.1 A Distributed Multi-process Vision System.

Image Acquisition and Processing

The image acquisition and processing module handles all image processing requirements for the other modules, thus minimizing the communication requirements. Three classes of image processing procedures are available in the image processing module

1) Edge Segment Extraction. On command, the module will transfer the pixels within a region of interest to a edge extraction card produced by TIRF. This card computes the gradient magnitude and orientation and detects pixels which are extrema in gradient magnitude. Detected pixels are grouped in single raster scan to construct edge segments. Gradient magnitudes are compared to two thresholds to provide a hysteresis based thresholding.

2) Edge Chain Extraction: In place of edge segments, another module may request edge chains. Edge points are are computed by the same algorithm as for edge segments. A one pass raster-chaining algorithm is used to construct a list of edge chains within the region of interest. The edge chaining code is computed on a co-processor card based on the Intel 680

3) Measures for Ocular Reflexes. In order to avoid communicating images, the measurements on which ocular-motor reflexes are based have been placed in this module. Measures include coarse to fine computation of phase for convergence, and gradient based measurements for aperture and focus.

This module is described in chapter 5. The module is based on the LTIRF Pyramid card, which is capable of acquiring and digitizing synchronised stereo image. The card is also capable of computing a 12 level binomial pyramid for the two images at video rates. The algorithms used in this card are described in chapter 6.

Image Tracking and Description

The image description is maintained by a tracking process which uses a first order Kalman filter to track image description primitives. This tracking process improves the stability of image primitives, permits the system to maintain correspondence of image features over time, and provides an estimate of the position and velocity of image primitives as well as the uncertainty of these estimates. A vocabulary of model access and grouping procedures give associative access to the 2D description modules. These procedures are used by a library of "demon" procedures which can be enabled in order to provide data driven interpretation of the image description.

The image description modules extract edges from a region of interest within a multi-resolution pyramid description of an image. Edges are then matched to predictions from a tracking process, and the correspondence is used to update a description of the image. Model access primitives use matching and grouping to interrogate the contents of the token model. A set of demons may be invoked by other modules to interrogate the description after each update using the model access primitives. When the image acquisition module signals that an image is available, the image is obtained over the high band-width channel. A binomial pyramid is then calculated for the image. Edges are extracted from within a multi-resolution "region of interest", which forms a "log-Cartesian" retina.

The edge extraction process begins by calculating the horizontal and vertical derivatives within the region of interest. These derivatives are then combined by table look-up to compute the gradient magnitude and orientation. Points which are extrema in magnitude are marked as potential edge points and compared against two thresholds. Hysteresis thresholding is applied so that only regions of edge points containing at least one point above the threshold are considered. Adjacent edge points with a similar orientation are grouped to form line segments. Edge segments are represented by a vector of parameters that includes the mid-point, orientation and half-length.

The 2D image descriptions are maintained by a tracking process that uses a first order Kalman filter to track image description primitives. This tracking process improves the stability of image primitives, permits the system to maintain correspondence of image features over time, and provides an estimate of the position and velocity of image primitives as well as the uncertainty of these estimates.

Access to the 2D model is provided by a large vocabulary of model access and grouping procedures. The four classes of model access procedures are:

Class 1 Procedures for extracting parametric primitives that are *close to an ideal geometric entity* such as a line or a point. Such primitives are easily implemented and have a complexity that is proportional to the number of primitives in the model.

Class 2 Procedures for extracting parametric primitives that are *similar to an ideal "prototype"*. The prototype is specified an estimated value and a standard deviation for each of the primitive parameters. Similarity is measured by a Mahalanobis

Distance. Parameters may be labelled as a "wildcard" by specifying a large (or infinite) standard deviation.

Class 3 Procedures for extracting *pairs of primitives* that satisfy some geometric relation, such as a junction, an alignment, or a parallelism. The geometric relation is defined as a set of parameters and specified as an estimate and a standard deviation.

Class 4 Procedures for extracting *grouping of groupings.* If performed by brute force, complex forms such as lines and contours composed of sets of line segments can require an exponential number of computations. By structuring the interrogation as a grouping, we can limit the theoretical complexity to $O(N^3)$ and maintain an average case cost that is nearly $O(N)$.

It is also possible to compose sequences of these grouping procedures, extracting, for example, all the junctions near an ideal line. These procedures may be called by other modules within the system, or they may be invoked by a set of interpretation demons. These demons are placed on an agenda by messages from other modules. After each update cycle the demon agenda is executed.

The tracking process on which the 2D and 3D modules are based is discussed in greater detail in this chapter. These procedures for model access, grouping and the interpretation demons are described in chapter 8 and 9.

3D Geometric Scene Description Module

In addition to a geometric description of a image, the system maintains a geometric description of the scene. This geometric description expresses the structure within a region of interest of the scene in terms of 3D parametric primitives. This module assumes that the phase based convergence reflex maintains the cameras converged on an object. Convergence maintains edge segments from a region of interest in the scene in the similar positions in the image. The image description access primitive "FindPrototypeSegment" is used to construct a list of possible matching segments in the left and right image. This list is sorted based on similarity of length, orientation and position. The most likely matches are selected for 3D reconstruction.

Reconstruction requires camera calibration. A novel procedure for dynamic auto-calibration of cameras has been developed. This procedure permits a reference frame for a pair of stereo cameras to be constructed for any scene objects. The projective transformation matrices from object centered coordinates can be obtained by direct observation (no matrix inversion) and can be maintained by a very simply operation. These matrices make it possible to reconstruct the 3D form of objects in an object centered reference frame. As with the image tracking and description module, the geometric description is maintained by a tracking process in order to provide stability and to maintain correspondence over time. The 3D description process is described in chapter 16 below. The process for maintenance of a 3D model is described in chapter 17.

Symbolic Scene Interpretation

The symbolic scene interpretation maintains a symbolic description of the scene in terms of known object categories (or classes) and qualitative relation. This description is built up and maintained by interrogating the contents of the image and scene description modules.

The symbolic scene interpretation maintains a symbolic description of the scene in terms of known object categories (or classes) and qualitative relation. This description is built up and maintained by interrogating the contents of the image and scene description modules. The SAVA II symbolic description process is implemented using the CLIPS rule interpreter system. Because CLIPS is C based, its integration within SAVA is relatively straightforward. Working memory of the production system serves as a blackboard into which recognition procedures can poste their results.

Interpretation is performed by recognition procedures. Each recognition procedure consists of a trigger, a condition part and an action part. Triggers typically correspond either to demons in the 2D or 3D description process, or to hypotheses currently on the hypothesis blackboard. Recognition procedures are organised into contexts. The procedures that make up the current context are placed on an agenda, their trigger conditions are tested. Whenever a trigger condition is detected, the condition part of the knowledge source performs further interrogation of the 2D model or the hypothesis blackboard. The action part of a knowledge source can update the working memory or can change the current context. The symbolic scene description process is described in chapters 19, 20, 21 and 22.

Process Supervisor

A major issue to be addressed by the consortium is the relative roles to be played by centralized and distributed control. Clearly some sort of centralized control structure is required for resource allocation and determination of highest level goals. The consortium plans experiments with different amounts of control placed in a centralized "supervisor" or distributed over the individual modules. The Process supervisor will serve as the centralized control module.

The University of Aalborg has recently developed a supervisor for SAVA II using the CLIPS rule based system. The Aalborg supervisor implements control of the camera control unit and the image description modules to execute commands from the set:

Find <object>
Watch <object>
Track <object>
Classify <Object>
Explore

The possible techniques to be used in system control and in the process supervisor are described in The supervisor for use with SAVA III is described in chapter 19.

2.2 A Framework for Continuous Modeling

The SAVA skeleton system modules are designed to maintain a description of the external environment based on observations guided by a-priori expectations. Modules are based on a framework for incremental modeling presented in this section. The section begins with a description of dynamic world modeling as an iterative process of integrating observations into an internal description. We then present a framework for incrementally modeling the contents of a scene by integrating successive observations. This framework is used in the design of the SAVA modules for description of images and for description of the 3D scene. The 2D and 3D modeling systems that have been constructed permit us to make several conclusions about the role of prediction and estimation in perception:

1) A perceptual system can be designed using a layered architecture of predict-match-update cycles (described below). The model maintained within each cycle serves as an observation for cycles in a more abstract coordinate space.

2) Tracking preserves correspondence. The correspondence between image features established by tracking can be preserved and propagated throughout a system, thus avoiding a more costly matching between 3D structures. This is particularly useful in real time stereo systems.

3) Real time tracking can be based on a very simple "nearest-neighbor" matching algorithm. A first-order predictive system makes possible a very simple matching algorithm.

These conclusions were observed as a result of building systems for dynamic modeling applying tools from estimation theory within a framework described in the next section.

In the following section we present world modeling as a continuously operating cyclic process composed of the phases: predict, match and update. We also describe techniques from estimation theory for each of these processes. In section 2.3 we present a "standard" module for scene description based on this framework.

2.2.1 A Framework for Perception: The Predict, Match and Update Cycle

Dynamic world modeling is a cyclic process in which a description of a scene is elaborated by fusing successive observations. The observations may come from multiple sensors, multiple modalities of sensor data processing, or from a-priori information. A framework for dynamic world modeling is illustrated in Figure 2.2. In this framework, independent observations are processed to extract information in a common coordinate space and vocabulary. This information is then used to update a model. This forms a cyclic process composed of three phases: Predict, Match and Update.

Predict: The current state of the model is used to predict the state of the external world at the time that the next observation is taken.

Match: The transformed observation in brought into correspondence with the predictions. Such matching requires that the observation and the prediction express information which is qualitatively similar. Matching requires that the predictions and observations be transformed to the same coordinate space and in a common vocabulary.

Update: The update phase integrates the observed information with the predicted state of the model to create an updated description of the environment composed of hypotheses.

The update phase serves both to add new information to the model as well as to remove "old" information. During the update phase, information which is no longer within the "focus of attention" of the system, as well as information which has been found transient or erroneous, is removed from the model. This process of "intelligent forgetting" is necessary to prevent the internal model from growing without limits.

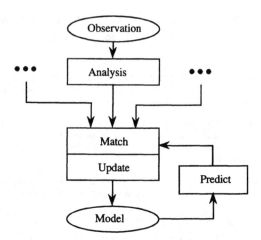

Fig. 2.2 A Framework for Dynamic World Modeling.

This framework can be applied at every level of abstraction within a perceptual system. In particular, such a process may be used in the 2D description of images, in the 3D scene modeling system, and in the symbolic scene description process.

2.2.2 Background for Dynamic World Modeling

The problem of dynamically modeling the environment is a form of "sensor fusion". Recent advances in sensor fusion from within the vision community have largely entailed the rediscovery and adaptation of techniques from estimation theory. These techniques have made their way to vision via the robotics community, with some push from military applications.

For instance, in the early 1980's, [Herman and Kanade 1986] combined passive stereo imagery from an aerial sensor. This early work characterized the problem as one of incremental combination of geometric information. A similar approach was employed by the author for incremental construction of world model of a mobile robot using a rotating ultrasonic sensor [Crowley 1985]. That work was generalized [Crowley 1984] to present fusion as a cyclic process of combining information from logical sensors. The importance of an explicit model of uncertainty was recognized, but the techniques were for the most part "ad-hoc". Driven by the needs of perception for mobile robotics, [Brooks 1985] and [Chatila 1985] also published ad-hoc techniques for manipulation of uncertainty.

In 1985, a pre-publication of a paper by Smith and Cheeseman was very widely circulated [Smith-Cheeseman-1987]. In this paper, the authors argue for the use of Bayesian estimation theory in vision and robotics. An optimal combination function was derived and shown to be equivalent to a simple form of Kalman filter. At the same period, Durrant-Whyte completed a thesis [Durrant-Whyte 1987] on the manipulation of uncertainty in robotics and perception. This thesis presents derivations of techniques for manipulating and integrating sensor information which are extensions of technique from estimation theory. Well versed in estimation theory, Faugeras and Ayache [Faugeras et al. 1986] contributed an adaptation of this theory to stereo and calibration. From 1987, a rapid paradigm shift occurred in the vision community, with techniques from estimation theory being aggressively adapted.

While most researchers applying estimation theory to perception can cite one of the references [Smith-Cheeseman 1987], [Durrant-Whyte 1987] or [Faugeras et al 1986] for their inspiration, the actual techniques were well known to some other scientific communities, in particular the community of control theory. The starting point for estimation theory is commonly thought to be the independent developments of [Kolmogorov 1941] and [Weiner 1949]. [Bucy 1959] showed that the method of calculating the optimal filter parameters by differential equation could also be applied to non-stationary processes. [Kalman 1960] published a recursive algorithm in the form of difference equations for recursive optimal estimation of linear systems. With time, it has been shown that these optimal estimation methods are closely related to Bayesian estimation, maximum likelihood methods, and least squares methods. These relationships are developed in textbooks by [Bucy-Joseph 1968], [Jazwinski 1970], and in particular by [Melsa-Sage 1971]. These relations are reviewed in a recent paper by [Brown et. al. 1989], as well as in a book by [Brammer-Siffling 1989].

These techniques from estimation theory provide a theoretical foundation for the processes which compose the proposed computational framework for fusion in the case of numerical data. An alternative approach for such a foundation is the use of minimum energy or minimum entropy criteria. An example of such a computation is provided by a Hopfield net [Hopfield 1982]. The idea is to minimize some sort of energy function that expresses quantitatively by how much each available measurement and each imposed constraint are violated [Li 1989]. This idea is related to regularization techniques for surface reconstruction employed by [Terzopoulos 1986]. The implementation of regularization algorithms using massively parallel neural nets has been discussed by [Koch et. al. 1985], [Poggio-Koch 1985] and [Blake-Zisserman 1987].

Estimation theory techniques may be applied to combining numerical parameters. In the case of symbolic information, the relevant computational mechanisms are inference techniques from artificial intelligence. In particular, fusion of symbolic information will require reasoning and inference in the presence of uncertainty using constraints.

The Artificial Intelligence community has developed a set of techniques for symbolic reasoning. In addition to brute force coding of inference procedures, rule based "inference engines" are widely used. Such inference may be backward chaining for diagnostic problems, consultation, or data base access as in the case of MYCIN [Buchanon-Shortliffe 1984]. Rule based inference may also be forward chaining for planning or process supervision, as is the case in OPS5 [Forgy 1982], [Brownston et al 1985]. Forward and backward chaining can be combined with object-oriented "inheritance" scheme as is the case in KEE and in SRL. Groups of "experts" using these techniques can be made to communicate using black-board systems, such as BB1 [Hayes-Roth 1985]. For perception, any of these inference techniques must be used in conjunction with techniques for applying constraint based reasoning to uncertain information.

Several competing families of techniques exist within the AI community for reasoning under uncertainty. Automated Truth Maintenance Systems [Doyle 1979] maintain chains of logical dependencies, when shifting between competing hypotheses. The MYCIN system [Buchanon-Shortliffe 1984] has made popular a set of ad-hoc formulae for maintaining the confidence factors of uncertain facts and inferences. Duda, Hart and Nilsson [Duda et al 1976] have attempted to place such reasoning on a formal basis by providing techniques for symbolic uncertainty management based on Bayesian theory. Shafer has also attempted to provide a formal basis for inference under uncertainty by providing techniques for combining evidence [Shafer 1976]. A large school of techniques known as "Fuzzy Logic" [Zadeh 1979] exist for combining imprecise assertions and inferences.

In this section we show how techniques from estimation theory may be applied to the predict-match-update cycle, in the case where the model if composed of parametric primitives.

2.2.3 Representation: Parametric Primitives

A dynamic world model, M(t), is a list of primitives which describe the "state" of a part of the world at an instant in time t.

Model: $M(t) \equiv \{ P_1(t), P_2(t), ..., P_m(t) \}$

A model may also include "grouping" primitives which assert relations between lower level primitives. Examples of such groupings include connectivity, co-parallelism, junctions and symmetries. Such groupings are an example of symbolic properties.

Each primitive P(t) in the world model, describes a local part of the world as a conjunction of estimated properties, $\hat{X}(t)$, plus a unique ID and a confidence factor, CF(t).

Primitive: $P(t) \equiv \{ID, \hat{X}(t), CF(t)\}$

The ID of a primitive acts as a name by which the primitive may be referred. The confidence factor, CF(t), permits the system to control the contents of the model. Newly observed segments enter the model with a low confidence. Successive observations permit the confidence to increase, where as if the segment is not observed in the succeeding cycles, it is considered as noise and removed from the model. Once the system has become confident in a segment, the confidence factor permits a segment to remain in existence for several cycles, even if it is obscured from observations. Experiments with have lead us to use a simple set of confidence "states" represented by integers. The number of confidence states depends on the application of the system.

The properties, $\hat{X}(t)$, may be either a numeric value or a symbolic label from a finite class of symbols.In the case of numeric properties, we may employ techniques from estimation theory. For symbolic properties we will need symbolic reasoning. In either case, a primitive represents the local state of a part of the world as an association of a set of N properties, represented by a vector , $\hat{X}(t)$.

$\hat{X}(t) \equiv \{ \hat{x}_1(t), \hat{x}_2(t),... \hat{x}_n(t)\}.$

The actual state of the external world, X(t), is estimated by an observation process which is assumed to be corrupted by random noise, N(t). The world state, X(t), is not directly knowable, and so our estimate is taken to be an expected value of X(t). For numeric properties, this expected value may be obtained by computing the expected (or average) value from a sample population.

$\hat{x}_m(t) = E\{x(t)\}$

At each cycle, the modeling system produces an estimate $\hat{X}(t)$ by combining a prediction, $Y^*(t)$, with an observation $Y(t)$. The difference between the predicted vector $Y^*(t)$ and the observed vector $Y(t)$ provides the basis for updating the estimate $\hat{X}(t)$, as described below.

In order for the modeling process to operate, both the primitive, $\hat{X}(t)$ and the observation, $Y(t)$ must be accompanied by an estimate of their uncertainty. This uncertainty may be seen as an expected deviation between the estimated vector, $\hat{X}(t)$, and the true vector, $X(t)$. In the case of numerical properties, an expected deviation may be approximated as a covariance matrix $\hat{C}(t)$ which represents the expected difference between the estimate and the actual world state. For numeric properties, the covariance is defined as :

$$\hat{C}(t) \equiv E\{[X(t) - \hat{X}(t)] [X(t) - \hat{X}(t)]^T\}$$

Modeling this precision as a covariance makes available a number of mathematical tools for matching and integrating observations. The uncertainty estimate is based on a model of the errors which corrupt the prediction and observation processes. Estimating these errors is both difficult and essential to the function of our system.

The uncertainty estimate provides two crucial roles in our system:

  1) It provides the tolerance bounds for matching observations to predictions, and
  2) It provides the relative strength of prediction and observation when calculating a new
     estimate.

Because $\hat{C}(t)$ determines the tolerance for matching, system performance will degrade rapidly if we under-estimate $\hat{C}(t)$. On the other hand, overestimating $\hat{C}(t)$ merely increases the computing time for finding a match.

2.2.4 Prediction: Discrete State Transition Equations

The prediction phase of the modeling process projects the estimated vector $\hat{X}(t)$ forward in time to a predicted value, $X^*(t+\Delta T)$. This phase also projects the estimated uncertainty $\hat{C}(t)$ forward to a predicted uncertainty $C^*(t+\Delta T)$. Such projection requires estimates of the temporal derivatives for the properties in $\hat{X}(t)$, as well as estimates of the covariances between the properties and their derivatives. These estimated derivatives can be included as properties in the vector $\hat{X}(t)$.

In the following, we will describe the case of a first order prediction; that is, only the first temporal derivative is estimated. Higher order predictions follow directly by estimating additional derivatives. We will illustrate the techniques for a primitive composed of two properties, $x_1(t)$ and $x_2(t)$. We employ a continuous time variable t to mark the fact that the prediction and estimation may be computed for a time interval, ΔT, which is not necessarily constant.

Temporal derivatives of a property are represented as additional components of the vector $X(t)$. It is not necessary that the observation vector, $Y(t)$, contain the derivatives of the properties to be estimated. The Kalman filter will iteratively estimate the derivatives of a property using only observations of its value. Furthermore, because these estimates are developed by integration, they are less sensitive to noise than instantaneous derivatives calculated by a simple difference.

Consider a property, $\hat{x}(t)$, of the vector $\hat{X}(t)$, having variance σ_x^2. A first order prediction of the value $x^*(t+\Delta T)$ requires an estimate of the first temporal derivatives, $\hat{x}'(t)$. To apply a first order prediction, all of the higher order terms are grouped into an unknown random vector $V(t)$, approximated by an estimate, $\hat{V}(t)$. The term $\hat{V}(t)$ models the effects of both higher order derivatives and other unpredicted phenomena. $V(t)$ is defined to have a variance (or energy) of $Q(t)$.

$$Q(t) = E\{V(t)\,V(t)^T\}$$

When $V(t)$ is unknown, it is assumed to have zero mean, and thus is estimated to be zero. In some situation it is possible to estimate the perturbation from knowledge of commands by an associated control system. In this case, an estimated driving vector $\hat{u}(t)$ and its uncertainty, $\hat{Q}(t)$ may be included in the prediction equations. Thus each term is predicted by:

$$x^*(t+\Delta T) = \hat{x}(t) + \frac{\partial \hat{x}(t)}{\partial t}\,\Delta T + \hat{u}(t) + \hat{v}(t)$$

In matrix form, the prediction can be written as:

$$X^*(t+\Delta T) := \varphi\,\hat{X}(t) + \hat{U}(t) + \hat{V}(t) \tag{1}$$

Predicting the uncertainty of $X^*(t+\Delta T)$ requires an estimate of the covariance between each property, $\hat{x}(t)$ and its derivative. An estimate of this uncertainty, $\hat{Q}_x(t)$, permits us to account for the effects of unmodeled derivatives when determining matching tolerances. This gives the second prediction equation:

$$C_x^*(t+\Delta T) := \varphi\,\hat{C}_x(t)\,\varphi^T + \hat{Q}_x(t) \tag{2}$$

2.2.5 Matching Observation to Prediction: The Mahalanobis Distance

The predict-match-update cycle presented in this paper simplifies the matching problem by applying the constraint of temporal continuity. That is, it is assumed that during the period ΔT between observations, the deviation between the predicted values and the observed values of the estimated primitives is small enough to permit a trivial "nearest neighbor" matching.

Let us define a matrix $^Y H_x$ which transforms the coordinate space of the estimated state, $X(t)$, to the coordinate space of the observation.

$$Y(t) = {}_X^Y H\ X(t)$$

The matrix $_X^Y\mathbf{H}$ constitutes a "model" of the sensing process[1] which predicts an observation, $Y(t)$ given knowledge of the properties $X(t)$. Estimating the model of the sensor is a crucial aspect of designing a world modeling system. The model of the observation process can not be assumed to be perfect. In machine vision, the observation process is typically perturbed by photo-optical, optical and electronic effects. Let us define this perturbation as $W(t)$. In most cases, $W(t)$ is unknown, leading us to estimate

$$\hat{W}(t) \equiv E\{W(t)\} = 0 \qquad \text{and} \qquad \hat{C}_y(t) \equiv E\{ W(t)\, W(t)^T\}$$

It is often the case that the observation Y is related to the state vector X by a nonlinear function, $F(X)$. In this case, $_X^Y\mathbf{H}$ is approximated by the derivative, or Jacobian, of the transformation, $_X^Y\mathbf{J}$.

$$_X^Y\mathbf{H} \approx {}_X^Y\mathbf{J} = \frac{\partial F(X)}{\partial X}$$

Let us assume a predicted model $M^*(t)$ composed of a list of primitives, $P^*_n(t)$, each containing a parameter vector, $X^*(t)$, and an observed model $O(t)$ composed of a list of observed primitives, $P_m(t)$, each containing the parameters $Y(t)$. The match phase determines the most likely association of observed and predicted primitives based on the similarity between the predicted and observed properties. The mathematical measure for such similarity is to determine the difference of the properties, normalized by their covariance. This distance, normalized by covariance, is a quadratic form known as the squared Mahalanobis distance.

The predicted parameter vector is given by:

$$Y^*_n := {}_X^Y\mathbf{H}\; X^*_n$$

with covariance

$$C^*_{yn} := {}_X^Y\mathbf{H}\; C^*_{xn}\; {}_X^Y\mathbf{H}^T$$

The observed properties are Y_m with covariance C_{ym}. The squared Mahalanobis distance between the predicted and observed properties is given by:

$$D^2_{nm} = \frac{1}{2}\{ (Y^*_n - Y_m)^T\, (C^*_{yn} + C_{ym})^{-1}\, (Y^*_n - Y_m)\}$$

In the predict-match-update cycles described below, matching involves minimizing the normalized distance between predicted and observed properties or verifying that the distance falls within a certain number of standard deviations.

2.2.6 Updating: The Kalman Filter Update Equations

Having determined that an observation corresponds to a prediction, the properties of the model can be updated. The extended Kalman filter permits us to estimate a set of properties and their derivatives,

---

[1] $_X^Y\mathbf{H}$ is sometimes known as the "Sensor Model". This use of the word "model" by the estimation theory community creates an unfortunate conflict of terms with the vision community.

$\hat{X}_n(t)$, from the association of a predicted set of properties, $Y_n^*(t)$, with an observed set of properties, $Y_m(t)$. It equally provides an estimate for the precision of the properties and their derivatives. This estimate is equivalent to a recursive least squares estimate for $X_n(t)$. The estimate and its precision will converge to a false value if the observation and the estimate are not independent.

The crucial element of the Kalman filter is a weighting matrix known as the Kalman Gain, $K(t)$. The Kalman Gain may be defined using the prediction uncertainty $C_y^*(t)$.

$$K(t) := C_x^*(t) \; {}_X^Y H^T [C_y^*(t) + C_y(t)]^{-1} \tag{3}$$

The Kalman gain provides a relative waiting between the prediction and observation, based on their relative uncertainties. The Kalman gain permits us to update the estimated set of properties and their derivatives from the difference between the predicted and observed properties:

$$\hat{X}(t) := X^*(t) + K(t) [Y(t) - Y^*(t)] \tag{4}$$

The precision of the estimate is determined by:

$$\hat{C}(t) := C^*(t) - K(t) {}_X^Y H \; C^*(t) \tag{5}$$

Equations (1) through (5) constitute the 5 equations of the Kalman Filter.

2.2.7 Fusion of Symbolic Properties

The framework for fusion of perceptual information described above can also be applied to symbolic properties. Fusion of symbolic information is a problem of associating (or grouping) symbols which represent perceptual phenomena. As with numerical properties, an internal "symbolic" model is composed of a vector of properties. In modern terms such an association of properties is called a "schema" or frame or a unit. We will use the term "schema".

A schema is a named association of numeric and symbolic properties. A schema may contain relations to other schema (such as ISA and Part-Of hierarchies), as well as procedures for operating on its properties. A dynamic world modeling system for symbolic information is concerned with instantiating and maintaining a collection of schema which describe and predict the external environment. The maintenance process can be formed using a predict-match-update cycle.

Primitives in the world model are expressed as a set of properties. Schema provide just such a representation. The properties may be symbolic labels or numerical measures. For numerical data, a common coordinate system serves as a basis for data association. In a symbolic description, such association may be on the basis of spatial or temporal coordinates, or it may be on the basis of a relationship between properties. The equivalent to a common coordinate system and common vocabulary at the symbolic level is a "context". A context is a collection of symbols and relations which are used to describe a situation. Knowing the context provides a set of symbols and relations which can be expected. This permits description to proceed by a process of prediction and verification.

As with numerical properties, symbolic properties have two kinds of uncertainties: precision and confidence. The classic AI method for representing precision is to provide a list of possible values. Such lists are used both for symbolic properties and for relations. Constraints are applied by intersecting lists of possible values.

A schema is an assertion about the external environment. This assertion may be more or less certain. The degree of confident in the truth of a schema can be represented by a confidence factor. As with numerical techniques, this confidence factor applies to the association of symbols, and increases or decreases as supporting or conflicting evidence is detected.

The phases of the predict-match-update each have their equivalent at the symbolic level. Let us consider the equivalence for each of these:

Prediction. The key to prediction is context. The prediction phase applies a-priori information about the context to predict the evolution of schemas in the model as well as the existence and location of new schema. An important role of the prediction phase is to select the perceptual actions which can detect the expected phenomena. The prediction phase can be used with both pre-attentive and post-attentive recognition. In pre-attentive recognition, the prediction phase can be used to "enable" or "arm" the set of pre-attentive cues which will be processed. In post-attentive recognition, prediction triggers procedures for detecting and locating instances of expected phenomena [Bossier-Demazure 1987].

Match. The match phase associates new perceptual phenomena with predictions from the internal model. As with numerical properties, a primary tool for matching is spatial location. Matching may also be an association based on similar properties.

Update. The update phase combines the results of prediction and observation in order to reconstruct the internal model. Production rules provide one method to express the world knowledge needed to update the internal model.

Control of Perception. The internal model may be thought of as a form of short term memory. The computational cost of the predict, match and update phases depends on the quantity of information in this model. Imposing a fixed cycle time on the model update process has the effect of determining a limit on the quantity of information (the number of schema) that can be placed in the model at any time. This problem of controlling the contents of the model is part of the problem of control of perception [Crowley 1991].

Prediction plays a crucial role in control of perception. Expectations generated by the context determine what objects and relations will be fed back into the match and update cycle. But context alone is not sufficient. The current task of the system determines what information is needed and thus which part of the context is attended to by prediction, matching and updating. Pre-attentive information must also be controlled based on prediction, or else the match and update phases would be flooded with too much information.

The SAVA skeleton system is designed to support processes for continuous world modeling based on theses principles. In the following section we show how these techniques have been used in the design of the SAVA skeleton system.

2.3 The Design of the SAVA Skeleton System

The SAVA system contains a generic module based on the predict-match-update cycle. This generic module used as the basis for each of the modules which maintain a description of the state of the environment. The core of the SAVA skeleton system contains the following components:

1) A *launcher* program that permits the user to assign modules to processors and to initiate operation.
2) A distributed *mailbox system* that is launched on the different processors to establish a communications system and to launch the component processes.
3) A library of *communication procedures* for modules. This library include procedures for communication by message as well as procedures for dedicated high band-width communication between processes.
4) A *generic module* structure built around a scheduler.
5) A set of graphical man-machine interfaces.

The SAVA system provides mailbox communication for data, control and acknowledgements, as well as a procedures for dedicated high-band-width channels between modules. Messages include formatting information that permits the message passing system to pack and unpack messages.

The heart of the generic module structure is the scheduler. Building a module consists of calling a procedure to place the addresses of the desired procedures in a table for the scheduler. The scheduler called each procedure in turn, interleaved with message handling. Modules in SAVA II contained a separate demons scheduler. Demons provide a cyclic computation on the model maintained by each module, and could be used to detect perceptual events in either a directed or data driven manner. Demons could be enabled and disabled by messages from other modules.

In the SAVA III system all of the procedures of a module are made explicitly available to an interpreter. This includes the original SAVA II scheduler, so that the system is upwards compatible. In SAVA III, the sequence of phases in the cycle of each module can be controlled by rules. Between phases, the message buffer is emptied into the working memory of CLIPS (the Facts-List) and immediately processed by the control rules. In the final phase of each cycle, demons use associative access procedures to tests the contents of the model for the presence of perceptual events to which the system is to react.

2.3.1 Implementing the Predict, Match and Update Cycle

We have developed a standard architecture for SAVA modules based on the predict-match-update cycle. Observations are obtained and "transformed" into the module's coordinate space and representation. Observations are matched to predictions and then integrated into the model. Prediction from the previous model serves to accelerate the description of additional observations, and to reduce the errors in the description. The update phase serves both to add and to remove older information. During the update phase, information that is no longer within the "focus of attention" of the system, and information that has been found transient or erroneous, is removed from the model. This serves to limit the size of the model

The contents of the model maintained by this process may be interrogated at any time by other modules in the system. This interrogation occurs by sending a message to a vocabulary of procedures for model access and perceptual grouping. These procedures may also be used by a set of demons for data driven reaction to events. We have defined a standard protocol for model interrogation based on the operations "Find", "Verify", and "Get". This protocol uses the model access and perceptual grouping procedures as associative access functions to interrogate the contents of the model, as illustrated in Figure 2.3. Such interrogation is described in greater detail in chapter 8.

2.3.2 Module Components

The architecture for the standard description module is shown in Figure 2.3. The module is composed of a number of procedures (shown as rectangles) that are called in sequence by a scheduler. Between each procedure call, the scheduler tests a mail box to see if any messages have arrived. Such messages may change the procedures that are used in the process, change the parameters that are used by the procedures, or interrogate the current contents of the description that is being maintained.

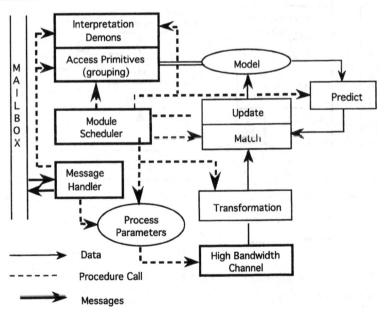

Fig. 2.3 Architecture of a Standard Module for SAVA II

At the end of each cycle, the scheduler calls a set of demon procedures. Demons are responsible for event detection, and play a critical role in the control of reasoning. Some of the demon procedures, such as motion detection, operate by default, and may be explicitly disabled. Most of the demons, however, are specifically designed for detecting certain types of structures. These demons are armed or disarmed by recognition procedures in the interpretation module according to the current interpretation context.

A description module repeatedly executes a cycle in which it:

  1) Gets new data from a lower level module
  2) Transforms this data into its own format
  3) Makes predictions from its internal model
  4) Matches the predictions with the transformed data
  5) Uses the match results to update the internal model
  6) Executes demons (to search for groupings in the data)

This cyclic process is executed by a scheduler, which cyclically calls the procedures within a schedule table. Between each procedure, the scheduler checks to see if any new messages have arrived. In the SAVA III, a rule based interpreter has been added to the standard module as shown in figure 2.4.

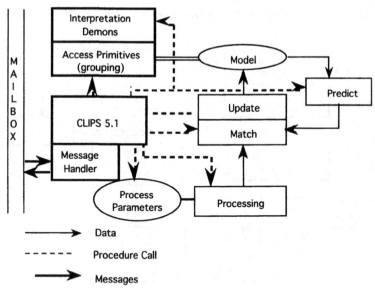

Fig. 2.4. Architecture of the standard module in SAVA III. In this module, the scheduler has been replaced by a rule based system.

2.3.4 Mail Box Communications

Control, data, and synchronisation information is exchanged between modules using a message passing system based on Unix Sockets. For heavy data traffic a facility exists to allocate dedicated sockets between modules. The SavaSend() function is used to send mailbox messages to other modules. A SavaSend command contains three fields:

Header: The destination and type of message. Format: An ascii description of the message format.

Body: The message including commands and parameters.

The destination is a symbolic name for another module. The types of message may be control, acknowledge or data. All message exchanges are initiated by a control message. The format string is transmitted with the message and is used both to encode and decode the message. In this way a change in message protocol may be made with a minimum of difficulty. This format string can contain conversion directives like %d, %f, and %c, based on the C language printf protocol. We have added conversion directives for sending arrays, structures and images.

Many SAVA functions accept a variable number of arguments. Furthermore, the type of these arguments is unspecified. These functions accept a fixed number of normal arguments, followed by an arbitrary number of arguments of unknown type. The last normal argument is a format string which describes the arguments following it.

Several classes of message functions exist. These can be divided into two groups:

Extract: The function uses the format string to extract data from arguments following it. These arguments are "normal" program variables.

Store: The function uses the format string to store data in the arguments following it. These arguments are pointers to program variables.

For instance, SavaSend() extracts data from program variables and sends this data across the mailbox to another module. SavaReceiveArgs() gets data from this mailbox and stores this in program variables.

2.3.5 High Band-Width Channels

Large data structures may be communicated between modules using the functions SavaRead and SavaWrite. The SavaWrite() function sends data to another module across a dedicated channel. Reception is made by the function SavaRead(). As with mail-box, messages, high bandwidth channel messages are encoded with an ascii format directive which is transmitted with the message. High bandwidth channels in SAVA are faster than message passing because the channels provide a direct connection. No intermediate routing is necessary.

2.3.6 Synchronisation of Modules

SAVA is supposed to be a test bed for real time processes. At the same time, SAVA has been built using UNIX , and is composed on modules which communicate over an EtherNet. The use of Unix and EtherNet guarantee that SAVA, as currently implemented, can not function as a true real time system. Each module is at the mercy of the UNIX scheduler, and communication delays depend on the load of the Ethernet. Thus SAVA, as currently implemented, is a "soft" real time system. That is, data is treated as it is acquired, and the system does model temporal processes, but there is not guarantee that any one process will give its result within a fixed delay.

SAVA modules require a common temporal reference, for two reasons. Time is necessary to predict and model dynamic events. Furthermore, time is necessary for controlling robotics processes. Thus all messages are time stamped with an estimated universal time stamp. However, the SAVA modules are distributed over several UNIX systems, and there is no guarantee that their clocks operate at the same frequency. Thus we have added a dedicated socket between the launching process and each of the SAVA modules. This channel provides, on request, the current temporal reference. The UNIX clock of the local system is used to time the delay between emission of the request and reception, and this delay is divided by two to estimate the delay from the time that the reference was transmitted and the time it was received. The result is used as the estimated universal time.

2.4 Demons: Active Interpretation Processes

CLIPS is a provides a forward chaining rule interpreter using the RETE matching algorithm. Rules are matched to working memory composed of numbered list of lists and structures. The cyclic process within a module is managed by a control token placed on the working memory. This control token is a simple list in which the first atom is the word "phase" and the second is the name of one of the phases: {get-data, predict, match, update, messages, demons}. The definition of theses phases is as follows:

get-data	Acquire a new observation
predict	predict the contents of the observation
match	match the prediction to the observation
update	Update the model using the correspondence of the prediction and the observation
demons	The phase in which demons are executed.

In each state, just before the state transition, the message queue is tested for new messages and any new messages are processed.

At the end of each update cycle, the scheduler executes the demon agenda. Demons may be invoked by other demons or by commands received from other modules, including from a human supervisor. A demon is instantiated by entering a demon token in working memory. A demon token is simply a list with three elements:

```
(demon   <name>   <id>)
```

where <name> is the name of the demon and <id> is a unique identity determined by the function "gensym". Multiple copies of the same demon can be instantiated, each having its own "id". Each demon can create its own state in working memory, indexed by <id>. A demon can be removed by removing the demon token from working memory.

The control part of a demon is a rule. As an example consider a demon to find ellipses in the image:

```
(defrule ellipse-finder    "The demon for an ellipse finder"
    (phase demons)
    (demon ellipse-finder ?id)
    (ellipse-demon-data  (id ?id) (parameters ?p))
=>
    (assert (get-ellipses ?p))
)
```

If we suppose that the function "get-ellipses" will instantiate a structure of type ellipse for each ellipse found, then a second rule can be written to treat each ellipse.

```
(defule  hypothesize-cylinder "The demon to generate cylinder hypotheses"
    (phase demons)
    (demon cylinder-finder)
    (ellipse (id ?id) (cx ?x) (cy ?y) (major ?ma) (minor ?mi) (angle ?angle))
    (test (< 5  (abs ?angle)))    ; ellipses should have horizontal major axes
=>
    (assert (cylinder  (cx ?x) (bottom ?y) (radius ?ma) (ellipse ?id)))
    (assert (cylinder  (cx ?x) (top ?y) (radius ?ma) (ellipse ?id)))
)
```

Other rules can be used to detect the existence of cylinders with the same axis and to reduce cylinder hypotheses to a minimum number, or to use the hypothesis of a cylinder with several ellipses to generate the hypothesis of a cup.

Goals for the module can be entered into working memory as a three element list:

```
(goal  <name>  <priority>)
```

Goals can then have the effect of activating and disactivating demons. An example of a goal invoking a demon is the rule "cup-demons".

```
(defule  cup-demons "invoke the cylinder finder"
    (phase demons)
    (goal find-cup ?p)
=>
    (assert (demon cylinder-finder))
)
```

An example of removal of demons is the rule "remove-non-cup-demons".

```
(defule remove-non-cup-demons "invoke the cylinder finder"
    (phase demons)
    (goal find-cup ?p)
    ?d<-(demon ?n ?id)
    (not (?n cylinder-finder))
=>
    (retract ?d)
)
```

Having a rule interpreter provides explicit control knowledge for demons and their control logic. It also permits the working memory to be used to create and free working memory for representing demons state. The result is a flexible, easy to use tool for experiments in control of perception.

References

[Ayache-Faugeras 1987] N. Ayache, O. Faugeras, "Maintaining Representation of the Environment of a Mobile Robot". *Proceedings of the International Symposium on Robotics Research*, Santa Cruz, California, USA, August 1987.

[Blake-Zisserman 1987] A. Blake, A. Zisserman, *Visual Reconstruction*, Cambridge MA, MIT Press, 1987.

[Bossier-Demazure 1987] O. Boissier and Y. Demazeau, "A DAI View on General Purpose Vision Systems, *Decentralized AI 3*, Werner & Demazeau, ed., Elsevier-North-Holland, June 1992.

[Brooks 1985] R. Brooks, "Visual Map Making for a Mobile Robot", *Proc. 1985 IEEE International Conference on Robotics and Automation*, 1985.

[Brown et al 1989] C. Brown, and H. Durrant Whyte, J. Leonard and B. Y. S. Rao, "Centralized and Decentralized Kalman Filter Techniques for Tracking, Navigation and Control", DARPA Image Understanding Workshop, May 1989.

[Brownston et al. 1985] L. Brownston, R. Farrell, E. Kant, N. Martin, *Programming Expert Systems in OPS-5*, Addison Wesley, Reading Mass, 1985.

[Buchanon-Shortliffe 1984] B. Buchanan, E. Shortliffe, *Rule Based Expert Systems*, Addison Wesley, Reading Mass, 1984.

[Bucy 1959] R. Bucy, "Optimum finite filters for a special non-stationary class of inputs", Internal Rep. BBD-600, Applied Physics Laboratory, Johns Hopkins University, 1959.

[Bucy-Joseph 1968] R. Bucy, P. Joseph, *Filtering for Stochastic Processes, with applications to Guidance*, Interscience New York, 1968.

[Chatila-Laumond 1985] R. Chatila, J. P. Laumond, "Position Referencing and Consistent World Modeling for Mobile Robots", *Proc 2nd IEEE International Conf. on Robotics and Automation*, St. Louis, March 1985.

[Chehikian et al. 1990] A. Chehikian, S. Depaoli, and P. Stelmaszyk, "Real Time Token Tracker", EUSIPCO, Barcelona, Sept. 1990 .

[Crowley 1984] J. L. Crowley "A Computational Paradigm for 3-D Scene Analysis", *IEEE Conf. on Computer Vision, Representation and Control*, Annapolis, March 1984.

[Crowley 1985] J. L. Crowley, "Navigation for an Intelligent Mobile Robot", *IEEE Journal on Robotics and Automation*, 1 (1), March 1985.

[Crowley 1986] J. L. Crowley, "Representation and Maintenance of a Composite Surface Model", *IEEE International Conference on Robotics and Automation*, San Francisco, Cal., April, 1986.

[Crowley 1987] J. L. Crowley, "Coordination of Action and Perception in a Surveillance Robot", IEEE Expert, Novembre 1987 (also appeared in the 1987 International Joint Conference on Artificial Intelligence).

[Crowley 1988] J. L. Crowley, P. Stelmaszyk, C. Discours, "Measuring Image Flow by Tracking Edge-Lines", *Proc. 2nd International Conference on Computer Vision*, Tarpon Springs, Fla. 1988.

[Crowley 1989a] J. L. Crowley, "Dynamic Modeling of Free-Space for a Mobile Robot", 1989 IEEE Conference on Robotics and Automation, Scottsdale, April 1989.

[Crowley 1989b] J. L. Crowley, "Knowledge, Symbolic Reasoning and Perception", Proceedings of the IAS-2 Conference, Amsterdam, December, 1989.

[Crowley 1991] J. L. Crowley, "Towards Continously Operating Integrated Vision Systems for Robotics Applications", Scandinavian Conference on Image Analysis, August 1991.

[Crowley et al. 1991] J. L. Crowley, P. Bobet et K. Sarachik , "Dynamic World Modeling Using Vertical Line Stereo", *Journal of Robotics and Autonomous Systems*, Elsevier Press, June, 1991.

[Crowley et al. 1992] J. L. Crowley, P. Stelmaszyk, T. Skordas et P. Puget, "Measurement and Integration of 3-D Structures By Tracking Edge Lines", *International Journal of Computer Vision*, July 1992.

[Crowley-Ramparany 1987] J.L. Crowley, F. Ramparany, "Mathematical Tools for Manipulating Uncertainty in Perception", *AAAI Workshop on Spatial Reasoning and Multi-Sensor Fusion*, Kaufmann Press, October, 1987.

[Doyle 1979] J. Doyle, "A Truth Maintenance Systems", *Artificial Intelligence*, Vol 12(3), 1979.

[Duda et al 1976] R. Duda, R. Hart, N. Nilsson, "Subjective Bayesian Methods for Rule Based Inference Systems", *Proc. 1976 Nat. Computer Conference*, AFIPS, Vol 45, 1976.

[Durrant-Whyte 1987] H. Durrant-Whyte, "Consistent Integration and Propagation of Disparate Sensor Observations", *Int. Journal of Robotics Research*, Spring, 1987.

[Faugeras et al. 1986] O. Faugeras, N. Ayache, B. Faverjon, "Building Visual Maps by Combining Noisey Stereo Measurements", *IEEE International Conference on Robotics and Automation*, San Francisco, Cal., April, 1986.

[Forgy 1982] C. Forgy, "RETE: A Fast Algorithm for the Many Pattern Many Object Pattern Match Problem", *Artificial Intelligence*, 19(1), Sept. 1982.

[Genesereth-Nilsson-1987] M. R. Genesereth and N. Nilsson, *Logical Foundations of Artificial Intelligence*, Morgan Kaufmann Publishers, 1987.

[Hayes-Roth 1985] B. Hayes-Roth, "A Blackboard Architecture for Control", *Artificial Intelligence*, Vol 26, 1985.

[Herman-Kanade 1986] M. Herman, T. Kanade, "Incremental reconstruction of 3D scenes from multiple complex images", *Artificial Intelligence* vol-30, 1986, pp.289

[Hopfield 1982] J. Hopfield, "Neural Networks and physical systems with emergent collective computational abilities", *Proc. Natl. Acad. Sci.*, vol-79, USA, 1982, pp 2554-2558.

[Jazwinski 1970] J. Jazwinski, *Stochastic Processes and Filtering Theory*, Academic Press, New York, 1970.

[Kalman 1960] R. Kalman, "A new approach to Linear Filtering and Prediction Problems", *Transactions of the ASME*, Series D. J. Basic Eng., Vol 82, 1960.

[Kalman-Bucy 1961] R. Kalman, R. Bucy, "New Results in LInear Filtering and Prediction Theory", *Transaction of the ASME*, Series D. J. Basic Eng., Vol 83, 1961.

[Kant 1781] E. Kant, *Critique of Pure Reason*, Translated by N. Kemp Smith, New York Random House, 1958 (original work published in 1781).

[Koch 1985] C. Koch, J. Marroquin, A. Yuille, "Analog neural networks in early vision", AI Lab. Memo, N° 751, MIT Cambridge, Mass, 1985.

[Kolmogorov 1941] A. Kolmogorov, "Interpolation and Extrapolation of Stationary Random Sequences", *Bulletin of the Academy of Sciences of the USSR* Math. Series, VOl 5., 1941.

[Krammer-Siffling 1989] K. Brammer, G. Siffling, *Kalman Bucy Filters*, Artech House Inc., Norwood MA, USA, 1989.

[Li 1989a] S. Li, "Invariant surface segmentation through energy minimization with discontinuities", submitted to Intl. J. of Computer Vision, 1989.

[Li 1989b] S. Li, "A curve analysis approach to surface feature extraction from range image", *Proc. International Workshop on Machine Intelligence and Vision*, Tokyo, 1989.

[Matthies et al. 1987] L. Matthies, R. Szeliski, T. Kanade, "Kalman Filter-based Algorithms for Estimating Depth from Image Sequences", CMU Tech. Report, CMU-CS-87-185, December 1987.

[Melsa-sage 1971] A. Melsa, J. Sage, *Estimation Theory, with Applications to Communications and Control*, McGraw-Hill, New York, 1971.

[Poggio-Koch 1985] T. Poggio, C. Koch, "Ill-posed problems in early vision: from computational theory to analog networks", *Proc. R. Soc. London*, B-226, 1985, pp.303-323.

[Shafer 1976] G. A. Shafer, *A Mathematical Theory of Evidence*, Princeton N. J., Princeton University Press. 1976

[Smith-Cheeseman 1987] R. Smith, P. Cheeseman, The Estimation and Representation of Spatial Uncertainty", *International Journal of Robotics Research* 5 (4), Winter, 1987.

[Terzopoulos 1986] D. Terzopoulos, "Regularization of inverse problems involving discontinuities", *IEEE Trans PAMI*, 8, 1986, pp.129-139.

[Weiner 1949] R. Weiner, *Extrapolation, Interpolation and Smoothing of Stationary Time Series*, John Wiley and Sons, New York., 1949.

[Zadeh 1979] L. Zadeh, "A Theory of Approximate Reasoning", *Machine Intelligence*, J. E. Haynes, D. Mitchie and L. I. Mikulich, eds, John Wiley and Sons, NY, 1979.

3. KTH Software Integration in the VAP Project *

Lars Olsson and Harald Winroth

CVAP, KTH

We describe two tools for building continuously operating and distributed systems, developed for computer vision applications.

MNT is a toolkit for building computational networks, which consist of connected but independently executing agents called modules. Modules pass data to each other via communications channels called ports. The MNT communication is based on *XOR*, which simplifies the movement of objects between different address spaces via files, pipes or shared memory buffers. XOR provides a protocol by which objects can be described independently of the implementation language.

3.1 Introduction

The technical annex for ESPRIT Basic Research Action 3038 "Vision As Process" states:

> The consortium aims to demonstrate that the paradigm of *vision as process* is basic to the function of a high level vision system. Such an hypothesis can only be demonstrated within the context of a complete vision system, in which the potential benefits of continuous control of perception and of associated temporal context are evident. The goal of this project is to adapt and refine existing vision techniques for integration, and to combine them in a first step towards a general purpose vision system.

This paragraph points to the need to *integrate* existing pieces of software and hardware to build a complete *continuously operating* vision system. Early on in the project it was decided that the system should be structured as a set of independently executing computational agents which communicate by message passing. A number of tools supporting the construction of such systems were therefore developed within the consortium.

This article describes two software integration tools developed at KTH. They are both general enough to be used in a variety of contexts, although one of them was developed specifically for the VAP project.

3.1.1 Encapsulation and Code Reuse

These tools are intended to facilitate the integration of disparate pieces of software and hardware into complete systems. They address the classical problem of software *encapsulation*, i.e. how to create computational modules with well-defined interfaces and hidden internal data structures and algorithms. In the context of the VAP project, the question of encapsulation was particularly important. First, system components were implemented in parallel by different research groups, and in several programming languages (C, C++ and LISP). Second, as more specialized hardware

* The work has been perfomed within the ESPRIT-BRA project "Vision as Process, VAP" (BR-3038). The support from the Swedish National Board for Technical and Industrial Development, NUTEK is also gratefully acknowledged.

such as camera heads and array processors were integrated into the system, it became increasingly important that software components could easily be replaced by functionally equivalent hardware components.

Good encapsulation mechanisms are also likely to simplify *code reuse*. Although some software components are highly application specific, many of them are of a more general nature and can be used in a variety of computer vision systems. This applies in particular to filters for low-level image processing. With standardized components readily available in module libraries, system prototypes can be assembled quickly and with minimal programming effort.

There are two ways of achieving this encapsulation: *flow-oriented* programming and *object-oriented* programming. While flow-based systems, which consist of filters operating on data streams, seem to be the natural choice for computer vision applications, the data structures forming the streams can still be described in object-oriented terms. This will be discussed further in the next section.

Fast system prototyping requires that the components are *loosely coupled*. Since components have to communicate with each other, they must obviously be connected in some way, but they should *not* be required to know each other by name. Instead, the interface of a component should be limited to a specification of the input streams accepted and the output streams generated from the input data. If a component assumes no more than that it will receive certain inputs and that it should produce certain outputs, it may be used in many contexts and be connected to any other component handling the same type of data.

3.1.2 The Data Representation Problem

If systems are to be composed of a set of standard components, these components must be able to communicate with each other using a set of standard data structures. Ideally, such a set of data types would be small, yet support the widest possible range of applications. In practice, however, it is very hard to predict the requirements of future systems. The set of types will probably develop over a period of time and must therefore be extendible in two ways: First, as new components are implemented it may be necessary to add completely new data types to the "vocabulary" used in the system. Second, it should be possible to extend existing data types by creating *sub-types*. For example, a simple geometrical object such as an edge may represent an intensity discontinuity, the border of a textured region, or the co-linear alignment of other features. Many computational modules will deal only with the geometrical properties (orientation, length etc.) of the edges, while more specialized modules might also need, for instance, the contrast of intensity edges, or the list of co-linear features forming apparent edges. The inclusion of such components requires more detailed information to be transferred between different parts of the system. If existing data structures can be extended through sub-typing, the communication interfaces of the new components will be backward compatible with the existing ones.

One obvious solution is to adopt an *object-oriented* communication scheme, which involves sending objects belonging to well-defined classes through the network. The standard mechanisms that can be found in most object-oriented systems and languages, such as classes, methods, inheritance and polymorphism, have long been used to support data encapsulation, sub-typing and code reuse[Bla91].

3.1.3 Message Passing Systems and Computational Networks

In many computer vision systems it is possible and desirable to process data in *parallel*. In binocular vision, for instance, images from the left and right cameras can be pre-processed independently. Furthermore, active vision systems will certainly contain *feedback loops*, for example where a high-level component controls the computation in low level layers based on data it has previously received from them. In general, it is necessary to allow for the coexistence of data driven low-level processing and goal driven high-level processing.

Synchronization is an important problem here. In a complex system with special hardware and external agents it is difficult to predict when data will arrive and when requests will be processed. A natural way to handle this is to construct systems made up of asynchronously executing components which transfer data and synchronize by sending *messages* to each other.

If components communicate by message passing it is feasible to implement components in different programming languages, provided that all messages are represented in the same way. Furthermore, by making use of the networking capabilities of modern workstations it is possible to distribute the components over a network of computers.

There are several ways in which components could communicate. One approach is to use a central *mailbox*, which is made available to all parts of the system. Any component can drop messages addressed to any other component into this mailbox. However, the sender usually has to know the name of the receiver, which violates our requirement that components should be loosely coupled.

Instead, we advocate the *computational network* approach in which components are equipped with a fixed set of input and output ports which may be connected to ports of other components. A computational network can be pictured as a graph where the nodes are asynchronously executing components and the arcs are the point-to-point connections between them. When a component sends a message, it is transferred to all connected components. Figure 3.1 shows a simple computational network for stereo matching. We have developed two software integration tools. XOR (eXternal

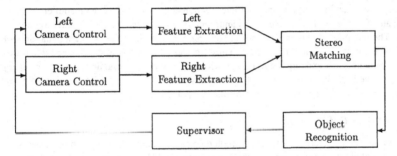

Fig. 3.1. A simple computational network for stereo matching. Image features, such as edges, are extracted from the input images. Features from the left and right channels are combined by the stereo matching module and are used for indexing a database of object descriptions. The supervisor then tries to track the target object using the camera controls.

Object Representation) is a package which addresses the data representation problem. It allows programmers to transfer a single object or a cluster of related objects from one component to another. Module Network Tool (MNT) addresses the problems of how to construct components and systems[Ols92]. MNT provides an API (Application Programming Interface) for constructing software components called modules, and programs for building computational networks from such modules.

3.2 XOR

XOR deals with the problem of moving objects between address spaces. The idea is to describe the information contained in the objects using a simple protocol. XOR goes beyond the more conventional approaches of implementing object persistence in that it is intended to make object descriptions *host language independent*.

3.2.1 XOR Protocol

The XOR protocol can be seen as an extension of XDR (External Data Representation), which is a DARPA standard for data encoding. The XDR protocol is defined as a data declaration language, similar in flavor to C. The primitives of this language are simple data types such as integers, floating point numbers and strings. With those primitives and some compound types (struct), more complex data structures can be described. XOR extends the XDR protocol with class references and references between objects. XDR defines a canonical file format, using e.g. IEEE floating point encoding and four-byte integers stored with the most significant byte first. This format is sufficient for XOR, which uses pure XDR format files.

By separating the XOR protocol, which defines *what* can be stored, from the physical *file format*, which determines *how* data is actually encoded, objects can be moved between architectures with different representation of integers and floating point numbers. More importantly, XOR's protocol based approach makes the external representation of objects independent of the specific mechanisms that implements them in a particular programming language. It is thus possible to transfer objects freely between components written in different languages.

3.2.2 Class Tags

The first and most important problem with persistent objects is to represent their class membership externally in an unambiguous and space-efficient way, using some sort of class identifiers. A program reading an XOR file has to map the class identifiers to suitable constructors in order to recreate the objects. The mapping of class tags to object constructors is implemented using a dynamic look-up table, which can be extended at run-time. Thus, no source code has to be changed when derived classes are created, and existing programs can without recompilation handle files containing objects of newly defined subclasses. Of course, the programs have to be linked with the code implementing the new classes, but this can be done at run-time using a dynamic linker.

3.2.3 Object References

Objects often contain references, typically implemented by address pointers, to other objects. These references contain important structural information and should be recreated automatically in the new address space. One can not assume that these structures are hierarchical or even acyclic.

Since pointers become invalid when objects are moved to a new address space, the pointers have to be translated into some sort of object identifiers externally. The problem is that it is impossible to give every object in the world a unique identity (even if it were feasible, it would be very hard to *find* an object, given its ID). XOR introduces the concept of *packets*, which limit the universe of object identifiers; only objects that are put into the the same packet (by the application program) are allowed to reference each other by addresses. Therefore, objects can be identified by a local index inside the packet. Packets are opaque on disk, and their contents cannot be changed unless they are read into workspace memory, so that all inter-object pointers can be resolved. XOR uses various compression techniques to reduce the space occupied by object references.

3.2.4 Streams

The actual communication mechanism is encapsulated in *stream* objects. The XOR system interacts with the stream objects through the interface defined in the stream base class. The actual stream operations are implemented differently for files, sockets, clipboards, etc. in a set of stream subclasses.

3.2.5   A Sample XOR Object Definition

To make things more concrete, this section demonstrates how XOR can be used in C++ programs. The C++ interface requires that all objects handled by XOR belong to a class derived from the abstract Xor base class. The objects to be transferred are described in a set of *data description function* calls made in special XOR constructors. Consider an abstract Picture class with two derived classes: Pixmap which contains actual pixels, and Window which is a logical subwindow of another Picture.

```
class Picture : public Xor
{
    XOR_DECLARE(Picture, Xor); // declarations needed by XOR
    unsigned int x_size, y_size;
public:
    Picture(XorPacket& p) : Xor(p) { p ^ x_size ^ y_size; }
};

class Pixmap: public Picture
{
    XOR_DECLARE(Pixmap, Picture);
    char *pixels;

public:
    Pixmap(XorPacket& p) : Xor(p)
    {
        unsigned int size = x_size * y_size;
        p ^ xor_buf(pixels, size, x_size * y_size);
    }
};

class Window : public Picture
{
    XOR_DECLARE(Window, Picture);
    Picture *father;
    unsigned int x_offset, y_offset;

public:
    Window(XorPacket& p) : Xor(p)
    {
        p ^ father ^ x_offset ^ y_offset; }
    }
};
```

The XOR constructors (the ones with XorPacket parameters) are used for both encoding and decoding objects. They contain calls to the C++ xor operator (^), which performs the actual I/O. Space for the pixels field in Pixmap will be allocated automatically by XOR during decoding. The last argument to xor_buf is a maximum size, which is used for error checking. When a window is decoded its father pointer will be resolved when the referenced Picture object has been recreated. Also note that if a Window object is transferred with XOR, the whole chain of sub-windows will be restored in the new address space. For simplicity, a few details have been left out in this example. In particular, we would need to declare virtual pixel access functions in the Picture base class.

3.3 MNT

MNT is a tool for constructing computational networks. It supports construction of the network building blocks, the modules, and allows them to be combined into networks.

We describe a SunOS implementation of MNT where modules are independent programs which are executed as separate processes, and where port connections are based on UNIX sockets. The implementation includes a C++ module construction API. A module consists of an application core surrounded by a system layer which handles module activity scheduling and inter-process communication.

3.3.1 Modules and Ports

A module is an independently executing process with a fixed (but arbitrary) number of typed input/output ports. A module consumes data items arriving on input ports and produces derived data items on its output ports. A module can be stateless, which means that the output stream is as a function of the current module inputs, or it can maintain an internal state over time. When a module is included in a network its ports are connected to ports of other modules, so that the network becomes a graph where nodes are modules, and arcs are port-to-port connections. Synchronization between modules is achieved through data transfers; a module is idle until data arrives on a port, at which time it wakes up, fetches and processes the data, and writes output data to the output ports before going back to sleep. A module may also execute code in the background when no input is pending.

A module program consists of a C++ *module object*, which belongs to a subclass of the MNT module base class, and a main(), which gives the program an entry point. The module base class represents an "empty" module and defines the basic properties shared by all modules. In particular, it declares virtual functions for handling input events and background processing. These functions are redefined in each module subclass to do application specific processing.

A module object always contain two system objects, a *scheduler* and a *server*. The scheduler object is responsible for module activity scheduling. It divides the available time between input handling and background processing. The server object handles inter-process communication and detects new input arriving to input ports.

In addition, each port is represented by a *port object* member of the module, see Figure 3.2. All communication with the outside world goes through these port objects, which hide most of the complexity of data transfer between modules.

Ports are typed, where the port type determines the data transfer protocol used by the port. MNT provides an port base class which contains the functionality common to all port objects, regardless of data transfer protocols. Specific protocols are supported by port subclasses. Besides the port base class, MntPort, two additional port subclasses are provided, one for raw binary data communication and another for object-oriented communication using the XOR protocol (see Figure 3.3).

Since each application specific module is defined as a new module subclass, the class declaration serves as a module specification as well. Usually, a class declaration is placed in an header file and it is often sufficient to look at this file to find out what the module is supposed to do.

3.3.2 A Sample MNT Module

Consider a simple filter module with ports *in* and *out*, which handle object transfer using the XOR protocol. Such a module could be declared:

```
class MyModule : public MntModule
{
```

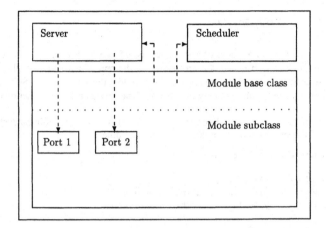

Fig. 3.2. Each module contains a server, a scheduler, and a set of port objects.

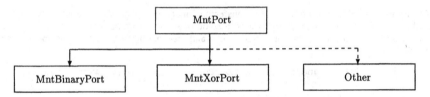

Fig. 3.3. Port class hierarchy.

```
private:
  MntXorPort in_port;
  MntXorPort out_port;
  // Other private data and functions

public:
            MyModule::MyModule(MntScheduler &,MntServer &);
  void      HandleInput(MntPort *);
  void      Execute(void);
};
```

Class `MyModule` is a subclass of the module base class `MntModule`. Its instances will contain two XOR ports and some private data. The constructor contains initialization code, which among other things registers the port objects with the server object. The `HandleInput` function is called when data have arrived on an input port. The `Execute()` function is called whenever the scheduler has decided it is appropriate to perform background computations.

The `MyModule` constructor should register its ports with the server object and perform other necessary initializations:

```
MyModule::MyModule(MntScheduler &scheduler,MntServer &server)
: MntModule(scheduler,server),
  in_port("in",this,this),
  out_port("out",this,this)
```

```
{
  // other initializations
}
```

The port constructor expects a number of arguments, including the external name of the port, the port input handler, and the port *owner*. Before returning, the port constructor asks the port owner, which in this case is the module object, to register the port with the server object. Subsequent input events will then be detected by the server, which will call the HandleInput function of the input handler object.

Finally, the application part (a module subclass instance) is combined with the system objects (the scheduler and the server) into a complete executable module:

```
int main(int argc,char *argv[])
{
  MntScheduler scheduler;
  MntServer    server(argc,argv,scheduler);
  MyModule     my_module(scheduler,server);

  scheduler.Dispatch();
  return 0;
}
```

MNT provides two port subclasses, one for transferring binary data, and one for transferring objects according to the XOR protocol. The preferred method of sending data between modules is to use the XOR port. This port class has functions to send and receive objects which are instances of an Xor subclass:

```
void MyModule::HandleInput(MntXorPort *in_port,MntXorPort *out_port)
{
  Xor *in_object  = in_port->Read();
  Xor *out_object = Compute(in_object);
  out_port->Write(out_object);
  delete in_object;
  delete out_object;
}
```

3.3.3   Networks

Once the modules have been implemented and placed in libraries, they can be used as building blocks in networks. A library module is actually a module *class*, and a network may contain any number of instances that class.

MNT provides a launcher program which given a network specification can build and launch a complete network. The network specification tells the launcher which module instances to launch, where to launch them, and how they should be connected. It may also specify other properties of the module instances in the network. The standard MNT launcher reads network specifications written in a simple specification language (see Figure 3.4) but MNT can also support other types of launchers, such as graphical network editors which allow users to create networks interactively.

In a network description, each module is characterized by its class, its instance name, and a set of ports. Each port also has a name, which is local to the module, and a type which determines the data transfer protocol supported by the port. Since ports are typed objects, the launcher can check that only ports with matching types are connected to each other.

A port may be connected to any number of ports of other module instances. Port connections allow for two-way data transfer and it is possible to send data to multiple receivers, as well as receiving data from multiple senders. A module need not be concerned with the number of modules

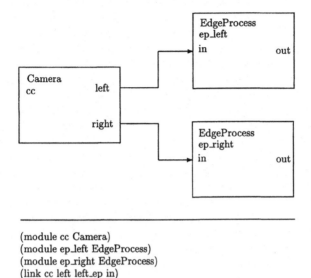

```
(module cc Camera)
(module ep_left EdgeProcess)
(module ep_right EdgeProcess)
(link cc left left_ep in)
(link cc right right_ep in)
```

Fig. 3.4. The launcher creates networks from specifications written in a simple description language.

its ports are connected to. When data are sent from an output port, the port object copies the data to every connected module, and when multiple senders send data to an input port, the data streams are merged before they are made available to the module. Exactly how data streams are merged is determined by the stream protocol.

When a module fetches input data from one of its input ports, it may optionally retrieve an associated *connection descriptor* identifying the source of the message. This descriptor can be used when replies are sent back from the same port. If such a connection descriptor is specified together with the reply, the output data will be sent only to the corresponding module instance, and not than to every connected module instance. Note that since the connection descriptor is an opaque handle, it will not give the modules specific knowledge about the existence and identity of other modules in the system. Connection descriptors are primarily intended for client/server applications.

References

[Bla91] Gordon Blair. *Object-oriented languages, systems and applications.* Pitman, 1991.
[Ols92] Lars Olsson. Mnt programmers guide. Forthcoming CVAP technical note, 1992.

4. A Vision Programmers Workbench

Niels O. S. Kirkeby and Henrik I. Christensen
LIA, AUC

4.1 Introduction

In construction of experimental machine vision systems two different approaches may be adopted a) a specific/dedicated approach or b) a modular approach. In the first approach each new system is designed basically from scratch, while the other approach assumes that systems may be constructed through integration of a set of standard modules.

In design of industrial systems demanding speed requirements are often present. To satisfy such constraints it is usually necessary to provide dedicated implementations to ensure a reasonable cost/performance ratio. For design and implementation of prototype systems it is, however, desirable to adopt a modular approach that allow recycling of modules already available from earlier work or from efforts elsewhere. This is by no means a new idea, the concept of re-use of modules in well known in both machine vision and other software efforts. Most efforts which have used such an approach have, however, not reported on the architectural requirements which has formed the basis for this effort. Notable exceptions are the Khoros system [Rasure, 1992] and the VISIONs system [Draper et al., 1988].

In the design and use of a modular system the component modules may be either dedicated or general purpose depending on the system being constructed. If several different systems are constructed, a set of general modules for e.g., acquisition of images and control of a binocular camera head may be used in all the systems. To accommodate re-use of modules it is necessary to have a standard for writing and use of such modules.

In 1988 a consortium consisting of three Danish laboratories (Aalborg University, Copenhagen University & The Danish Technical University) decided to initiate a joint effort to specify, design and implement an architecture that would allow sets of modules to be interconnected to form machine vision and image processing systems. The objectives for the effort were:

Design of a testbed that would allow simple construction of application systems

The design should allow reuse of as many modules as possible

The architecture should allow use of existing software to the extent possible

Distributed processing using a cluster of interconnected workstations should be possible (interconnection is achieved through use of standard Ethernet with associated protocols)

Easy reconfiguration should be possible

From these objectives a software architecture has been designed and implemented. the system has evolved from an initial revision 1.0 which only implemented the basic functionality of communication between modules to the present version 4.0 which is described in this paper. The effort has been carried out in parallel to several other efforts and there has been a continued communication with the SAVA and MNT efforts reported in the volume. The exchange of information between these efforts has enhanced the design in several ways.

4.2. Outline Of Vipwob Facilities

4.2.1 System Architectures

In the construction of image analysis and computer vision systems are large variety of different system architectures is employed. The most tradition architectures are: pipeline and blackboard. The pipeline structure is well known as it corresponds tot he situation where a user applied one operator after another either manually or in a script file. The blackboard structure where a number of different modules may perform analysis based on the same sets of data is also well known, in particular after the VISIONS group in Amhurst, Mass presented its utility for construction of large systems [Draper, 1988]. Recent development in commercial systems have indicated the utility of construction of systems as a directed graph structure, where nodes are processing modules while arcs represent communication links between such modules. Examples of such commercial systems are AVS™ [AVS, 1990] and Explorer™ [SGI, 1990]. The methods presented above emphasise the breakdown of systems into processing modules interconnected by communication media. The architecture of the system is consequently made explicit. Only in the blackboard architecture is the data representation made explicit, in the remaining models there is no indication of use of common data-representations. The advantage or disadvantage of this may vary depending on the approaches adopted. In the initial design of the VIPWOB system the primary focus was interconnection of both existing and new data processing modules and use of standard representations to be designed with the system was thus not possible and consequently the directed graph approach was adopted as the kind of architectures that it should be possible to implement in VIPWOB. This choice was made as it imposes a minimum of structure on application programmes; i.e., it allow implementation of standard architectures like pipelines, and the blackboard (through use of a common database server). The different architectures are shown in figure 4.1.

In practise the system should thus be able to handle multiple inputs and outputs any output should be able to replicate it data to an arbitrary number of inputs. The distribution of data to from multiple channels is a task which should be transparent to the application programmer and it must consequently be part of the basis software.

For some applications the computing power of a single workstation is insufficient and it is thus desirable to be able to distribute the set of modules over a number of workstations. In other applications special hardware (i.e., a computer with dedicated image processing hardware) is needed and it is here desirable to run modules which exploit such hardware on that particular machine while other processes may be executed on other computers. The system should thus support execution of modules on multiple computers. Initially it has been decided that such computers should be interconnected by an Ethernet network, but facilities has been provided to allow communication using shared memory if needed.

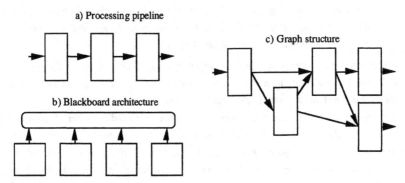

Fig. 4.1. Various vision system architectures that should be supported by standard environments for experimental research

The approach adopted here is thus similar to that used in systems such as Khoros (Rasure, 1989), AVS (AVS, 1989), Explorer (SGI, 1991), and MNT (Olsson, 1992). In the approach used here it is, however, possible to have multiple connections to one channel, a feature which is not present in any of the other systems. The adopted approach also different from the other category of approaches where fixed system architectures are used. An example is the SAVA system described by Crowley et al. (1991). Such dedicated systems are often more efficient, but through careful design it is here shown that comparable performance may be obtained with general purpose architectures, at least when performance is measured in terms of communication bandwidth.

In system architectures such as the one shown in figure 1c a disadvantage is that all the necessary communication links must be defined explicitly, which in practice may results in an almost fully connected graph. To avoid this problem an approach as shown in figure 1b (the blackboard) may be adopted. In practise speed considerations will often argue for a mixed architecture. It is thus desirable to have an architecture which support both explicit (highly efficient) communication channels and use of centralised (possibly low bandwidth) communication channels that may be established dynamically. VIPWOB support such mixed architectures through use of a variety of different communication facilities. The variety of these are described in the next section.

4.2.2 System configurations

The set of tasks that may be carried out in a computer vision system is vast and it is consequently necessary to provide a rich set of communication facilities. Communication between sets of modules may be divided into the following categories:

Point to point data flow communication

Client-server communication

Explicit module synchronisation

Sharing of data structures between modules (blackboard type of structures)

The point to point communication is the most frequently used communication type. It provides a channel for sequential transfer of data from one module to another. The channel is memoryless and data are transferred in the order the are written to the channel.

In the client server communication a module (the client) requests data from another module (the server) and once data are available a reply with the data is returned from the server to the client. Communication is thus two-way and might be asynchronous. The term asynchronous indicates here that several requests might be submitted before replies are returned and the order of replies may be different from the order of requests. The client server communication models may for example be used for inquires about module parameters, setting of thresholds in response to the modality of the histogram, selection of groupings for verification of the presence of an object, etc. In this type of communication the request is typically a rather limited, in size, specification, while the response may be messages of arbitrary size.

The explicit synchronisation of modules is included here as it often is needed in systems which does analysis of sequences of images. Explicit synchronisation may for example be used for communication of the fact that a new image has been acquired, that processing of something has been completed or simply to synchronise for example the left and the right modules in a stereo reconstruction task. Another example of synchronisation is in the use of special hardware such a motorised lenses, where the end of motion may be signalled to a module. In principle signals may be viewed as messages with a one byte message (the type of the signal), but in contrast to messages which typically are queued for later processing, the signal is attended to as soon as it is received (unless the signal handler is disabled) and it has consequently a higher priority in processing than regular inter-module communication. A problem is distributed computing is signalling across different processors. In section 4.4 it is outlined how this problem has been solved in VIPWOB.

The last category of communication methods is used of shared data structures. In some tasks it is inconvenient to use a sequential communication method, as the receiver might need to access data in an order which is different from the one produced by the transmitter. For such situations it is convenient to have a common area which allow temporary storage of data and facilities that allow random access to this area. An example where such access is used is in labelling of images, in grouping of line segments, matching of information over time etc. In traditional multi-process systems such data sharing is achieved through use of shared memory. Use of this techniques does however assume that all processes have access to the same physical memory, which rarely is the case in a distributed system. It is consequently necessary to provide special facilities that allow simulation of shared memory (in computer science this is often referred to a distributed shared memory). In section 4.4 it is described how this facility has been provided in VIPWOB.

4.3. The Module Interface

4.3.1 The Module Context

A module is in principle any part of a image processing or computer vision system, ranging from a single algorithm for convolution of an image to a set of methods for knowledge based interpretation of a set of features. As a modular approach is used it is desirable to be able to use modules in a variety of systems. This implies that the interface to a module should be general.

For some modules it might be preferable that they may be executed both as individual modules and as part of the VIPWOB system. When a module is executed as an individual programme (i.e., a programme for thresholding, line extraction, geometric matching etc.) the communication channels should be mapped to regular files. I.e., it should appear as a regular programme.

For an application programmer it is important that issues related to communication to other modules is hidden. Preferably the communication should be emulated through use of functions which are similar to those used for traditional file based I/O. From a programmers point of view an interface as shown in figure 4.2 is ideal.

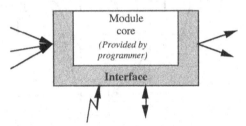

Fig. 4.2. The relation between module (core) software and the rest of the system.

Using this approach all interfacing involving set-up of the system, distribution of data to channels, handling of signals, access to shared information etc. is handled by the interface, and the programmer uses functions calls from and application programmers interface (API).

To enable use of existing source code, to the extent possible, it has been chosen to overload the systems calls for *read, write, open* and *close*. This implies that programmes may use the stdio library for I/O and the underlying operations will be hidden to the user.

For initiation of processing and selection of a mode of operation (in VIPWOB or outside) a single function call is included (VWinit). This function call must be performed before any I/O is performed. The call will return true is the module is executed as part of a VIPWOB application and it will return false otherwise. If different actions are needed in the two modes of operation such changes may be implemented through use of the return value from VWinit. For existing source code which only exploits the STDIO functions for I/O is thus only necessary to include the VWinit call and recompile to allow the module to be used together with VIPWOB. For more advanced application detailed control of the methods of communication may be utilised but this will of course imply the module will not be able to operate as a regular programme outside VIPWOB. Use of such facilities may be implemented both explicitly in the source code or at run-time. The explicit control of communication channels is described in the next section.

4.3.2 Communication channels

As mentioned in section 4.2.2 communication between modules may be performed using both a point-to-point and client-server models of operation. For both kinds of communication the link between a transmitting and a receiving module may implemented as an explicit channel (similar to sockets under UNIX) or it may be implemented as an implicit channel where data is transferred by an unknown, to the application programmer, method. Consider again the application as a set of modules where explicit communication may be described in terms of a directed graph, with a single component. As all modules in the application are connected it is possible to transfer data, in implicit channels, through routing through modules interconnected by explicit modules. Consider the application in figure 4.3.

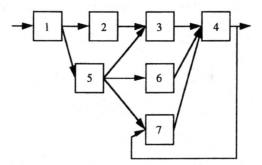

Fig. 4.3. A sample architecture used to explain the difference between explicit and implicit communication. (See text for details)

In the architecture in figure 4.3 it is possible to send messages from module 1 to module 7 by using module 5 as a routing node, which simply used module 5 as a communication agent. In practical applications it is of course necessary to consider issues such as distribution onto different computers in the routing of messages. The specific implementation of this method is described in section 4.4.

In VIPWOB terminology the two types of communication channels (explicit and implicit) are termed *high bandwidth* and *low bandwidth* channels. The names indicate also that care has been taken to implement the explicit channels with a maximum of efficiency while the implicit channels provide a means of communication which is inexpensive but also less efficient. The implicit channel is applicable for non critical communication.

The interface used for memory-less channels (point-to-point and client-server) are, as already mentioned, implemented through overloading of the function calls *open, read, write,* and *close* . The syntax adopted for read, write and close is the same as the one used in standard Kernighan - Ritche C [Kernighan-Richie 1986]. In a similar fashion the user might use buffer library calls, i.e., the syntax for fopen is *FILE* fopen(%vw%output, r)* where the name refers to a vipwob channel through use of the identifier %vw%. There are also decicated i/o library cals in VIPWOB. The library call for opening of a channel is VWOpenChannel(name,type, dir,fp), where name denotes a symbolic reference to the channel, type indicates the underlying massge passing protocol (f is a file, i an internet socket, and u a UNIX socket) and dir is the direction of the channel (r is read, w is write and b is bidirectional).

For the application programmer channels will thus appear as file descriptors or file pointers depending on the set of functions used for the interface (i.e., raw or buffer based communication).

In practice the communication between modules is implemented in terms of UNIX or TCP/IP sockets, depending on the location of the modules at each end of a channel. If both modules connected to a channel reside on the same computer a UNIX socket is used, while communication between modules that are executed on different computer are interconnected by TCP/IP sockets. All communication over channels are, in principle, full duplex, but an application may choose to exploit the channels in simplex mode only.

4.3.3 Access to Common Information

In VIPWOB it is possible to share information between modules using two different mechanisms. If the modules reside within the same physical memory space they may exploit ordinary shared memory facilities provided by the operating system. Such use of shared memory is for example convenient when accessing a frame grabber. If the modules may reside on different processors a simple sharing of information is available in VIPWOB. This sharing of information is based on use of named variables.

In VIPWOB it is possible to define variables. Such variables may be used for storage for ASCII information in the form of a string. I.e., if the variable should contain a number, say the present frame number, then it must be converted to string form before it is stored. The choice of ASCII as the basis for storage of information was chosen to avoid conversion of number according to the format used by different computers (BIG-ENDIAN or SMALL-ENDIAN) in a multi-processor set-up.

For variables it is possible to create, delete and monitor access. A variables is created through a *set* system call, while the value is retrieved using a *get* system call. For a variable it is possible to lock it is that no other module may change its value. Used in this fashion it might act as a semaphore. In addition it is possible for a module to monitor access (reading, change, deletion, or creation) of a variable. These monitoring facilities may be used for simple inter process communication, which among other things can be used for synchronisation.

There is presently no limit to the allowed size of the information that may be stored with any variable. The only limit imposed is the one imposed by the memory available on the processor where the module is executed.

4.3.4 Signalling

For synchronisation of processors it is common to use signals, at least in real-time programming. Use of signals does, however, represent a problem in a multi-processor environment as signals typically are implemented in term of interrupting facilities associated with individual processors.

In VIPWOB use of signals across processors has been implemented in terms of communication links. To the individual modules the use of signals is similar to the facilities provided by the UNIX operating system.

This implies that there is a system call for assertion of a signal. The function call has two parameters: a destination (the address of another module) and the type of the signal (a number). In addition there is a system call for installation of signal handlers. Installed signal handlers are activated whenever a signal, of the type specified at the time of installation, is received.

The implementation of signalling is carried out within a processor using the signalling facilities provided the operating system, while signalling across processors uses message passing. The implementation is described in [Kirkeby-Christensen, 1993].

4.4. Putting it All Together

4.4.1 Building Systems

Systems may be viewed as a collection of modules that are interconnected by channels. The architecture may be described as a directed graph, where nodes represent modules and arcs represent communication channels. As the *open* system call is available the graph structure may be changed at run-time. At the time when the system was designed it was imaged that the system initially would be constructed as a minimum graph where each module only is connected to one common module. Once modules begin to communicate the statistics for the data flow is monitored and if the data flow above a certain threshold the implicit communication between two modules is replaced by an explicit channels. This implies that the architecture will evolve from a minimum architecture to an architecture where all high bandwidth channels are implemented as explicit links while all other channels are implemented as implicit channels. In the implemented system it has however be left for the application programmer to choose a model for communication and if statistics should be used to change the architecture at run time such changes must be implemented by software provided by the application programmer.

The definition of a system in VIPWOB is performed through a configuration file. One might argue that the graphical interface provided by systems such as Khoros [Rasure, 1989], AVS [AVS, 1989] and Explorer [SGI, 1991] are better as the provide an interface which is easier to understand, but from our point of view such interfaces may be provided as an extension to VIPWOB, and in particular for industrial applications such interfaces are a luxury.

In the definition of an application both nodes and arcs in the graph structure must be provided. Nodes, which represent modules, are specified as shown below:

```
module <name> {
        prog = <path to binary file>
        host = <user>@<host>
        options=option string
}
```

A number of other specifiers may also be provided, as described in (Kirkeby-Christensen, 1993). Of the specifiers outlined above only the field for *name* and *prog* must be provided the other fields are optional. If no host is specified it is assumed the module should be executed on the local host.

For the specification of communications channels (arcs in the graph) a structure as shown below is used.

```
channel { <module name>:<fid> -> <module name>:<fid>}
```

The <module name> specifies the nodes which the channel is connected to, while <fid> specifies the file descriptors the channel is associated with. An optional field *type* specify the type of the channel (unix-socket, internet, ...). The channel specification is, of course, only used for specification of channels to be provided when the system is initialised. The architecture, may as already outlined, be changed at run-time.

Examples of configuration files for full systems are provided in section 4.5 to illustrate use of the system.

4.4.2 The Shell

The configuration file outlined above must be interpreted before it is possible to execute a set of modules as a complete system. To handle this a special module called a *vwshell* has been implemented. The shell uses a configuration file as inputs and executes all the modules specified in the file and set-up the set of communication channels specified. Once these operations have been performed the shell accept input from a user terminal. This allow interactive change of the set up, which in particular is useful during debugging and in interactive applications, where it might be necessary to change options/parameters for individual modules. The shell provides also a convenient mechanism for shutdown of a running system as it not only will terminate all modules but also perform garbage collection and clean-up of various administrative system data.

In distributed applications the redirection of error output is typically a problem. Often there are no unique location for storage of such output. in VIPWOB the systems has been designed so that all output from the *stderr* out is directed to the shell. The shell will thus collect all error output and present it to the user. Alternatively the user might redirect error output from the shell and be assured that all information output by other modules to *stderr* will also be redirected to the same location.

4.4.3 The Master and Slave Servers

For execution of modules and coordination of information about processes and shared information a set of server processes are used. When the system is activated the information in the configuration file is read by the shell and requests for activation of modules is passed onto a master server. The master server determines which processor the module should be executed on. If the module is to be executed on a machine where there no modules have been activated on earlier a slave server is initially activated. Slave servers and the master server communicated through a communication channel. Over this channel the slave servers are requested to fork modules and monitor their execution. I.e. slave servers are notified when a module terminates. To distinguish between different modules all modules are assigned an address. The master server has address 0.0 while slave servers have address which is the number of the host (enumerated from 1 and up) while the minor number always is 0 (i.e., they are numbered x.0). A module on a particular processor is assigned the next free number from 1 and up. By using such a scheme each module is assigned a unique address which and there is a one-to-one mapping from symbolic names to numeric addresses.

Storage of shared information by means of named variables is performed through allocation of space in the master server process. This does of course imply that all queries to shared information must be routed to the master server. This is, however, transparent to modules as a query simply is forwarded to the local slave server which in turn routes the request to the master server. The central storage of shared information does, however, represent a bottle neck in terms of bandwidth and it might also be slow. For applications where response time is highly critical it is thus recommended that special modules that allow dedicated sharing of information are used. Alternative one might use distributed shared memory. An example of use of shared memory has been described by [Kristensen, 1993].

The routing of messages between different modules is also handled by slave servers. The set of slave servers represent a set of fully connected processes. In an earlier version of the system all messages were routed through the master server, but this is clearly a non-optimal architecture and has consequently been changed to allow routing using direct connections. Communication through slave servers are, however, only used for low bandwidth channels, as all explicitly defined channels are implemented as UNIX or TCP sockets.

Signalling is implemented through the slave servers. When a module wants to signal another module it forwards the request to the local slave server. If the module to be signals is on the same processor the slave server simply signals the process using function calls in the operating system. If the module is on another processor the signal is routed to the slave server on that particular processor. For the mapping of names to addresses the slave server may have to query the master server, but if the destination is given as a numeric address rather than a symbolic module name the routing is carried out directly.

4.5. Experimental Evaluation

The preeceding sections have outlined the facilities available in the VIPWOB system. To appreciate the facilities available an sample application will be outlined. The system architecture is shown in figure 4.4.

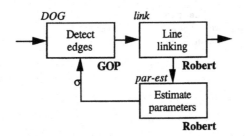

Fig. 4.4. Example system used to demonstrate the operation of VIPWOB

The system in figure 4.4 is used for adaptive line detection. In the sense that images initially are subjected to the a Difference of Gaussian filter with subsequent line linking. depending on the quality of the extracted lines the size of the filter may be changed by the parameter estimation module. The modules are executed with the edge detection module on a special computer named GOP while the two other modules are executed on the computer named Robert (a computer for mobile robotics). The name for the executeable programmes are indicated in italics in figure 4.4. For the above system the configuration file is shown below.

```
module detect_edges {
    prog = /home/staff/hic/bin/gop/DOG
    host = hic@gop.vision.auc.dk
}

module line_linking {
    prog = /home/staff/hic/bin/m68k/link
    host = robert.vision.auc.dk
}

module estimate_parameters {
    prog = /home/staff/hic/bin/m68k/par-est
    host = robert.vision.auc.dk
    options = -dog -vw
}

channel{ detect_edges:3 -> line_linking:4 }

channel{ line_linking:5 -> estimate_parameters:0 }
```

The communication channel between the parameter estimation module and the edge detection module is not specified in the configuration file as it is implemented as a massgae passing channel, where a new massge is sent from one module to the other whenever the is a need for change of the filtering parameter σ.

The modules named in figure 4.4 all have the statement VWInit() at the beginning of the programme and knows that I/O will be performed on the file descriptors mentioned in the configuration file. In the edge detection module (DOG) a message handler is set-up when the program is initiated. This is performed using software as outlined below

```
/* the function which takes care of input */
int MsgHndl(Messageheader *mesg, byte *data){
    /* read msg and change s */
    ......
    return(1);
}

main(){
    .......
    VWInit();
    /* Specify that messages should be routed through MsgHndl() */
    VWIntercept(MsgHndl,SIGUSR2);
    .........
}
```

In the module *par-est* the function VWResolve is used to get hold of the address of the module *detect_edges* and whenever a change in σ is needed the function VWSendMessage() is called which will transfer the message and notify the receiver that a message has arrived.

To determine the performance of the VIPWOB system for usage with problems as outlined above a set of measurements has been carried out. All tests were carried out on SiliconGraphics Indigo workstations interconnected on an ethernet segment.

The experiments included time for transfer of data over a dedicated channel between local and remote modules, and use of message passing for doing the same. In addition polling of the common database was measured. The results are shown below in table 4.1. For eahc situation a linear regression over transmission time has been estimated. For all circumstances the fit is better the 0.95. The measurements was carried out for packages ranging from $\frac{1}{8}$kB to 64 kB. The results are averages for transfer of 1000 packages and the timing includes only system and user time as the real time is depending on the load of the actual machines used.

The above mentioned performance figures indicate that processes in theory might transfer 4MB/s but due to the bandwidth limitation of the Ethernet (10Mb/s) such a performance is not obtainable. By monitoring the ethernet a transfer rate of 800kB/s can be observed corresponding to $1.2 \cdot 10^{-3}$s per kB, which still is considered acceptable. In local communication a speed of about 5MB per second may be obtained.

4.6. Summary

The VIPWOB system has been designed to facilitate construction of image analysis and computer vision systems and particular attention has been paid to implementation and design of facilities that allow use of the system for diverse kinds of system, ranging for being a simple execution environment to a facility for detailed control of the interaction between modules in a real-time application.

The manuscript has outlined the facilities available in VIPWOB and described the performance parameters for the system when executed on Sun and Silicon Graphics computers.

Table 4.1. Timing results for the transfer of data between modules in a VIPWOB system.

Test setup	Initiation time	Time per kB
Transfer of data using a dedicated channel between two modules on the same computer	$3.3 \cdot 10^{-4}$s	$9 \cdot 10^{-5}$s
Transfer of data using a dedicated channel between two modules on different computers	$3.4 \cdot 10^{-4}$s	$2.5 \cdot 10^{-4}$s
Message passing between two modules on the same computer	$1.9 \cdot 10^{-3}$s	$2.5 \cdot 10^{-4}$s
Message passing between modules on different computers	$3.1 \cdot 10^{-3}$s	$2.4 \cdot 10^{-4}$s
Access to shared variable on the same computer	$2.1 \cdot 10^{-3}$s	
Access to a shared variable on another computer	$3.3 \cdot 10^{-3}$s	

The system has been implemented on a wide range of UNIX platforms, including Digital VAX computers running Ultrix, HP series 300 and 700 running HPUX, Motorola 680x0 computers running SVR5r3 and SVR5r4, SGI computers running IRIX 3.3.2 and IRIX 4.0.5A-F, and Sun Sparc running SunOS 4.0.1-3. The system is auto-configuring and installation is thus trivial. Adaptation to new platforms may typically be carried out in less that a day. The only requirement, is presently that the computers runs UNIX and that BSD sockets are available. Change to use of AT&T streams has been considered but not implemented.

The software system is available in public domain. It is distributed under the GNU license agreement. It is possible to obtain a copy of the software through anonymous ftp from the computer danablue.vision.auc.dk. The package includes extensive documentation (>100 pages) in terms of a users guide, a users reference, and an internal guide manual.

Acknowledgement

The VIPWOB project was conceived in 1989 by Erik Granum, Knut Conradsen and Peter Johansen, The Danish Image Processing Syndicate. The initial design was performed by Henrik I Christensen, Carsten Trant, Bjarne Ersbøll, Jens Michael Carstensen, Mikael Kjærulf, and Mads Nielsen. Initial implementations were made by Peter Melsen.

The project has in addition gained substantially through interactions with members of the CEC-Vision as Process consortium, in particular James L. Crowley and Christophe Discours, LIFIA, and Lars Olsson and Harald Winroth, KTH.

The VIPWOB project has been sponsored by the Danish Technical Research Council under the MOBS programme and the CEC through the Basic Research Action BR-3038 Vision as Process. This funding is gratefully acknowledged.

References

[AVS, 1989] Advanced Visualisation Systems Inc., The AVS System 4.0, Tech Report, Los Angeles, CA, 1989.

[Draper 1988] B. Draper, B. Collins, E. Riseman, A. Hanson, The Schema System, IJCV, Vol. 3, No. 1, 1988.

[K & R, 1986] B. Kernighan & D. Ritche, The C Programming Language, Prentice-Hall Inc., 1986.

[Kirkeby-Christensen, 1993a] N.O.S. Kirkeby & H.I. Christensen, The VIPWOB 4.0 Documentation Suite, LIA-Tech Report 93-02, Institute of Electronic Systems, Aalborg University, February 1993.

[Kristensen, 1993] S. Kristensen, Implementation of HELIX facilities under VIPWOB 4.0, LIA-Tech Report 93-10, Institute of Electronic Systems, Aalborg University, March 1993.

[Rasure 1989] J. Rasure, The Khoros System, in *Scientific Visualisation*, Springer Verlag, 1989.

[SGI, 1991] Silicon Graphics Inc., The Explorer System, Tech Report, Mountain View, CA, 1991

Part II

Image Processing and Description

The partners in the consortium were initially united concerning the importance of multiple resolution processing for image description. However, as the VAP project was getting underway, an extended discussion developed concerning the form of image description appropriate for the common test-bed system. One line of reasoning argued that the traditional approach of extracting edge-segments was woefully inadequate as a general image description. This group argued for novel image description techniques based on such things as peaks and ridges from a Laplacian pyramid or measurements made with quadrature Gabor filters in spatio-temporal pyramids. A counter argument was that while such approaches may show promise they are both unproven and quite expensive to compute. It was further argued that the project goal was not to develop new image processing techniques, but to explore techniques for integration and control. This group argued for the use of real time hardware to compute image descriptions based widely accepted image description techniques.

This issue was resolved by adopting a two track approach. In order to have fast reliable image description on which to base experiments on integration and control, the TIRF and LIFIA laboratories would develop techniques and hardware based on edge points calculated from derivatives from a binomial (Gaussian) pyramid. In parallel the LITH group would develop more advanced techniques using spatio-temporal filtering to be integrated into the common system at a later phase.

This section describes the results of the edge-based approach. Chapter 5 describes the image processing module. The image processing module responds to requests from other modules and provides

1)   Acquisition of synchronised stereo images

2)   Real time (40 ms) computation of binomial pyramids for the two stereo images

3)   Real time (100 ms) computation of a gradient magnitude and orientation within a multi-resolution region of interest (log Cartesian fovea) extracted from the pyramids.

4)   Real time (100 ms) detection and linking of edge segments.

5)   Fast chaining of edge points

6)   Extraction of ellipses from edge chains

7)   Real time (100 ms) computations of measures for reactively controlling camera focus, camera aperture, and stereo vergence.

Chapter 6 describes the design of a fast optimum algorithm for computation of low-pass pyramids. A number of algorithms are compared, and a new fast form of the cascade convolution algorithm is presented. It is shown the binomial coefficients provide an optimum signal-to-noise ratio. An algorithm is presented which produces a pyramid in which the resolution levels are spaced at exactly half octave intervals. With this algorithm, the ratio of sample density to impulse response is identical at all levels of the pyramid, which means that image description procedures may be applied in the same manner at any level of resolution.

Hardware has been constructed for fast computation of edge segments using gradient magnitude and orientation. Chapter 7 describes the algorithm for one pass extraction of edge segments. This algorithm applies two tests for edge segmentation:

  • Similarity of gradient orientation, and
  • A simple recursive measure of the surface under an edge chain.

These two tests are complementary. Gradient orientation provides a clean segmentation at corners, while the surface under an edge provides a close approximation for curved edges.

Edge segments are provided to modules for tracking and description of edges from the two stereo cameras. Chapter 8 described the module architecture for tracking and description. Tracking provides temporal consistency, while description is performed by a vocabulary of primitives for grouping. Tracking is performed by a first order Kalman filter organised as a cyclic process composed of the phases: Predict, Match and Update. The result is a model of the contents of an image sequence in which edges parameters and their temporal derivatives are estimated. Other modules may send message to interrogate the edge model using a vocabulary of perceptual grouping procedures. These perceptual grouping procedures provide a form of associative access to the model. Between each phase, the module scheduler processes interrogation messages. At the end of each cycle the scheduler executes an agenda of "demon" procedures.

Demons use the grouping primitives to locate higher level structures and to watch for perceptual events. Demons may be added or removed from the agenda by messages from other modules, or by other other demons. When a demon detects the specified structure, it generates a message to the module which scheduled it. In this way demons provide a sort of "pre-attentive" detection of image structure which may then serve as perceptual events for the system. Demons can be seen as observing "agents".

Chapter 9 describes a more sophisticated technique for detecting junctions based on context. Contextual rules are used to recover junctions in terms of physical properties of 3D edges. A probabilistic relaxation scheme is used to detect and extract contextual edges for the active scene interpretation processes described in Part V.

5. Image Processing in the SAVA System

James L. Crowley
LIFIA, INPG

This chapter describes the image acquisition and processing module of the VAP SAVA-II integrated vision system. This first section reviews the objectives of this module and then provide an overview of the architecture. It then describes the hardware and software versions of this module. In the second section, a more detailed description in given of the stereo image acquisition and processing system. Stereo image acquisition in the TIRF VME pyramid card is described. The third section describes technique for detecting edge points which is used in the various VAP edge detection procedures. The TIRF edge segment extractor will be described as well as an edge chain extractor which has been constructed to aid the recognition module. In the final section, we describe the measures which are performed for the motor-ocular reflexes of iris, focus and convergence.

5.1 Overview of the Module Components

In early versions of the SAVA skeleton system, image processing and tracking were integrated into a common module. Efforts to speed up the image processing and edge extraction code taught an unexpected lesson. Progress in algorithm implementation coupled with rises in the computational power of work-station technology were such that the bottleneck was not image processing but image communication. This conclusion had a major impact in the design of the SAVA II skeleton system and in the month 33 demo. An effort has been made to minimise image communication where ever it occurs.

The binomial pyramid hardware has proved to be a powerful tool for reducing image transmission. It is well known that image processing and interpretation algorithms can exploit multiple resolution pyramids to dramatically improve computational costs. Equally important has been the reduction of costs for communication to man machine interfaces. Communication of the current stereo images from the digitizer to the computer screen of the camera control unit is currently the biggest bottleneck in the system. To minimise this communication time, the user may select the smallest level of the pyramid which is necessary for him to control the camera head. In particular, in normal operation a very low resolution image (64 by 64) is transmitted, resulting in a significant speed up in the tracking cycle.

The image acquisition and processing module has been designed to minimise image communication. In theory, all image processing is to be performed in this module. Thus the module may be seen as responding to requests from any of the other modules for information about the scene via stereo images. The information which is currently provided includes:

1) Measurements for the ocular-motor reflexes of iris, focus and convergence
2) Edge segment extraction for the 2D tracking and description modules
3) Edge chain extraction for detection for object recognition
4) Low resolution images for the man machine interfaces

The architecture of the this module is shown in figure 5.1

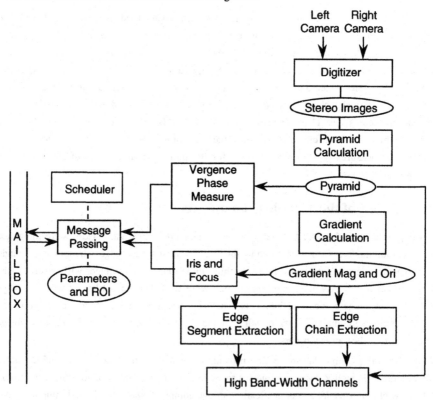

Fig. 5.1 The architecture of the Image Acquisition and Processing Module

All processing is in response to requests received as messages from other modules. Stereo images are acquired from the left and right cameras and a 12 level half octave pyramid is calculated. Images from any level of this pyramid may be transmitted over a high bandwidth channel to the camera control unit. The pyramid may be used to compute a coarse to fine vergence measure using phase.

On command from the 2D modules, the gradient magnitude and orientation is computed within a multiple resolution region of interest from the pyramid, and edge points are extracted and fused to form edge segments. These edge segments are then communicated to the 2D modules over a high band-width channel. A similar command has been made available to extract edge chains. The edge chain extractor computes the gradient magnitude and orientation, and then detects and chains edge points to form chains which are sent over a high band-width channel to the object recognition module. In addition, the measurements based on the gradient may be computed and sent to the camera control unit in order to servo the ocular reflexes of iris and focus.

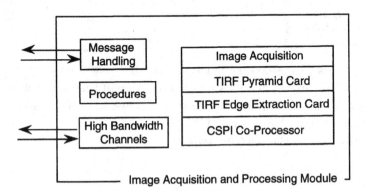

Fig. 5.2 The hardware and software components of the module

The image acquisition and processing module is constructed using a several specialised computer cards. Synchronised stereo images are acquired at 6 bits per pixel at 256 by 256 pixels using a digitizer provided by TIRF. This digitizer is on the pyramid card. Acquired images are immediately passed through the pyramid calculation hardware constructed by TIRF to be placed in a pyramid memory. The pyramid memory is mapped into the address space of a Unix work-station. The work-station may transfer a multi-resolution region of interest to either the TIRF edge extraction card or to the CSPI co-processor based on an Intel i860 RISC processor. At the current time, the system is limited by the amount of time it takes the sun to transfer this the region of interest from the pyramid memory to the memory of the edge extractor card or the CSPI co-processor. Display of images may be performed either by copying images to the memory of a frame-buffer or copying by high bandwidth channel to a display window on the colour screen of a workstation.

5.2 Hardware and Software for the Binomial Pyramid

In this section we review the use of cascaded convolution with resampling, using binomial coefficients, to computer a low pass pyramid.

5.2.1 Low Pass filtering by Cascade of Binomials

The binomial coefficients provide a finite size integer coefficient form of Gaussian filter, with a number of interesting signal processing properties [Chehikian-Crowley 91]. A family of binomial coefficients may be defined by "Pascal's triangle" as shown in Figure 5.3.

Pascal's triangle	Level (n)	Sum	Variance (σ_n^2)	Std.
1	0	1	0	0
1 1	1	2	$1/4$	$1/2$
1 2 1	2	4	$1/2$	$\sqrt{2}/2$
1 3 3 1	3	8	$3/4$	$\sqrt{3}/2$
1 4 6 4 1	4	16	1	1
1 5 10 10 5 1	5	32	$5/4$	$\sqrt{5}/2$
1 6 15 20 15 6 1	6	64	$6/4$	$\sqrt{6}/2$
1 7 21 35 35 21 7 1	7	128	$7/4$	$\sqrt{7}/2$
1 8 29 56 70 56 29 8 1	8	256	2	$\sqrt{2}$

Fig. 5.3 Pascal's triangle for generating a family of binomial coefficients

Let us define the nth level of Pascal's triangle as the binomial filter as b_n. This filter can be expressed as n "auto-convolutions" with the filter [1,1]. We denote auto-convolution by the superscript "*n". $b_n(i) = b_0(i)^{*n} = [1, 1]^{*n}$

The gain of b_n is given by the sum of the coefficients:

$$\sum_{i=-n/2}^{n/2} b_n(i) = 2^n$$

The variance of the filter is given by n times the variance of the kernel filter, which is $\frac{1}{4}$.

$$\sigma_n^2 = n \ \sigma_0^2 = \frac{n}{4} \ .$$

5.2.2 Scale Space Pyramids by Cascaded Convolution

Cascade convolution with resampling can be employed to derive a fast algorithm to compute a multiple resolution pyramid. The classic version of this algorithm was developed in the late 70's and published by Crowley and Stern in 1984 [Crowley-Stern 84]. A similar algorithm has been developed by Burt [Burt 84]. This representation is equivalent to a discretely sampled scale space with scale samples at multiples of $\sqrt{2}$. This algorithm is illustrated by the flow diagram in figure 5.4.

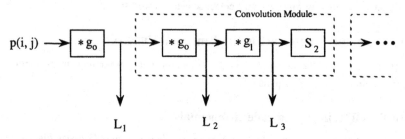

Fig. 5.4 Algorithm Binomial Pyramid by Cascade Convolution with Resampling

In [Chehikian-Crowley 91] it is shown that the optimum signal to noise ratio and the minimum height side lobe occur for this algorithm when the kernel filter $g_0(i) = \begin{bmatrix} 1 & 2 & 1 \end{bmatrix}$ and the filter $g_1(i) = \begin{bmatrix} 1 & 4 & 6 & 4 & 1 \end{bmatrix}$. This algorithm is the basis of the real time VME hardware constructed by the consortium for calculating binomial pyramids. This algorithm produces images with an exact exponential sequence of variances. The variance of each level of the pyramid, k, is given by:

$$\sigma_k^2 = 2^{k-2}$$

That is, the standard deviations of the resulting filters are from the set:

$$\{\frac{\sqrt{2}}{2}, 1, \sqrt{2}, 2, 2\sqrt{2}, 4, 4\sqrt{2}, 8, 8\sqrt{2}, 16, 16\sqrt{2}, 32, 32\sqrt{2}\}$$

5.3 Edge Chain and Segment Extraction

A low pass pyramid provides the smoothing that is necessary for calculation of derivatives. In this section we examine how the gradient magnitude and orientation are computed within a multiple resolution region of interest within the pyramid. This region of interest forms a log-Cartesian "fovea".

5.3.1 A Multiple Resolution Region of Interest

Processing within the image processing module is performed over a multiple-resolution region of interests (ROI). The multi-resolution ROI is defined as a set of overlapping windows in the pyramid such that the window at each level contains the same number of sample. This gives the effect of an log-Cartesian fovea, as shown in figure 5.5.

A multiple resolution ROI is defined as the samples within a pyramid centred on a point at column i, row j and level k. The ROI includes a fixed number of samples, distributed about the center point at Δi, Δj, and Δk. For each image, the Gradient and Laplacian are computed and described only within the ROI. The number of points is chosen to assure that the description can be computed within a specified "frame time". The ROI may be moved at each new frame in response to commands from other modules in the system.

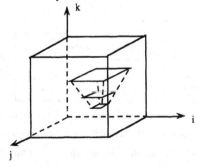

Fig. 5.5 The Region of Interest in Scale Space

A scale space "Region of Interest " is defined with the following parameters:

i, j, k;   The Center Point of the highest resolution (col, row, level) of the ROI

Δi, Δj, Δk;   The extent of the ROI in row, col and level

NumPoints;   The number of samples contained in the ROI

The center point of the ROI is defined by three parameters i, j, k. The parameter k identifies the level of the center of the ROI. The parameters i and j are in coordinates of pixels from the original image, and not samples at the level k. The parameter Δk specifies how many levels are covered by the ROI. The parameters Δi and Δj specify a rectangle in terms of samples (not pixels) at each level.

The ROI has the same number of samples at each level. Because the sample density decreases with increasing scale, the result is a set of concentric rectangles centered on the point i, j, k. Although the ROI has a fixed number of points, its geometry may be altered to adapt to the goals of the system. For determining correspondence, it may be appropriate to use an ROI which spans a small region in many levels. For making precision measurements, it may be appropriate to cover a large region in only one level. In either case, the fact that the ROI has a fixed number of points helps to assure that the calculation will be accomplished in the fixed time delay imposed by the constraint of real time processing.

5.3.2 Calculating Discrete Derivatives with a First Difference Operator

The process for computing an edge map of the contents of a multi-resolution region of interest is shown in figure 5.6. Image samples have been smoothed during the pyramid computation. A window of pixels from the region of interest (ROI) is transmitted to a process which computes the first derivative along rows and columns. The two first derivative images are then used to compute the gradient magnitude and orientation by table lookup. The magnitude and orientation are passed to a process which detects local extrema above either a lower or an upper threshold. Extrema points are then passed to one of the chaining processes, either to extract straight edge segments or edge chains. Hysteresis thresholding is applied during chaining based on the flags set during the peak detection process.

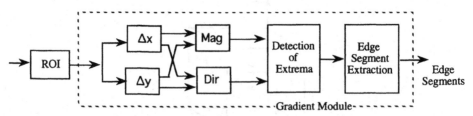

Fig. 5.6 The Process for extracting edges from the Gradient magnitude

As classically defined, derivatives can only be measured for continuous analytic functions. It is not possible to exactly measure the first derivative a signal represented by discrete samples. For discrete signals we can approximate the derivative by linear operations based on the first difference. In the case of sampled signals, derivatives are measured by a filter which is convolved (or correlated) with the signal. The design of such operations is based on analysis from the field of digital signal processing. Such analysis shows that a good signal to noise ratio is given by the filter $d_2 = [1 \quad 0 \quad -1]$.

The gradient magnitude is computed by convolution within the ROI with the first difference operator $[1 \quad 0 \quad -1]$ in the row and column direction. Row and column derivatives are converted to gradient magnitude and orientation by table look-up. Points which are local extrema in gradient and have the same orientation are detected by hysteresis thresholding.

The maximum image, max(i,j) and the direction image dir(i, j) are defined as:

$$\max(i, j) = \sqrt{d_x(i, j)^2 + d_y(i, j)^2}$$

$$\text{dir}(i, j) = \frac{180}{\pi} \, \text{Tan}^{-1}(\frac{-D \, d_y(i,j)}{d_x(i,j)})$$

The direction image provides gradient direction as a short integer (SARY) with a value of 0 to 359. The factor -D accounts for the fact that pixels are not square and that the image is provided in a left handed coordinate system. More interesting for maximum detection and chaining is the orientation image ori(i, j). The orientation image is a direction image in which gradient orientation is expressed as a value between 1 and 32.

$$\text{ori}(i, j) = \frac{16}{\pi} \, \text{Tan}^{-1}(\frac{-D \, d_y(i,j)}{d_x(i,j)}) + 16$$

To avoid excessive calls to the square root and arc tangent functions, each of these images is computed by table lookup. For the gradient magnitude, the possible values of first derivative range from -128 to 127. By adding a bias of 128, we obtain a range from 0 to 255 for both the x and y directions. An image of 255 by 255 is used to store the possible gradient magnitudes. Conversion magnitude is then accomplished by reading the value in this image.

Orientation is quantized to 32 orientation bins. A table of orientation bins is also built for all possible pairs of values in vertical and horizontal gradient. Edge information is may be extracted by detecting points within the gradient magnitude image which are local maxima in the direction of maximum gradient.

5.3.3 Extrema Detection

Edge point detection is performed by detecting extrema in the maximum gradient direction. Extrema are marked with a 1 if the gradient magnitude is greater than a lower threshold, and 2, if the magnitude is greater than a second, higher, threshold.

An edge point is an extrema if is larger than or equal to its immediate neighbours on either side in the direction of maximum gradient. Let us define the gradient magnitude at pixel (i, j) as the value

$m(i, j)$. Suppose that n_1 and n_2 are the neighbours on either side of (i, j) in the direction of maximum gradient, and the n_3 and n_4 are the pixels at a distance of 2 pixels. The point (i, j) is labelled as an extrema if:

$$E(i, j) = \begin{cases} 1 & \text{if} \quad |n_1| \le |M(i, j)| \ge |n_2| \text{ and} \\ & \qquad g|n_3| < |M(i, j)| > g|n_4| \text{ and} \\ & \qquad M(i, j) > \text{threshold1} \\ 0 & \text{otherwise} \end{cases}$$

where g is a gain, typically set to 1.1. If $E(i, j)$ is 1 and $M(i, j)$ is greater than the second threshold, then $E(i, j)$ is set to 2. A connected set of edge points must contain at least one edge point marked with a 2 to be considered. This provides a hysteresis thresholding effect for edge extraction. Extrema points are passed to a chaining process described in chapter 7.

5.3.4 Edge Segment Extraction Hardware

A number of edge segment and edge chain extraction algorithms have been written by the VAP consortium. These algorithms differ mostly in the way that the edge points are merged. TIRF has constructed an edge segment extraction card using an AMD 2000 signal processing chip. This chip must be programmed in a form of assembler language, making modifications to the chaining algorithm very delicate. In addition, the chip contains only limited program memory. On the other hand, the card is optimized for edge segment extraction. The month 33 demo system has available both a the TIRF edge extraction hardware, and the possibility of using recently improved edge extraction software running on the CSPI co-processor. The primary difference between these procedures is the criteria which is used to break edge chains to form edge segments. The modifications were subsequently adopted by a later version of the software for the TIRF edge extraction board.

Edge segments are extracted by a raster-scan of the edge image in which a linked list of "active" edges are maintained. Each active edge is composed of an orientation bin, a start point, a stop point, as well as pointers to the next and previous edges in the list. During the scan, segments which belong to edge bin are chained to neighbours belonging to the same bin or an adjacent bin. For each chain a prediction of the next pixel is made based on the average orientation bin number of the chain. If an edge point of the same bin or an adjacent bin is found within one pixel of the predicted point, the edge is extended to that point. If an edge point is detected which is not adjacent to the prediction from an active chain, a new chain is started. If two segments of the same orientation bin meet, then they are merged. Each time that a point is added to a segment, the total surface under the segment is updated by adding an increment in surface. If the total surface passes a threshold, then the segment is closed and a new segment is opened. This process is described in greater detail in chapter 6.

In addition to providing edge segments, a procedure has been written for the CSPI co-processor to extract edge chains within the ROI. This edge chain extraction algorithm is essentially the same as the edge segment algorithm, except that the pixel locations of each edge point are stored, and that the edge chains are only segmented when meeting other edge chains.

5.4 Image Measurements for Camera Control

This section describes measurements which are made within the image acquisition module for ocular reflexes. We first describe measures for iris and focus. We then describe measures for convergence.

5.4.1 Measure for Iris and Focus

We have found that the maximizing the variance of pixels in the region of interest provides a robust estimator for aperture. This calculation is performed by the image acquisition and processing module and the result communicated to the camera control unit.

It is well known that focus can be controlled by the "sharpness" of contrast. The problem is how to measure such "sharpness". In [Krotkov 87] we can find a description of several methods for measuring image sharpness. Horn [Horn 65] proposes to maximize the high-frequency energy in the power spectrum. Jarvis proposes to sum the magnitude of the first derivative of neighbouring pixels along a scan line [Jarvis 83]. Schlag [Schlag et al. 83] and Krotkov [Krotkov 87] propose to sum the squared gradient magnitude. Tenenbaum and Schlag compare gradient magnitude to a threshold and sum uniquely those pixels which are above a threshold. The problem is then the choice of such a threshold. We have found that such a measure performs poorly. After experiments with several measures, we have found our best results with the sum of gradient magnitude, without the use of the threshold.

We measure image gradient at the level five or our low-pass pyramid, providing a binomial smoothing window with a standard deviation of $4\sqrt{2}$. Gradient is calculated using compositions of the filter [1 0 −1] in the row and column directions. By default, the "region of interest" is at the center of the image, but this region may be placed anywhere in the image by a message from another software module. Local extrema in the gradient magnitude are summed within the region of interest. An initialize command causes focus to look for a global maximum in this sum. Subsequently, the reflex action seeks to keep the focus at a local maximum. Note that this measure exhibits a plateau around the proper focal value. This region corresponds to the "depth of field". Reducing the aperture will enlarge the depth of field and thus enlarge this plateau.

5.4.2 Measure for Convergence

As described above, when we specify the 3D interest point, the system computes the necessary camera angles so as to bring the optical axes to intersect at the specified gaze point. This corresponds to a simple perceptual action which might be called "Look-at <3D point>". Our collaborators at University of Linköping have found that a very robust measure for convergence is provided by the difference in phase of the correlation of an even and odd filters with the image

[Westelius et al 91]. They have demonstrated vergence control of a simulated head using even and odd Gabor-like filters. In collaboration with them, we have determined that a reasonable approximation of the phase may be obtained using first and second derivative filters [1 0 -1] and [1 -2 1]. While the phase measured in this way is not linear with position, it does seem to be monotonic.

We exploit the multiple resolution pyramid to converge on an object in a coarse to fine manner. An image row and an initial column positions in the two cameras are selected for convergence. We measure the phase at this row in the two cameras at level 9 of our pyramid. The phase provides a shift in each image. This shift is then used to compute the column for the next higher resolution level. The process is repeated at each level. The final shift in each image is converted from pixels to encoder counts for the vergence motors and pan motors. The sum of the shift is used to compute a pan motion for axe 6. The difference is used to compute a symmetric vergence angle for the two vergence motors. The result is that the head stays symmetrically verged on an edge of the largest object which covers the selected scan line. The process repeats for each pair of images which are taken by the image acquisition module.

5.5 Conclusions

If there is one lesson that is clear from our work in image description it is that edge segments are a very poor representation for an image. They are unstable in the presence of non-polyhedral objects. They are un-reliable in the presence of highlights and shadows. They are ambiguous in position for stereo matching. It is essential that a more robust and reliable image description be found.

In the first year of the project we performed experiments with detection functions for bars, bare-ends, corners and spots. Work in this area was suspended in order to devote man-power to developing the VAP skeleton system. It is now time to return to this work, in order to provide reliable point primitives for stereo and to provide the pre-attentive features that are needed for recognition. It is also apparent that at the least the token tracker should track and group arcs as well as segments. Such an extension is relatively easy and will be carried out in the near future.

It is also apparent that much remains to be done in the area of measures for ocular-motor reflexes. Work will continue in this area using the SAVA integrated system as an experimental tool.

References

[Burt 84] P Burt, The Laplacian Pyramid as a Compact Image Code, IEEE trans. on Communication, COM-31, Vol. 4, pp. 532-540, 1984.

[Califano 90] A. Califano, R. Kjeldsen, and R.M. Bolle, Data and Model Driven Foviation, 10 ICPR, Atlantic City, IEEE CS Press. June 1990.

[Canny 86] Canny, J. A Computational Approach to Edge Detection, IEEE Trans. on P.A.M.I., Vol. 8, No. 6, Nov. 1986.

[Chehikian 91] Chehikian A., A One Pass Edge Extraction Algorithm, 7th Scandinavian Conference on Image Analysis, Aalborg, 1991.

[Chehikian-Crowley 91] Chehikian A., Crowley J.L., Fast Computation of Optimal Semi-Octave Pyramids, 7th Scandinavian Conference on Image Analysis, Aalborg, 1991.

[Clark and Ferrier 88] Clark, J. and Ferrier, N., Modal Control of an Attentive Vision System, IEEE 2nd Internat.. Conf. on Comp. Vision, Tarpon Springs, Fla., pp. 514-523, Dec. 1988.

[Crowley 91] Crowley, J. L. Towards Continuously Operating Integrated Vision Systems for Robotics Applications, SCIA-91, Seventh Scandinavian Conference on Image Analysis, Aalborg, August 91.

[Crowley-Stern 84] Crowley J. L., Stern R. M. Fast computation of the difference of low-pass transform , IEEE Trans. Pattern Anal. Mach. Intell., Vol. 6, 212- 222, 1984.

[Discours 89] Discours, Christophe, Analyse du Mouvement par Mise en Correspondance d'Indices Visuels, Thèse de Doctorat, INPG, February 1990.

[Eklundh 92] J.O. Eklundh and K. Pahlavan, Head, Eye and Head-Eye System, *Internat. Jour. Patt. Rec. and Art. Intl.*, Vol. 7, No. 1. 1992.

[Horn 68] Horn, B. P. K., Focusing, MIT Artificial Intelligence Lab Memo No. 160, May 1968.

[Jarvis 83]. Jarvis, R. A., A Perspective on Range Finding techniques for Computer Vision, IEEE Trans. on PAMI Vol. 3, No. 2, pp 122-139, March 1983.

[Krotkov 87] Krotkov, E., Focusing, International Journal of Computer Vision, 1, p223-237, 1987.

[Krotkov 90] Krotkow, E., Henriksen, K. and Kories, R., Stereo Ranging from Verging Cameras, IEEE Transactions on P.A.M.I, Vol. 12, No. 12, pp. 1200-1205, December 1990.

[Lindberg 90] Lindberg , T. Scale Space for Discrete Signals, IEEE Transactions on P.A.M.I., Vol. 12, No. 3, March 1990.

[Schlag et. al. 83] Schlag, J., A. C. Sanderson, C. P. Neumann, and F. C. Wimberly, Implementation of Automatic Focusing Algorithms for a Computer Vision System with Camera Control, CMU-RI-TR-83-14, August, 1983.

[Westelius et. al. 91] Westelius, C. J., H. Knutsson, and G. H. Granlund, Focus of Attention Control, SCIA-91, Seventh Scandinavian Conference on Image Analysis, Aalborg, August 91.

6. Building a Pyramid of Low-Pass Images

Alain Chehikian
LTIRF, INPG

Abstract

Multiple resolution representations are often used in computer vision, as they provide a natural way for describing an image by a hierarchy of structures. However, because of the resampling process, the images they provide are corrupted by an aliasing noise which makes difficult the detection of structures, specially when the detection process implies the computation of derivatives. Choosing the filtering kernel is thus essential. Paradoxically, the importance of proper signal-to-noise analysis have been widely neglected by the vision community. In this paper, we study two commonly used algorithms from the point of view of the aliasing noise they create. Then we propose an optimum filtering kernel which: minimizes the aliasing noise, does not create new structures, has interesting properties of rotational symmetry and reduced computation cost.

6.1 Introduction

Pyramid image representations derive their name from an alternation of low-pass filtering and resampling resulting in a set of NxN, N/2xN/2, N/4xN/4... images, where NxN is the size of the original image. Many authors have proposed such algorithms which differ from the low-pass filtering kernel as well as from the resampling sequence. [Burt 1981,1984], and [Crowley-Stern 1984] are the more often cited authors in the vision literature as the computational cost resulting from their algorithm is noticeably lower than other proposed algorithms including wavelet representation algorithms[Mallat 1989]. One can notice that all these authors, except Meer [Meer et al 1987]], have neglected that resampling produces aliasing noise, thus corrupting the images. One can also notice that all these authors, except Crowley, have neglected that using an improper filtering kernel can produce new structures e.g. ghost edges, in the filtered image.

In this chapter we define a set of constraints which must be satisfied by a filtering kernel, emphasizing on a constraint which, when satisfied, avoids the production of new (ghost) structures. We also define a measure for the corruption of the image by the aliasing noise. Then we analyse the previously mentioned algorithms from the point of the aliasing noise. From that analysis, we propose an algorithm which satisfies that constraints, is optimal from the point of the aliasing noise, and can be used to produce octave-spaced as well as half-octave-spaced low-pass pyramids.

6.2 A Family of Scale-Space Low-Pass Filtering Kernels

A family of low-pass filtering kernels may be determined analytically by listing a set of constraints which such filters should satisfy. Such constraints are the following:

Constraint 1: In order to reduce the computation cost, kernels must be separable.

Constraint 2: In order to process the image independently of the orientation of the structures it contains, the impulse response as well as the frequency response must be rotationally symmetric.

Constraint 3: In order to avoid phase shifting, kernels must be symmetric and odd-sized.

Constraint 4: In order to perform low-pass filtering, kernels must be normalized.

Constraint 5: In order to avoid new structure creation, the impulse response must be unimodal and non-negative, and the frequency response must be unimodal and non-negative inside the Nyquist domain referred to the current sampling rate[6].

The first constraint aims to reduce the computation cost, other constraints result from the use of the pyramid in vision processes. Let be $2p+1$ the size of the mono-dimensional filtering Kernel. Such a kernel can be expressed as:

$$f = [w_p \, ... \, w_1 \, w_0 \, w_1 \, ... \, w_p]$$

while the two dimensional kernel is:

$$F(x,y) = f(x)*f(y)$$

whose Fourier transform is:

$$F(u,v) = (\, w_0 + 2w_1.\cos(u) + ... + 2w_p.\cos(pu))(\, w_0 + 2w_1.\cos(v) + ... + 2w_p.\cos(pv))$$

When used at the level k of the pyramid, where k identifies the number of resampling performed to reach this level, this frequency response becomes:

$$F_k(u,v) = (\, w_0 + w_1 \cos(2^k u) + ... + 2w_p \cos(2^k pu))(\, w_0 + w_1 \cos(2^k v) + ... + 2w_p \cos(2^k pv))$$

At level k, the total frequency response $R_k(u,v)$ will be obtained by the product of a set of frequency responses of the form $F_m(u,v)$ where the subscript m identifies the pyramid levels before that level k. This total frequency response cannot be defined before the resampling sequence is defined and therefore depends on the considered pyramid algorithm. Nevertheless, the Nyquist domain can be defined as:

$$u_{Nk} \in \, [-\pi/2^k , +\pi/2^k]$$

$$v_{Nk} \in \, [-\pi/2^k , +\pi/2^k]$$

And the corruption of the image by the aliasing noise can be characterized by the signal-to-noise ratio SNR defined by:

$$SNR = \sqrt{\frac{\int_0^{u_{Nk}} \int_0^{v_{Nk}} R_k^2 \, dv \, du}{\int_{u_{Nk}}^{2u_{Nk}} \int_{v_{Nk}}^{2v_{Nk}} R_k^2 \, dv \, du}}$$

where $k > 0$. We can now analyse the results obtained from Burt's algorithm and Crowley's algorithm.

6.3 Burt's Algorithm

Burt uses the sequence of filtering and resampling described in figure 6.1. We have represented on this figure two sets I_{sl} of output images, where l from the set $\{0,1\}$ identifies whether or not the sub-sampling has been performed. The two sets have the same bandwidth for a given step s, but they differ from the resampling index: $k = s+l$. In his original algorithm, Burt uses the set I_{s0} whose computation cost is $O(3/4 \ N^2.(2p+1))$ while for the set I_{s1} the computation cost is $O(2 \ N^2.(2p+1))$. But as we will see later, the two sets differ noticeably by their SNR figure.

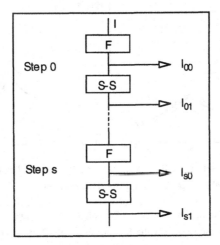

Fig. 6.1. Burt's algorithm

Burt considers 5-sized and 4-sized separable, symmetric and normalized kernels. We will only consider the first one which satisfies the constraints 1, 2, 3 and 4 listed above. Burt uses an unimodality constraint which partly satisfies our constraint 5. His 5-sized kernel is of the form:

$$f = [w_2 \ w_1 \ w_0 \ w_1 \ w_2]$$

He defines also the equal contribution constraint which results in:

$$w_0 + 2w_2 = 2w_1$$

When all the Burt's constraints are satisfied, only one free variable $a \in [0.25 ; 0.5]$, defines the kernel:

$$w_0 = a$$
$$w_1 = 0.25$$
$$w_2 = (0.5-a)/2$$

The Burt's pyramid is characterized by the frequency response at the step s:

$$R_s(u,v) = \prod_{m=0}^{s} F(2^m u, 2^m v)$$

where:

$$F(u,v) = [w_0 + 2w_1\cos(u) + 2w_2\cos(2u)][w_0 + 2w_1\cos(v) + 2w_2\cos(2v)]$$

We have represented on figure 6.2 the resulting SNR figure for the two sets I_{sl}:

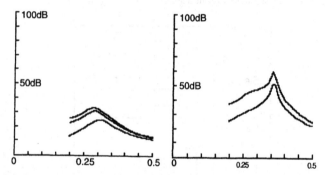

Fig. 6.2. The SNR figures versus a of the Burt's pyramids. On left: the original pyramid. Images I_{s0}, s = 1, 2, 3 from bottom to top. On right: Images I_{s1}, s= 0, 1 from bottom to top.

Figure 6.2 shows that, from the point of view of the aliasing noise, there is an optimum value for a. This optimum value is close to 0.3 for the I_{s0} images, while Burt suggests the value a= 0.4 to obtain a close to Gaussian response. For that suggested value the SNR is very low, close to 20 dB so that the images produced by Burt's algorithm are of very poor quality. When using the I_{s1} images, the optimum value of a is close to 0.375, and the SNR is greatly enhanced.

Moreover, the Burt's algorithm does not produce a true octave-spaced pyramid, as, if σ_0^2 is the variance associated with the basic filtering kernel:

$$\sigma_0^2 = 2(w_1 + 4w_2) = 2.5 - 4a$$

the variance associated to the total frequency response is:

$$\sigma_k^2 = \sigma_0^2 \sum_{m=0}^{k} 2^{2m} = \sigma_0^2, 5\sigma_0^2, 21\sigma_0^2, 85\sigma_0^2,...$$

which is not an exact exponential sequence of the form 2^k but tends towards such an exponential when k increases.

6.4. Crowley's Algorithm

This algorithm produces an exact half-octave spaced pyramid of Gaussian. The algorithm is illustrated on figure 6.3.

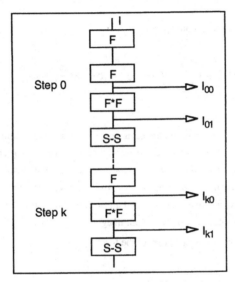

Fig 6.3 Crowley's algorithm

During an initial step, the input image is convolved with a filter F, then the sequence of convolution with F, and F*F is applied before resampling.

The filtering kernel F is seven coefficients wide and results from a cascade of three convolutions with a 3-sized binomial kernel:

$$f = [w_1 \ w_0 \ w_1 \ ]^{*3}$$

with $w_0 = 0.5$ and $w_1 = 0.25$.

This kernel provides a good approximation of a Gaussian kernel with $\sigma_0^2 = 1.5$. It satisfies all the constraints listed formerly. Moreover, it reduces the computation cost as it needs only shift-and-add operations. At a level k of the pyramid, the frequency response of the bi-dimensional kernel is:

$$F_k(u,v) = (w_0 + 2w_1 \cos(2^k u))^3 (w_0 + 2w_1 \cos(2^k v))^3$$

While the total frequency response associated with an output image I_{kl} is:

$$R_{kl} = F(u,v) \ F_k(u,v)^{(2l+1)} \prod_{m=0}^{k-1} F_m(u,v)^3$$

It turns out that the signal-to-noise ratio of an output image I_{k0} is close to 170dB, while for an output I_{k1} SNR is close to 240 dB. However, these high signal-to-noise ratios are due to an excess of filtering, so that even if output images are not corrupted by the aliasing noise, they are corrupted by a lack of information.

However, Crowley's algorithm produces a true half-octave spaced pyramid, as, if σ_0^2 is the variance associated with the basic filtering kernel:

$$\sigma_0^2 = 3w_1 = 1.5$$

the variance associated with an output I_{kl} is:

$$\sigma_{kl}^2 = \sigma_0^2(1 + (2l+1)2^{2k} + 3\sum_{m=0}^{k-1} 2^{2m}) = 2\sigma_0^2, 4\sigma_0^2, 8\sigma_0^2, 16\sigma_0^2, 32\sigma_0^2, \ldots$$

6.5 Our Proposal

From Crowley's algorithm we can retain the sequence of filtering and resampling, as it provides an exact exponential sequence of half-octave spaced pyramid and a sequence of octave-spaced pyramid as well. However, the basic filtering kernel used by Crowley results in over-filtering and thus in a loss of information. In our proposal, the basic kernel is defined so that it satisfies the constraints listed formerly, the size is not defined a priori, nor the coefficients, but will result from the study of the signal-to-noise ratio of the output images.

Let be the uni-dimensional filtering kernel of the form:

$$f = [w_1 \ w_0 \ w_1]^{*n}$$

This kernel satisfies our constraint 3. It will satisfy our constraint 4 if:

$$w_1 = (1-w_0)/2$$

It will satisfy our constraint 5 if:

$$0.5 \le w_0 < 1$$

So that this kernel is defined by two free variables w_0 and n ($w_0 = 0.5$ and $n = 3$, correspond to Crowley's kernel).

At a level k of the pyramid, the frequency response of the bi-dimensional kernel will be:

$$F_k(u,v) = (w_0 + (1-w_0)\cos(2^k u))^n (w_0 + (1-w_0)\cos(2^k v))^n$$

As for crowley's algorithm, the total frequency response associated with an output image I_{kl} will be:

$$R_{kl} = F(u,v) F_k(u,v)^{(2l+1)} \prod_{m=0}^{k-1} F_m(u,v)^3$$

Figure 6.4 shows the variation of SNR when w_0 varies from 0.4 to 0.9 and $n = 1, 2, 3$.

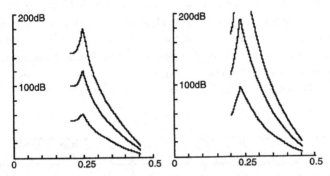

Fig. 6.4. Variation of SNR versus w0. On left I_{k0} images, on right I_{k1} images. n = 1,2,3 from bottom to top

We can see that the highest SNR is always obtained when w_0 is close to its lowest permitted value i.e. 0.5. Moreover, choosing n = 1 results in a signal-to-noise ratio noticeably higher than the SNR due to image noise (36 dB for a camera providing 64 distinct grey levels), higher also than the SNR due to quantification (48 dB if the pixels are coded on 8 bits). Choosing n higher than 1 will only result in additional computation cost and in a loss of information due to over-filtering. These remarks prevail on us to choose the free variables: $w_0 = 0.5$ and n = 1. Thus the basic kernel will be:

$$F(x,y) = \frac{1}{16}\begin{pmatrix} 1 & 2 & 1 \\ 2 & 4 & 2 \\ 1 & 2 & 1 \end{pmatrix}$$

The variance associated with that kernel being $\sigma^2_0 = 0.5$, the variance associated with an output image I_{kl} will be:

$$\sigma^2_{kl} = 1, 2, 4, 8, 16...$$

Moreover, such a kernel, because of its small size and the value of the coefficients, will be very of easy use in a hardware implementation. We should before compare the results obtained by our algorithm with those of Burt and Crowley. We also need to study how our constraint 2, i.e. rotational symmetry, is satisfied.

6.6 Comparing the Algorithms

To compare these algorithms, we need to define some criteria. One criterion should be the corruption produced by the aliasing noise. This will be characterized by the SNR as formerly formulated. Another important criterion will be the adequacy between the bandwidth of an output image and the sampling rate. This adequacy will be expressed by the ratio $\sigma_k/2^k$. For a given sampling rate 2^k, the greater is this ratio, the greater is the over-filtering. The rotational symmetry of the frequency response $R_{kl}(u,v)$ will be another important criterion. We will express this rotational symmetry by the Root Mean Square distance between the 0° and the 45° profiles of

$R_{kl}(u,v)$. For a true rotationally symmetric frequency response this RMS distance would be zero. By the same way, we can also characterize the likeness to a Gaussian, by the RMS distance between the frequency response $R_{kl}(u,v)$ and a true Gaussian with the same variance. Table I summarizes these features for an octave-spaced pyramid provided by Burt's original algorithm (I_{s0} images), Burt's modified algorithm (I_{s1} images), Crowley's algorithm (I_{k0} images), and our proposed algorithm(I_{k0} images). Burt's algorithm is considered with the free variable a = 0.375 which satisfies our constraint 5.

Table 6.1

	Burt I_{01}	Burt I_{10}	Crow. I_{10}	Prop. I_{10}	Burt I_{11}	Burt I_{20}	Crow. I_{20}	Prop. I_{20}
k	1	1	1	1	2	2	2	2
σ^2_k	1	5	12	4	5	21	48	16
SNR	18 dB	45 dB	174 dB	58 dB	21 dB	51 dB	171 dB	57 dB
$\sigma^2_k/2k$	0.5	1.12	1.73	1	0.56	1.15	1.73	1
Dissymetry x10-2	1.5	1	ε	1	1	0.9	ε	1
Gaussian Likeness x10-2	2.9	1.9	ε	2	1.9	1.7	ε	1.9

We observe in table 6.1 that Burt's original algorithm provides very noisy images due to a lack of filtering before resampling. At the opposite, Crowley's algorithm provides high SNR over-filtered images. From the point of view of information content, the effect is similar. From the point of view of rotational symmetry and Gaussian likeness, all these algorithms point out good features. This is due to the fact that they use filtering kernels commonly used to transform equiprobable random data into Gaussian random data. Moreover, our proposed algorithm provides an exact octave-spaced (half-octave spaced as well) pyramid which points out the best fit between the image bandwidths and the sampling rates.

Another important criterion will be the computation costs, we list below. These computation costs have been calculated so that they optimize the criterion for each algorithm.

Burt's original:	$10N^2$	multiply-and-adds.
Burt's modified:	$40/3N^2$	multiply-and-adds.
Crowley's half-octave:	$70N^2$	multiply-and-adds.
Crowley's octave:	$140/3N^2$	multiply-and-adds.
Proposed half-octave:	$30N^2$	multiply-and-adds.
Proposed octave:	$20N^2$	multiply-and-adds.

We observe that Burt's original algorithm is the fastest, but it points out the drawbacks mentioned above. Our proposed algorithm needs twice this computation cost but provides obviously better quality images. Compared with the modified Burt's algorithm, our proposed algorithm needs 50% more computation but provides better results.

6.7 Hardware implementation

The proposed algorithm have been implemented in hardware to produce a pyramid representation of a stereo pair, at frame rate. The internal architecture of this hardware is illustrated in figure 6.5.

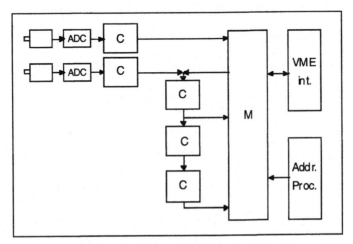

Fig. 6.5: The hardware architecture. C: Binomial 3x3 convolver. M: multiple access memory

The hardware includes a stereo frame-grabber, filtering operations are achieved by a set 3x3 binomial convolving proprietary chips while resamplings are achieved by an address processor. After completion of the process the dual half-octave spaced pyramids are available in a memory mapped in the address space of the host computer and directly accessed by the VME bus. The hardware is able to acquire and process a stereo pair every 40 milli-seconds.

References

[Burt 1981] Burt P.J. Fast Filter Transform for Image Processing, CGIP 16, 20-51, 1981.

[Burt 1984] Burt P.J. The Laplacian Pyramid as a Compact Image Code, IEEE Trans. on Communications, COM Vol. 31, No 4, 532-540, 1984.

[Crowley-Stern 1984] Crowley J. L., Stern R. M. Fast Computation Of The Difference Of Low-Pass Transform, IEEE Trans. Pattern Anal. Mach. Intell., Vol. 4., No. 6, 212- 222, 1984.

[Mallat 1989] Mallat S. G. A Theory For Multiresolution Signal Decomposition: The Wavelet Representation, IEEE Trans. Pattern Anal. Mach. Intell., Vol. 11, 674-693, 1989.

[Meer et al. 1987] Meer P., Baugher E. S., Rosenfield A. Frequency Domain Analysis And Synthesis Of Image Pyramid Generating Kernels, IEEE Trans. Pattern Anal. Mach. Intell., Vol. 9, 1987.

[Lindeberg 1990] Lindeberg , T., Scale Space for Discrete Signals, IEEE Transactions on P.A.M.I., Vol. 12, No. 3, March 1990.

7. One-pass Polygonal Approximation of a Contour Image

Alain Chehikian
LTIRF, INPG

Abstract

We present a new method for describing a contour image by straight line segments. During a first step, not described here, edge pixels, and their orientation, are extracted using a variation of Canny's algorithm. The orientation code is used by the algorithm for chaining and polygonal approximation, first to guide the chaining process, and then to interrupt this process when the current pixel should not be merged into the current chain. By this use of orientation information, chaining and polygonal approximation can be easily performed during one single scan of the contour image. We show that this algorithm provides fair results on images of polyhedral scenes. However, results are not satisfactory when curves are present in the contour image. This leads to an improved algorithm using an additional splitting criterion based on the distance between the pixels merged into a chain and the straight line segment representing the chain.

7.1 Introduction

A grey level image contains a large amount of information, and low-level image processing has to extract a restricted amount of primitive features which are significant for a given application. These primitive features may be regions, contours approximated by straight line segments, by curve segments, etc. In vision for robotics, where scenes usually are highly structured, contours may often be described by straight line segments, and this is the kind of description we are interested in. An algorithm for the extraction of straight line segments takes an image composed of contour pixels, obtained by previous operations, and transforms it into a list structure. Given the amount of input data, such an algorithm tends to be expensive in computing time; parallel implementation is not efficient, because of the very nature of the problem. In this paper, we present an algorithm for the extraction of straight line segments achieving contour chaining and polygonal approximation during a single scan of a contour image, and easily implementable on signal processing hardware.

There is a large amount of literature about methods to extract straight line segments from a contour image. Usually, these methods work on contour chains, where links between connected contour pixels have been made explicit. [Giraudon 1987] proposes an efficient algorithm for chaining in a single scan, and then applies two further steps of fusion and elimination.

We classify methods for polygonal approximation into two large categories. Splitting methods proceed in a recursive way, starting from the segment linking the end-points of the chain. Then, a characteristic point is searched for in the chain. If such a point is found, the chain is split, and its two parts are processed recursively. The characteristic point most often is the point at the greatest distance from the line ([Pavlidis 1977], [Nevatia-Babu 1980]). For fusion type methods, each chain is scanned, and pixels are accumulated into one portion as long as they satisfy some criterion. [Wall-Danielson 1984] incrementally computes the surface between the chain and the segment joining the first pixel with the current pixel. The process is interrupted when the surface grows above a threshold proportional to the length of the segment. [Sklansky-Gonzales 1980] considers a region of uncertainty around each pixel, due to noise. The best segment approximating a sequence of pixels goes through these regions and minimises the pixel-segment distance. [Dettori 1982] constructs a fixed-width band along each chain of pixels. A chain is interrupted when the current pixel falls outside this band, and the approximating segment is defined by the chain's endpoints.

In the algorithm proposed by [Burns et al. 1986], chaining and polygonal approximation are based on gradient orientation of contour pixels. Pixels with the same orientation are grouped together; the gradient is approximated by a plane, taking into account grey-levels, and the approximating segment is given by the intersection of this surface and the horizontal surface defined by the mean intensity.

In fact, the extraction of straight line segments should be considered as two distinct operations: chaining of contour points, and segmentation of chains. For some researchers, chaining and segmentation are handled separately, and it is possible to approximate the chain by any geometrical entity: straight lines, circular arcs, etc. Both operations may be combined when the kind of approximation is fixed. In this paper, we deliberately take the second standpoint; in the sequel, we use the term "chaining" to designate the combination of connected pixels which may be on a straight line segment. The segment may be defined either by the co-ordinates of the endpoints, or by a least squares approximation.

For the chaining operation, the methods found in the literature use a continuity criterion based on distance, or on orientation, or on curvature. The good choice of a threshold for this criterion always is a major problem. In the proposed algorithm we use two criteria: an orientation based continuity criterion to allow and guide the chaining, and a distance based criterion for splitting. As these two criteria are of completely different nature, the choice of thresholds becomes much less important.

The article is organised as follows. Section 7.2 describes the algorithm for chaining and approximation based on gradient orientation of contour pixels. Section 7.3 describes the algorithm for chaining and polygonal approximation based on the surface between the chain and the segment. Section 7.4 describes the new algorithm using the two criteria: gradient orientation to guide the chaining, surface to split.

In a preliminary phase, the grey level image has been processed to extract contour pixels. Four test images are used in this paper. Two images are synthesised. "Rings" represents concentric rings which alternatively are bright and dark; radius and thickness are growing by powers of 2. "Chessboard" has the same grey levels as "Rings". Two natural images contain rich rectilinear contours: the aerial image "Aquitaine", and the office image "Bureau". Figure 7.1 shows the contour images, computed with a variation of Canny's algorithm [Canny 1986].

(a)   (b)

(c)   (d)

Fig. 7.1. Contour images.

7.2 Orientation Based Chaining and Polygonal Approximation

The basic idea of the algorithm is straightforward: if a set of connected contour pixels describes a single straight line segment, then the pixels' gradients have a unique orientation. As the image and the gradient computation are altered by noise, a tolerance for gradient orientation is necessary. If the tolerance is too small, the chain is split into an excessive number of segments. If the tolerance is too large, there is excessive chaining.

The tolerance can be directly introduced by coding the orientation on a finite number of channels, in which case the quantisation step defines the tolerance for orientation variation. However, there is a problem for pixels having an orientation close to a quantisation limit. Figure 7.2 illustrates the good case and the bad case. This figure represents the plane (G_x, G_y) of the gradient vector components.

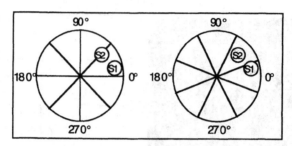

Fig. 7.2. Coding the orientation on n=8 channels

The plane is divided into 8 (more generally n) sectors representing the channels for encoding. In the partition on the left, the channels are centred on multiples of the value $2\pi/8$. On the right, the channels are rotated by half a channel with respect to the left. We indicate in these figures two sets of gradient values for pixels contained in two segments S1 and S2. The partition on the left allows to code correctly the orientation of the gradient vector associated with S1, whereas the pixels of segment S2 are spread over two channels, causing the segment to split. The partition on the right handles the problem correctly for S2, but not for S1.

Burns et al. [1987] solves this difficulty by the means of a double coding described in figure (2). Each pixel is described by two orientations, and two types of chains are considered. The final result is taken from the chaining assuring the best local continuity.

In this case, the tolerance on orientation is:

$$\pm\frac{1}{2}\frac{2\pi}{n}, \text{ centred on } \frac{2k\pi}{2n}, \ k \in [0, 2n-1].$$

This method gives good results, but uses a fairly complex algorithm, and cannot be applied in a single scan. We propose a method giving the same result using a very simple algorithm.

7.2.1 Implementing the Orientation Criterion

Our method encodes orientation with 3n channels which are centred on multiples of the value $\frac{2\pi}{3n}$. The orientation of a contour pixel is coded as C_k ($C_k \in [0, 3n-1]$).

If C_k is the orientation of the first pixel of a chain, subsequent pixels must have an orientation in one of the following sets:

$$\{C_{k+2}, C_{k+1}, C_k\}$$
$$\{C_{k+1}, C_k, C_{k-1}\}$$
$$\{C_k, C_{k-1}, C_{k-2}\}$$

The resulting tolerance on pixel orientation is then:

$$\pm\frac{1}{2}.3.\frac{2\pi}{3n} = \pm\frac{1}{2}\frac{2\pi}{n}, \text{ centred on } \frac{2k\pi}{3n}, \text{ k} \in [0, 3n-1],$$

i.e., it is identical to the tolerance obtained by Burns.

However, this method has a simpler implementation. Each chain is characterized by an orientation code which is an incremental function of the orientation codes of the pixels it incorporates. When a new chain is created, its orientation code is the same than the only pixel it incorporates. A look-up table, addressed by the pair (pixel-orientation-code, chain-orientation-code), indicates wether the current pixel can be incorporated or not to a chain. If yes, the look-up table contains the resulting chain-code, otherwise it contains a negative value.

In our implementation the look-up table is constructed as follow:

- If the chain-code and the pixel-code are equal, the resulting chain-code is unchanged. (All the pixels in the updated chain, including the current pixel, have a unique orientation.)
- If the chain-code and the pixel-code differ from one unit, the resulting chain-code is the lowest code plus 3n. (All the pixels in the updated chain, including the current pixel, have two adjacent orientations.)
- If the chain-code and the pixel-code differ from two units, the resulting chain-code is the intermediate code plus 6n. (All the pixels in the updated chain, including the current pixel, have two non-adjacent but permitted orientations.)
- If the chain-code is equal or greater than 3n, e.g. C_k+3n and,
    - if the pixel-code is C_k or $C_k +1$, the resulting chain-code is unchanged,
    - if the pixel-code is $C_k -1$, the resulting chain-code is $C_k +6n$,
    - if the pixel-code is $C_k +2$, the resulting chain-code is $C_k +1+6n$,
    - else the look-up table contains a negative value.
- If the chain-code is equal or greater than 6n, e.g. C_k+6n and,
    - if the pixel-code is C_k or $C_k +1$ or $C_k -1$, the resulting chain-code is unchanged,
    - else the look-up table contains a negative value.
- In any other case, the look-up table contains a negative value.

Thus, such a look-up table will contain (9n)(3n) values.

Moreover, a single scan chaining algorithm should be able to handle the additional problem of merging two existing chains. The look-up table is thus extended to handle this problem:

- if one chain-code at least, is lower than 3n, the former conditions are applicable.
- If the two chain-codes are of the forms C_k+3n and C_l+3n:
    - if k = l+1, the resulting chain-code is C_k+6n
    - if k = l-1, the resulting chain-code is C_l+6n
    - if k = l, the resulting chain-code is C_l+6n
    - else the look-up table contains a negative value.

- If the two chain-codes are of the forms C_k+3n and C_l+6n:

  if $k = l$, the resulting chain-code is C_k+6n

  if $k = l-1$, the resulting chain-code is C_l+6n

  else the look-up table contains a negative value.

- If the two chain-codes are of the form C_k+6n and are equal, the resulting chain-code is unchanged.

- else the look-up table contains a negative value.

Given the necessity of merging chains, the look-up table must have the dimension $(9n)^2$. Our experience with real images obtained from the camera head of a mobile robot in a structured environment gives n=8 as a good compromise. In this case, the size of the compatibility table (5184 bytes) is very reasonable, and allows easy implementation on a DSP.

7.2.2 Orientation based chaining algorithm

The chaining algorithm we propose scans the pixels of a contour image in top-bottom/left-right order. It needs information about the pixel's 8-neighbourhood, restricted to the past. Figure 7.3 shows this neighbourhood.

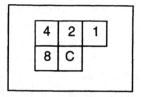

Fig. 7.3. The current pixel C and its neighbourhood

For convenience, the pixels of the neighbourhood are assigned a binary weight. Each pixel in the neighbourhood is characterized by a data structure "past" composed of the set {id,p}, where id is an identifier, p a pointer at a "segment" type data structure.

The identifier id represents one of four different states:

- id=0 means the neighbourhood pixel is not the endpoint of a chain.
- id=1 indicates the neighbourhood pixel currently is the first and unique pixel of a chain.
- id=2 means the neighbourhood pixel is (or descends from) the "head" of a chain. At any moment, every chain has two endpoints. During the scan, the first pixel is labelled "head", the second "tail"; later on, these labels are transmitted to the new endpoints.
- id=3 means the pixel is (or descends from) a chain's "tail". This chain has two endpoints.

The data structure "segment" pointed by p contains the orientation code of the chain and the co-ordinates of the pixels labelled "head" and "tail". It may contain other data, e.g. if a least square approximation of the segment is desired.

If, during the scan, the current pixel is a contour pixel, its neighbourhood is characterized by a value N computed as follows

> for each neighbour,
>> if id ≠ 0 and a chain orientation is compatible with the pixel orientation,
>>> add to N the weight (1,2,4, or 8) of the neighbour.

In this way, the neighbourhood is characterized by N∈ [0, 15]. Figure 7.4 shows the operations executed by the finite state automaton for chaining.

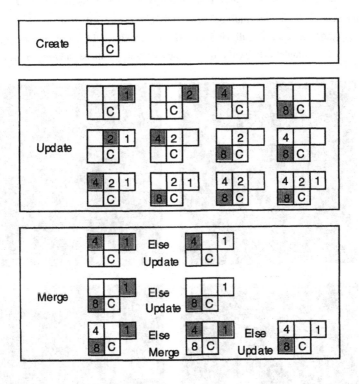

Fig. 7.4. Procedures to carry out according to the neibourhood. C: current pixel . X end-point of a compatible chain

These operations depend on N:

N= 0: (The current pixel is the first of a chain.)
- ♦ *Create* a chain

N = 1, 2, 4 or 8: (The current pixel is connected to only one existing chain.)
- ♦ *Update* the chain.

N = 3, 6, 7, 10, 11, 12, 14, or 15: (The current pixel is connected with several existing chains.)
- ♦ *Update* the most recent chain in the past.

This choice is arbitrary - there is no reason to privilege any chain in particular.

N = 5 or 9: (The current pixel is connected with two non-connected chains.)

♦ if the chains are compatible, *Merge*.

♦ else *Update* most recent chain.

N = 13: (The pixel is connected with three non connected chains.)

♦ if the most distant chains are compatible, *Merge* these.

♦ else if the chains that are closest to each other are compatible, *Merge* these.

♦ else *Update* most recent chain.

Thus, the chaining operation makes use of three auxiliary procedures: *Create*, *Update*, *Merge*.

A fourth procedure, *Eliminate*, must be used to take care of isolated points. It is applied when the north-east neighbour (weight 4) still has id=1 after the current pixel has been processed. All these procedures are very simple; they handle all possible data configurations and need no heuristics.

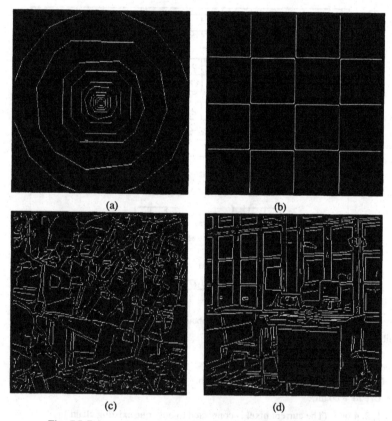

(a)   (b)

(c)   (d)

Fig. 7.5. Polygonal approximation based on the gradient orientation

Experimental results

Figure 7.5 shows the results obtained by this algorithm on the contour images of figure 7.1. Let us first consider the office scene figure.7.5d and the chessboard figure.7.5.b. The results look very convincing, and they effectively are very good because these images contain numerous rectangular features. The orientation tolerance of $2\pi/8$ preserves angles in the presence of connected perpendicular segments. Second, let us take a look at the aerial view with its fields (figure.7.5.c). Here also the result is convincing, there is no excessive fusion of connected segments having similar orientations. One may suspect, however, that some noise in pixel orientation has helped to separate the segments. On the contrary, the synthetic, noise-free image "rings" (figure.7.5.a), clearly shows the drawback due to the tolerance on gradient orientation. One could reduce this tolerance to obtain satisfactory results for this image, but only at the price of a bad segmentation of the natural images, which would be unacceptable.

7.3 Polygonal approximation based on surface error

This algorithm is based on the idea that a straight line segment constitutes a good approximation of a chain of connected pixels if the surface between the segment and the chain is less than a certain threshold. Figure 7.6 illustrates three interesting cases.

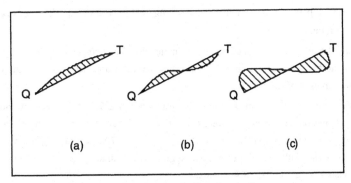

Fig. 7.6. Polygonal approximation based on surface error

In this figure, chaining goes from T towards Q; the area has positive sign when the segment is on the left-hand side of the chain. In (a), the area is positive; in (b) and (c), it is successively negative and positive, with final value 0. So this is a sound error criterion for (a), but unsatisfying for (b) and (c), where each of the positive and negative areas is too big. Wall [4] proposes an elegant solution by incrementally calculating the area during chaining. Chaining is interrupted when the computed surface passes over a threshold which is proportional to the length of the segment. Figure 7.7 makes explicit this incremental surface calculation.

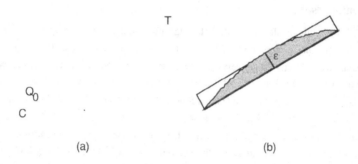

Fig. 7.7. Incremental calculation of the surface between the chain and the segment.

In (7.a), Q_0 is the last point in the chain, the vertically striped surface represents the previous surface A_0. C is the current point. When it is added to the chain, the error surface must be augmented by the horizontally striped surface ΔA. The increment ΔA is approximated by the triangle TQ_0C. Chaining is interrupted when

$$| A_0 + \Delta A | > k.length(TC)$$

It can be shown that the maximum distance ε (cf. figure 7.b) is less than k for a convex chain, less than 2k for an arbitrary chain.

Concerning the algorithm, the method can be implemented in a single scan. It needs no complex computation, and is very rapid. Figure 7.8 illustrates the result obtained by this method, with k=2, on the contour images of figure 7.1.

First consider figure 7.8.a. Compared with the results of figure 7.5, the surface criterion seems to be superior. However, this criterion breaks down for high curvature chains, as e.g. in the inner circles. This is particularly evident in the figures 7.8.b and 7.8.d containing a large number of angles, which he algorithm has a tendency to agglutinate. A smaller value of k might give better results, but also a larger number of segments.

7.4 Chaining and polygonal approximation using a double criterion

The preceding sections have shown the advantages and the limits of each of the two methods. In short, the orientation criterion is well adapted to the segmentation of chains including angles. Due to high curvature in angles, gradient orientation changes rapidly, causing the chaining to split. The criterion breaks down on chains with low curvature. One also has to decide on the threshold which defines when a change in orientation becomes significant for segmentation. A high value for the threshold may lead to excessive chaining, a low value to excessive segmentation, especially in the presence of noise. However, the orientation criterion is interesting from the algorithmic viewpoint. In particular, in the presence of a junction of any type, the orientation determines very simply the chain the current pixel should be included in. Moreover, using the orientation criterion one can restrict the processing to the neighbours in the "past", and not consider the neighbours in the current pixel's future. This greatly reduces the number of cases the finite state automaton has to handle, allowing a completely systematic commutation (no heuristics).

The surface criterion, which we consider as a way to limit the maximum distance between a chain of pixels and its approximating segment, is well adapted to segment low curvature chains. As is well known, it breaks down in the presence of angles. The algorithm we propose combines the advantages of the two preceding methods. It uses the orientation criterion to decide on the action to take for each contour point. However, this action is only carried out when the surface test is positive.

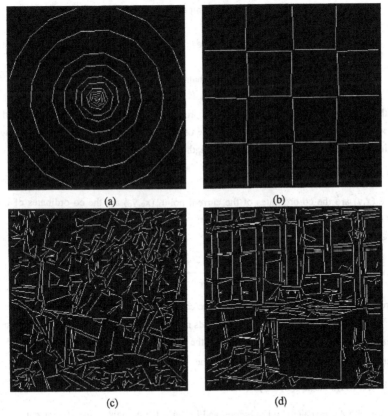

Fig. 7.8. Polygonal approximation based on the surface error.

7.4.1 Chaining algorithm

The algorithm follows the principal steps of the chaining algorithm using the orientation criterion.

During the scan, whenever the current pixel is a contour point:

calculate N

if N=0: (no connected chain)

    ♦ *Create*.

if N = 1, 2, 4, or 8: (unique connected chain)

    ♦ if surface_test then *Update*.

if N = 3, 6, 7, 10, 11, 12, 14, or 15: (several disjoint and connected chains)

    ♦ if surface_test then *Update* chain with highest weight.

    ♦ else if surface_test then *Update* chain with next highest weight.

    ♦ else if ...

if N = 5,9, or 13: (several disjoint and non-neighbouring chains)

    ♦ if the most distant chains have compatible orientations, and if surface_test
then *Merge*.

    ♦ else if the least distant chains have compatible orientations, and if surface_test
then *Merge*.

    ♦ else if surface_test then *Update* chain with highest weight.

    ♦ else if ...

With respect to the preceding algorithm, the procedures *Update* and *Merge* are modified, taking into account a surface test. This test makes use of the length of the new approximating segment, including the current point. In our case, this length is computed with the formula

$$L = |x - x_e| + |y - y_e|$$

where (x,y) are the co-ordinates of the current point, (x_e, y_e) are the co-ordinates of the most distant endpoint of the chain under consideration.

The surface is computed with the formula

$$A = A_0 + \Delta A = A_0 + \frac{(y - y_q)(x - x_t) - (y - y_t)(x - x_q)}{2}$$

where (x_t, y_t) and (x_q, y_q) are, respectively, the co-ordinates of the head and the tail of the current chain. Notice that the calculation of A just needs multiplications by 0, -1, or +1. The division by 2 is not necessary, as it can be taken care of with the coefficient k.

When two chains I and J are merged, the surface is calculated according to the formula:

$$A = A_I + A_J + surface(E_I C E_J)$$

where A_I and A_J are calculated as before, and $E_I C E_J$ is the triangle formed by the current point C and the most distant endpoints E_I and E_J of the chains I and J.

7.4.2 Experimental results

Figure 7.9 presents the experimental results for the test images. These results have been obtained using 24 channels to encode gradient orientation; just as with figure 7.5). The surface test has been implemented with k=2, as for figure 7.8). In figure 7.9.a one clearly sees that the low-curvature chains of the exterior circles have been cut by the surface criterion: the corresponding results in figure 7.8.a and 7.9.a are just the same. High-curvature chains (circles close to the centre) are cut according to gradient orientation. This observation also holds in the presence of angles, as can be seen in figures 7.9.b and 7.9.d. Finally, the comparison between the contour images (figure 1) and the polygonal approximations (figure 7.9) shows that the contours are very closely matched by the approximation. Also note that the asymmetry introduced by the scanning direction has become unperceptible.

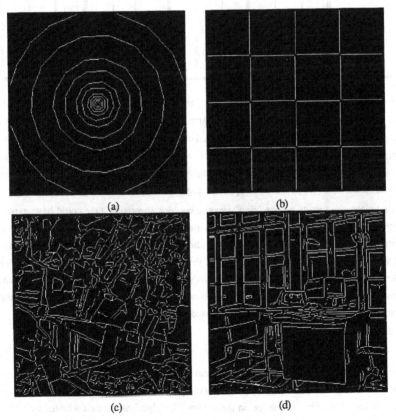

(a) (b)

(c) (d)

Fig. 7.9. Polygonal approximation by the proposed algorithm

Table 7.1 presents the results obtained with the three different chaining criteria. For each test image, the table indicates: the number of segments (in parentheses the number of segments of length 2 or 3), the mean length of the segments, the number of segments of mean length or more.

Table 7.1 Results obtained with the three chaining criteria.

	Rings			Chessboard			Aquitaine			Bureau		
Orientation crit.	9	2	3	4	3	2	176	3	1	7	7	2
	0	3	3	2	4	4	0 (958)		171	68		80
	(12)			(18)						(341)		
Surface crit.	9	2	4	2	5	2	977	7	4	4	1	1
	1 (0)	3	8	4 (0)	9	4	(213)		36	21	5	82
										(54)		
Proposed algor.	1	1	6	4	3	2	176	3	1	7	7	2
	13	8	2	2	4	4	6 (959)		178	69		81
	(12)			(18)						(340)		

One may note that the images "Chessboard", "Aquitaine"; and "Bureau" give very similar results for the orientation criterion only and our algorithm. This is due to the fact that these images feature a large number of angles causing segmentation. The angles themselves are represented by segments of length 2 or 3. On the opposite, the number of segments obtained using the surface criterion only is inferior for the three images. This is easily explained by the tendency of the algorithm to include angles in the chains. This produces an apparent augmentation in the average segment length.

At first sight, the results for the "ring" image may look surprising. One could have expected the surface-based algorithm to produce a larger number of segments than the orientation based algorithm. This is not the case, the number is almost the same. This comes from the fact that the orientation criterion produces few segments for the outer circles, and comparatively more segments for the inner circles. The surface criterion evolves in an opposite way, so the total number of segments is about the same for each of these criteria. As can be seen from the last line in the table, this number goes up when the two criteria are used simultaneously.

7.5 Conclusion

We have presented an algorithm for chaining and polygonal approximation of a contour image in a single scan. Using simultaneously two segmentation criteria of complementary nature, the value of thresholds have very little influence on the results. The algorithm is well adapted to images with straight contour lines, and gives equally good results in the presence of curved contours.

Concerning the algorithm, it is easily implemented and highly efficient. On a SPARC processor, we obtain 0.2 seconds of computing time. We also have implemented it on a specialised signal processor ADSP2100, where we obtain a computing time of roughly 40 milliseconds [Rungsunseri-Chehikian 1991].

References

[Burns et al 1986] J.B. Burns, A.R. Hanson, E.M. Riseman, Extracting Straight Lines, IEEE Trans. on Pattern Analysis and Machine Intelligence, IEEE Trans on PAMI, Vol. 8, pp. 425-455, 1986.

[Canny 1986] J. F. Canny, A Computational Approach to Edge Detection, IEEE Trans. on Pattern Analysis and Machine Intelligence, IEEE Trans on PAMI, Vol. 8, pp. 679-698, 1986.

[Detori 1982] G. Detori, An On-line Algorithm for Polygonal Approximation of Digitized Plane Curves, Proc. 6th Int. Conf. Pattern Recognition, Vol 2, pp. 840-842, 1982.

[Giraudon 1987] G. Giraudon, Chaînage efficace de contour, Rapport de Recherche INRIA N° 605, 1987.

[Nevatia-Buba 1980] R. Nevatia, K.R. Babu, Linear Feature Extraction and Description, Computer Graphics and Image processing, Vol. 13, pp. 257-269, 1980.

[Pavlidis 1977] T. Pavlidis, Polygonal Approximation by Newton's method, IEEE Trans. on Computer, Vol. 26, pp 800, 807, 1977.

[Rungsunseri-Chehikian 1991] Y. Rungsunseri, A. Chehikian, A Real Time System for Extracting Edges and Lines in Images, 7th Scandinavian Conference in Image Analysis, pp 839-846, Aalborg, Denmark, 1991.

[Slanski-Gonzales 1980] J. Slansky, V. Gonzales, Fast Polygonal Approximation of Digitized Curves, Pattern Recognition, Vol. 12, pp. 327-331, 1980.

[Wall-Danielson 1984] K. Wall, P.E. Danielson, A Fast Sequential Method for Polygonal Approximation of Digitized Curves, CVGIP Vol. 28, pp. 220-227, 1984.

8. Tracking and Description of Image Structures

James L. Crowley, Christophe Discours, and Bruno Zoppis
LIFIA, INPG

This chapter presents the image description and tracking module developed using the SAVA skeleton system. Copies of this module exist for the right and left stereo cameras. The module tracks edge segments using the predict-match-update cycle described in chapter 2. During tracking the module may be interrogated by other modules using a large vocabulary of preocedures for perceptual grouping.

In section 1 we describe the components of the module and their control. This description provides an example for the module components described in chapter 2. We include examples of the CLIPS rules which implement the control of the module, and describe the mathematics of the tracking process. In section 2 we present a vocabulary of perceptual grouping procedures which can be used to interrogate the module. These procedures are organised as a taxonomy composed of 4 classes. Each class is presented, and the algorithm for each procedures is briefly described. The image description and tracking module described in this chapter represents an well developed example of how the architecture presented in chapter 2 can be used to construct an integrated vision system.

8.1 Design of the Image Tracking and Description Module

In this section we will show how the SAVA Standard module has been used to construct a process for tracking perceptual organisation of image structures.

8.1.1 The SAVA Standard Module

The components of the the SAVA standard module is shown in Figure 8.1. The standard module is composed of a number of procedures (shown as rectangles) that are called in sequence by a scheduling process. Between each procedure call, the mail box is tested to see if any messages have arrived. Such messages may change the procedures that are used in the process, change the parameters that are used by the procedures, or interrogate the current contents of the description that is being maintained.

At the end of each cycle, the scheduler calls a set of demon procedures. Demons are responsible for event detection, and play a critical role in the control of reasoning. Some of the demon procedures, such as motion detection, operate by default, and may be explicitly disabled. Most of the demons, however, are specifically designed for detecting certain types of structures. These demons are armed or disarmed by recognition procedures in the interpretation module according to the current interpretation context.

In the 2D Description module, the model is composed of tokens that represent edge. Each segment is represented by a set of parameters and their temporal derivatives. Newly observed edge segments are obtained from the image acquisition and processing module described in chapter 5. These edges are computed within a log cartesian region of interest. extracted from the binomial pyramid described in chapter 6. Edge points are detected using the gradient magnitude and orientation. Extrema in the dgradient magnitude are detected and chained as described in chapter 7. The list of segements is then transmitted over the high bandwidth channel in response to a message get-edge-segments. In this section we describe each of the steps in this process.

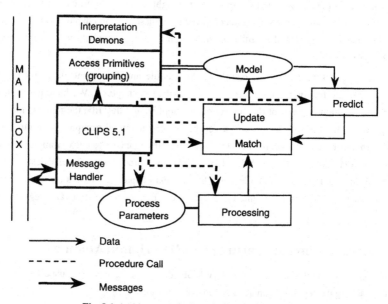

Fig. 8.1 Architecture of a Description Module in SAVA

In SAVA III, the role of the scheduler has been replaced by CLIPS 5.1. Each of the procedures which perform processing are declared to the CLIPS interpreter so that they can be called from within rules or functions.

8.1.2 Scheduler and The Control States

This section illustrates the scheduler for the 2D description module with a simple subset of the rules which control the operation. As much as possible, rules are restricted to control and functions are used for processing data. The scheduler in the image description module is implemented as a cylic sequence of states represented by a control element in CLIPS working memory. States are represented by a token in working memory of the form:

```
(phase <state> <cycle>)
```

where <cycle> is an integer number which records the cycle number, and <state> can be one of the following:

get-data: Acquire a new set of edge segments from image aquisition and processing module.

predict :   Using the time stamp from the newly acquired edge segments, predict the position and orientation of the segements in the model

match: Determine the list of correspondances between the segments.

update: update the model using the correspondence of the prediction and the observation.

demons: Execute any active demons.

In each state, just before the state transition, the message queue is tested and any new messages are treated. For example, at the end of the match phase, the following rule will transform the state to update, after testing for arrival of any new messages.

```
(defrule update-phase
   (declare (salience -100))
   ?p <- (phase match ?c)
=>
   (check-message)
   (retract ?p)
   (assert (phase update ?c))
)
```

Check messages places any message received as an ascii string in working memory, preceeded by the word "message" and the name of the sender. A high priority rule will interpret the string.

```
(defrule interpret-message
   (declare (salience 100))
   ?m <- (message ?sender ?str)
=>
   (eval ?str)
   (remove ?m)
)
```

The CLIPS function "eval" interprets an ascii string. If the interpretation is not sucessful, eval returns FALSE. In any case, the result is sent to the ?sender.

Each value of phase corresponds to a context in which a set of rules become activated. Let us illustrate how the control works by looking at the state to acquire edges.

8.1.3 Parametric Representation for Line Segments

A minimal representation for a line segment requires four parameters. The classic representation is the Cartesian coordinates of the two end-points. An alternative representation is to express a line segment as a center point, a half-length and an orientation. This second representation provides advantages for matching and for perceptual grouping. The model in the image description and tracking module is composed of line segments represented by 4 independently estimated parameters: (x_c, y_c, θ, h) as shown in Figure 8.2, as well as their temporal derivatives and covariances.

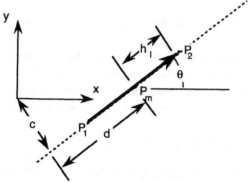

Fig. 8.2 Midpoint-Distance-Length Representation for Edge Lines

Thus the primary representation for segments is given by the vector $S = \{x_c, y_c, \theta, h\}$, where:

x_c The horizontal position of the center point.

y_c The vertical position of the center point.

θ The orientation of the segment.

h The half-length of the segment.

We also compute a number of "dependent" parameters which are useful for matching and grouping, as described below. These include the segment end-points P_1 and P_2 and the line equation coefficients, a, b and c, as well as the distance from the perpendicular projection of the origin, d. These redundant parameters are computed from the orientation using the formulas:

$a = \sin(\theta)$

$b = -\cos(\theta)$.

$c = x \sin(\theta) + y \cos(\theta)$

$d = x \cos(\theta) - y \sin(\theta)$

In the flow model it is necessary to estimate the temporal derivative of each parameter along with its covariance. A direct approach would lead us to a Kalman filter which estimates each line segment with an 8 dimensional vector (4 parameters and 4 derivatives) and an an 8 by 8 covariance matrix. We have greatly simplified the computational load by separating the estimation into 4 independent estimations. For each parameter $x \in S$ we maintain a first order estimate vector $\hat{X}(t)$.

$$\hat{X}(t) \equiv \begin{bmatrix} \hat{x}(t) \\ \hat{x}'(t) \end{bmatrix}$$

where $\hat{x}'(t) = \dfrac{\partial \hat{x}(t)}{\partial t}$.

In addition, for each property $x \in S$, we represent its variance, the variance of the temporal derivative, and a covariance between the estimate and its temporal derivative

$$\hat{C}_{x}(t) \equiv \begin{bmatrix} \sigma_x^2 & \sigma_{x'x} \\ \sigma_{xx'} & \sigma_{x'}^2 \end{bmatrix}$$

Segments also include a confidence factor, CF, a recency R, and a unique identity, ID (an integer). The confidence factor is expressed as an integer state, from the set $\{1, 2, 3, 4, 5\}$. CF = 1 represents a model segment which is very uncertain. CF = 5 represents a segment which has been reliably tracked for at least 3 frames. When the token for a segment is first created, it is added to the model with a confidence factor of CF = 3. During each update cycle, if a correspondence is found for a token, the CF is incremented by 1, up to a maximum value of 5. If no correspondence is found, the CF is decremented by 1, to a minimum value of 0. If the CF passes to 0, the token is removed from the model. The use of a confidence factor gives the flow model a degree of immunity to the temporary loss of edge lines in as many as four successive observation images. The ID is a unique index which makes it possible to identify a segment at different times. The recency is the cycle number where the segment was last updated or matched with perceptual grouping.

In summary, each line segment is represented in the model with the following parameters:

Horizontal Position:	$\{ x_c, x_c', \sigma_x^2, \sigma_{xx'}, \sigma_{x'}^2 \}$
Tangential Position:	$\{ y_c, y_c', \sigma_y^2, \sigma_{yy'}, \sigma_{y'}^2 \}$
Orientation:	$\{ \theta, \theta', \sigma_\theta^2, \sigma_{\theta\theta'}, \sigma_{\theta'}^2 \}$
Half Length:	$\{ h, h', \sigma_h^2, \sigma_{hh'}, \sigma_{h'}^2 \}$
Confidence Factor:	CF
Identity:	ID
Recency	R
Line coefficients:	(such that $ax + by + c = 0$)
projection of origin	d
End-points	$\{P_1 = (x_1, y_1), P_2 = (x_2, y_2)\}$

8.1.4 Data Acquisition

Segments are provided within a multi-resolution region of interest by the image acquisition and processing (IAP) module by chaining the extremal points in the gradient as described in chapters 6 and 7. A list of observed edge segments is acquired by sending a message to the IAP module. The IAP module acquires a new image (or pair of stereo images) and processes them to extract edge segments within the specified ROI using the specified parameters. The state "get-data" triggers a rule:

```
(define get-edge-segements "Rule to request edge segments"
    ?p <- (phase get-data ?c)
    (camera ?c)
=>
    (get-head-position)
    (get-new-edges)
    (remove ?p))
    (assert (phase wait-edges ?c)))
)
```

The function "get-head-position" sends a message which obtains the position and orientation of the head, and the vergence angle of the two cameras. The return message is preceded by a string "camera-position" which is interpreted by the 2D tracking and description module as an update to the camera-position data struction. This function also creates a token of the form (new-camera-position) on the working memory to indicate that it has responded. This token is deleted by the next change in phase.

The function get-new-edges sends a message to the IAP module to acquire a new image using the specified function, extract edge segments from the specified region of interest, using the specified edge segment thresholds. These parameters are contained in a global structure and included in the function call. These arguments are composed into a string for transmission in a message to the IAP module. The character-string is interpreted as a function by the IAP module.

The IAP module responds by acquiring a new image from the specified camera, calculating a binomial pyramid, and extracting the edge segments. During this time, the 2D description module polls the message queue, waiting for the message that the edges are available on the high bandwidth channel. Polling is performed by the rule "wait-for-edges"

```
(defrule wait-for-edges  "test if edges arrive"
  ?p <- (phase wait-edges ?c)
=>
  (check-messages)
)
```

When the IAP module has produced its edges, it responds with a message of the form:

```
"get-edges-from-hbwc number %d time %d"
```

The message is interpreted by the rule "interpret-message" described above. The function "get-edges-from-hbwc" reads the number of edges and the time-stamp, creates a time stamp in working memory, and then extracts the edges from the high bandwidth channel. The resulting edges are placed in a list of structures, with a pointer in a working memory element of the form (new-data ?pointer). This token is used to signal a state change. The state is then changed to the "predict" by the rule "wait-edges".

```
(defrule wait-edges  "test if edges arrived"
  ?p <- (phase wait-edges ?c)
  (new-data ?edges)
  ?pos <- (new-camera-position)
=>
  (remove ?p)
  (remove ?pos)
  (assert (?p <- (phase predict ?c)
)
```

8.1.5 Predict Match and Update

The predict, match and update functions are performed by function calls from CLIPS. Predict uses the time stamp, the change in camera position and the current model to predict a new model. The model is represented in working memory as a pointer to a data structure that was allocated by update. The predicted model is represented as a pointer to a data structure allocated by the function predict.

```
(defrule predict
   ?p <- (phase predict ?c)
   (model ?m)
   (update-time ?ut)
   (observe-time ?ot)
=>
   (bind ?dt (- ot ut))
   (assert (predicted-edges (predict ?m ?dt)))
   (remove ?p)
   (assert (phase match ?c))
)
```

Camera angle is the sum of the head angle and the vergence angle. The change in angle is translated to pixels by a calibrated coefficient for pixels/degree. This then forms the commanded displacement value v_x for the segment parameter x.

Prediction uses the Kalman Filter Prediction equations (equations 1 and 2 in chapter 2) to predict the evolution of the parameters of the edge segments and their covariances. The results is a list of predicted edge segments in a working memory element of the form (predicted-edges ?pointer). Given an estimate set of parameters, $\hat{S}(t)$, the prediction follows directly from the prediction equations (1) and (2) of chapter 2. For each $x \in S \equiv \{x_c, y_c, \theta, h\}$:

$$X^*(t+\Delta T) := \varphi \, \hat{X}(t) + \hat{V}(t)$$

and

$$C_x^*(t+\Delta T) := \varphi \, \hat{C}_x(t) \, \varphi^T + \hat{Q}_x(t) \, \Delta T^4$$

where

$$\varphi \equiv \begin{bmatrix} 1 & \Delta T \\ 0 & 1 \end{bmatrix}$$

Unmodeled acceleration between update cycles are a source of uncertainty. To account for the possibility of accelerations, the uncertainty of each attribute is increased by a constant term, σ_{acc}^2, multiplied by the time interval to the fourth power. For each of the parameters, this term is included in the term $\hat{Q}_x(t)$:

$$\hat{Q}_x(t) \equiv \begin{bmatrix} \sigma_{acc}^2 & 0 \\ 0 & 0 \end{bmatrix}$$

8.1.6 Demons

At the end of each update cycle, the scheduler executes an agenda of interpretation procedures named "demons". Demons provide a mechanism for dynamically interpreting the contents of the model. Demons are invoked by other demons or by messages received from other modules. A demon is instantiated by entering a demon token in working memory. A demon token is a list with four elements:

```
(demon   <name>  <id>)
```

where <name> is the name of the demon and <id> is a unique identity determined by the function "gensym". Demons are based on a combination of CLIPS rules, CLIPS functions and C procedures.Any state information is savec in structures in CLIPS 5.1 working memory.

8.2 Procedures for Model Interrogation and Perceptual Grouping

This section presents the procedures that have been implemented for interpreting the 2D image descriptions from the left and right camera. Similar procedures have been developed for access and grouping for the 3D description module. These procedures are organised in 4 classes according to their computational complexity, and the nature of the interrogation. This library of procedures continues to grow as we develop additional recognition procedures.

The perceptual grouping procedures are based on calculations that use the numerical attributes of edges maintained by tracking. For each model interrogation procedure we briefly describe the algorithm, the quality factor that is used, the properties that describe the grouping, and we show its computational cost.

We have defined a standard protocol for model interrogation based on the procedures "Find", "Verify", and "Get". In the 2D model, this protocol uses the model access and perceptual grouping procedures described below in section 8.4. The role of the standard access functions is illustrated by figure 8.3.

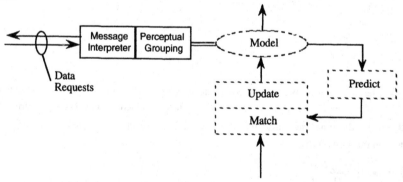

Fig.8.3 The model access primitives: Find, Get and Verify, from chapter 2

The interrogation procedures have the following form:

<List of IDs>   ← Find <Grouping Primitive> <Parameters> <ROI>
<CF>   ← Verify <Grouping Primitive> <Parameters>
<List of Parameters>   ← Get <ID>

The procedure "Find" searches for class of structures in the module using a set of parameters and a region of interest. "Find" returns a set of identities. "Get" permits access to the parameters of a token based on its ID. "Verify" returns a confidence factor for the existence of a primitive or grouping with a set of parameters.

8.2.1 A Taxonomy of Interrogation and Grouping Procedures

Model interrogation procedures may be organized into a taxonomy according to the nature of the interrogation that they make. We have identified four classes of interrogation procedures:

Class 1 Procedures for extracting parametric primitives that are *close to an ideal geometric entity* such as a line or a point. Such primitives are easily implemented and have a complexity that is proportional to the number fo primitives in the model.

Class 2 Procedures for extracting parametric primitives that are *similar to an ideal "prototype"*. The prototype is specified an estimated value and a standard deviation for each of the primitive parameters. Similarity is measured by a Mahalanobis Distance. Parameters may be labeled as a "wild-card" by specifying a large (or infinite) standard deviation.

Class 3 Procedures for extracting *pairs of primitives* that satisfy some geometric relation, such as a junction, an alignment, or a parallelism. The geometric relation is defined as a set of parameters and specified as a estimation and a standard deviation. These procedures have theoretical complexity of $O(N^2)$.

Class 4 Procedures for extracting *grouping of groupings*. If performed by brute force, complex forms such as lines and contours composed of sets of line segments can require an exponential number of computations. By structuring the interrogation as a grouping, we can limit the theoretical complexity to $O(N^2)$ or $O(N^3)$.

In addition to these four classes it is also possible to interrogate the model using a composition of groupings. That is we can apply the grouping operations in sequence, for example, extracting all junctions of a certain form that are near an ideal line. Thus the $O(N)$ primitive can be used to restrict the number of entities interrogated (the value of N) by an $O(N^2)$ or $O(N^3)$ procedure.

The result of an interrogation is a list of segments or groups of segments that satisfy the specified geometric property. These segments or groups are accompanied by a "quality factor" based on the similarity between the requested and observed entity, the list of ID's for the entities that satisfy the relation is returned to the calling module, and a small set of properties of the grouping. This list is sorted based on the quality factor.

For each procedure, we will present its computational cost in the following terms:
A   Number of additions or subtractions per operation
M   Number of multiplies or divisions per operation.
C   Number of comparisons per operation
N   Number of segments in the 2D model.

All computational costs are worst case. The actual code contains a number of optimizations that permit us to minimize the cost.

8.2.2 Class 1 - Grouping with Respect to a Point or a Line

The first class of procedures for extracting parametric primitives that are *close to an ideal geometric entity* such as a line or a point. Such primitives are easily implemented and have a complexity that is proportional to the number of primitives in the model. At the current time, procedures exist to retrieve a list of segments near a point in the image, or near to a line equation.

Segments near a Point:

Detection of segments near a point is a typical case for model access. The parameters for this procedure are the positions of the point, $P = (x, y)$, and the tolerance, σ_d.

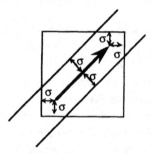

Fig. 8.4. A Segment is near a point if the point passes two tests

A segment is close to a point if the point is within a bounding box defined by the segment, and if the point has a distance of less than σ_d to the segment. The bounding box is defined by adding and subtracting the tolerance σ_d to the end-points of the segment. This operation requires 4 additions and 2 comparisons. The point is then tested for inclusion within this rectangle, requiring at most 4 comparisons. If the point is in the box, its perpendicular distance to the segment is computed using:

$$d_p = |\, A\, x + B\, y + C\,|$$

In the worst case, where the point is within the bounding box of every segment, the computational cost for N segments is:

$$\text{Cost} = (4\, A + 2\, M + 6\, C)\, N$$

Where A is the cost of an add, M the cost of a multiply and C the cost of a comparison.

The edge segments that meet the test are assembled into a list sorted based on the quality factor. The quality factor is given by the formula:

$$Q = \frac{d_{min}}{\sigma_d}.$$

A segment with a quality factor of zero is ideal. Segments that pass the test are inserted in the list based on their quality factor. The sorted list of segment ID's with their quality factors is returned.

8.2.3 Class 2 - Grouping with Respect to a Prototype

The second class of model interrogation procedures are for edges that are *similar to an ideal "prototype"*. The prototype is specified an estimated value and a standard deviation for each of the primitive parameters. Similarity is measured by a Mahalanobis Distance. Parameters may be labeled as a "wild-card" by specifying a large (or infinite) standard deviation. This test is illustrated in figure 8.5.

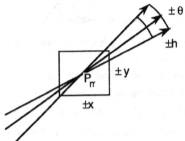

Fig. 8.5. Similarity with respect to a prototype segment

The prototype segment is given by parameters $(x_p, y_p, h_p, \theta_p)$. A tolerance vector for the parameters is given by $(\sigma_x, \sigma_y, \sigma_\theta, \sigma_h)$. The Mahalanobis distance for each parameter is given by square of the difference divided by the square of the tolerance. We can perform an equivalent test by simply comparing the absolute value of the difference to the tolerance, where abs() has a cost of one compare. In the worst case, for N segments, this test can costs

$$Cost = (4 \, A + 8 \, C) \, N$$

Where A is the cost of an addition or subtraction and C is the cost of a compare. For each segment that passes all four tests, the quality is given by:

$$Q = \left(\frac{x_p - x}{\sigma_x}\right)^2 + \left(\frac{y_p - y}{\sigma_y}\right)^2 + \left(\frac{h_p - h}{\sigma_h}\right)^2 + \left(\frac{\theta_p - \theta}{\sigma_\theta}\right)^2$$

The ID's of the segments that satisfy all four tests are inserted in a list based on their quality factor. The sorted list of segment ID's with their quality factors is returned.

8.2.4 Class 3 - Grouping of Pairs of Segments

The simplest grouping of segments are pair-wise groupings. Procedures for extracting *pairs of primitives* that satisfy some geometric relation, such as a junction, an alignment, or a parallelism constitute our third class of model access procedures. The geometric relation is defined as a set of parameters and specified as an estimation and a standard deviation. Detection of pair-wise groupings has an inherent complexity of $O(N^2)$.

Junctions of Two Segments:

A widely used pair-wise grouping are junctions of two segments. Junctions are defined with three parameters, θ, ψ, and d, as shown in figure 8.6.

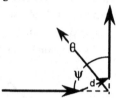

Fig. 8.6 A Pair-wise Junction of Two Segments

A prototype junction is specified as:

ψ, σ_ψ: The internal angle of the junction and its tolerance.

θ, σ_θ: The orientation of the junction and its tolerance.

σ_d: The tolerance of the distance between end-points.

For each segment having its point P_2 within the region of interest, the procedure tests every segment (other than itself) and computes the square of the distance from the point P_2 to the point P_1 of the other segment. This operation costs $3A + 2M$. If the squared distance is less than the square of the tolerance, then the internal angle and external angle for segments 1 and 2 are computed as

$$\psi_s = \theta_1 - \theta_2 .$$
$$\theta_s = \theta_1 + \theta_2 .$$

These values are subtracted from the prototype values, and the absolute value of the difference is compared to the tolerance, for a cost of 4 additions and 2 compares. Thus in the worst case, this operation costs

$$\text{Cost} = (7A + 2M + 3C)\, N(N-1)$$

The quality factor is based on the similarity with the prototype:

$$Q = \left(\frac{d}{\sigma_d}\right)^2 + \left(\frac{\psi_p - \psi_s}{\sigma_\psi}\right)^2 + \left(\frac{\theta_p - \theta_s}{\sigma_\theta}\right)^2$$

The sorted list of junctions is returned with the quality factor and the properties d, θ, and ψ.

T-Junction

A T-junction is another common class of pair-wise junction. A T-junction has the same parameters as a pair-wise junction. An prototype T junction is specified as:

ψ, σ_ψ: The internal angle of the junction and its tolerance.

θ, σ_θ: The orientation of the junction and its tolerance.

σ_d: The tolerance of the distance between end-points.

As with a junction, the internal angle and external angle are computed as

$\psi_s = \theta_1 - \theta_2$.

$\theta_s = \theta_1 + \theta_2$.

There are two types of T junctions: Those formed with point P_1 and those formed with point P_2, as shown in figure 8.7.

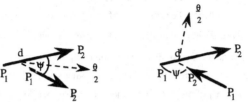

Case 1: Interior T Junction   Case 2: Exterior T Junction

Fig. 8.7 Two types of T Junctions and their parameters

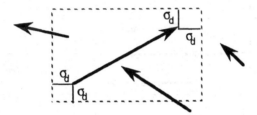

Fig. 8.8. Candidate T Junctions have end-points in the bounding box

For each segment, a list of candidate T junctions is computed by testing the end-points of all other segments to see if they fall within the bounding box defined by adding σ_d to the segment end-points. For any end point found within the box, the perpendicular distance to the support line of the edge segment is computed, and compared to the tolerance distance. Because this must be computed between all pairs of segments, the cost is:

Cost = (4 A + 2 M + 6 C)N(N-1)

If the distance is less than the tolerance the edge is kept in the list of T junctions. The quality is given by:

$$Q = \left(\frac{d}{\sigma_d}\right)^2 + \left(\frac{\theta_s - \theta_k - \theta}{\sigma_\theta}\right)^2 + \left(\frac{\theta_s + \theta_k - \psi}{\sigma_\psi}\right)^2$$

Where d is the perpendicular distance, and θ_s is the orientation of the segment "s" that defined the box, and θ_k is the orientation of the segment whose end-point was in the bounding box. If one of the end-points of the segment "s" is closer than σ_d to the end point being tested, then the pair of segments forms a normal junction, and not a T junction.

Aligned Segments

Alignment, parallel and anti-parallel are a mutually exclusive relations between two segments as indicated in figure 8.9. These relations are based on three measurements:

1)   The similarity of the orientation, σ_θ,

2)   The perpendicular distance of midpoints, σ_c, and

3)   The distance between the projection of end-points to the supporting line, σ_d

The parameters that define alignment are C, D, and θ. Two segments are said to be aligned if:

1)   The two segments have similar orientation ($|\theta_1 - \theta_2| < \sigma_\theta$)

2)   Each segment is near the support line of the other segment ($|C_1 - C_2| < \sigma_c$), and

3)   The end-point P_1 of one segment projects to the support line of the second segment a distance less than σ_d beyond the end-point P_2 of the second segment.

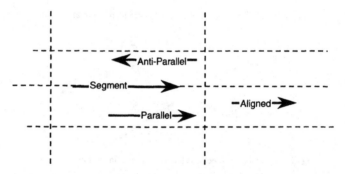

Fig. 8.9 Parallel, Anti-Parallel, and aligned segments

For each segment, s, the tests are applied to every other segment, k. The first test is easily made by comparing the parameters θ. If the difference in orientations is less than the test continue. The second test is based on the perpendicular distance from of the end points of segment k to the support line of segment s, as computed by the line equations:

$$c_1 = A\,X_1 + B\,Y_1 + C$$

$$c_2 = A\,X_2 + B\,Y_2 + C$$

C_1 and C_2 have opposite signs, then the segment crosses the supporting line, and the distance is $C_{min} = 0$. Otherwise, the test is based on the smaller absolute value of C.

$$C_{min} = \min\{\, |c_1|, |c_2| \,\}$$

If $C_{min} < \sigma_c$ then the test is passed. This test requires 4M + 4A + 3C for each segment.

The final test is based on where the end-points of the second segment project onto the support line. This projection can be expressed in terms of a parameter, D, that is the distance from the projection of the origin onto the support line. For a line equation, $Ax + By + C = 0$, the D parameter is computed by:

$D = Bx + Ay$

As each segment, s, is considered, we compute D_s for the midpoint of the segment s. To determine the overlap of a segment k with the segment s, we use the parameters (A, B) of s to compute the parameter D_k for the midpoint of segment k. Three cases are possible based on the half-length, h of the segments s and k.

Case 1:	$D_s + h_s < D_k - h_k$	Segment k is aligned after segment s.
Case 2:	$D_s - h_s > D_k + h_k$	Segment k is aligned before segment s.
Case 3:	Otherwise	Segment k and segment s overlap.

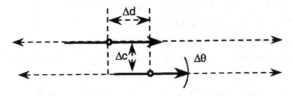

Fig. 8.10 The parameters that define parallelism, anti-parallelism and alignment

For the procedure to detect alignment, only case 1 of test 3 is applied. This gives a cost for the third test of $(2M + A) N^2$ for a total cost of: The total cost of detecting alignments is

$$Cost = (6M + 5A + 4C)N^2$$

In case 1, the segments are marked as aligned. In case 2, the relationship is not marked. It will be detected when segment k is compared to all other segments. To detect parallel segments, we apply tests 1 and 2 and case 3 of test test.

Anti-parallel segments are similar to parallel segments, except the the difference in orientation should be close to 180°. The quality factor for aligned segments is

$$Q = \left(\frac{\theta_s - \theta_k}{\sigma_\theta}\right)^2 + \left(\frac{C_{min}}{\sigma_c}\right)^2 + \left(\frac{D_s - D_k + h_s + h_k}{\sigma_D}\right)^2.$$

The quality factor for parallel segments is

$$Q = \left(\frac{\theta_s - \theta_k}{\sigma_\theta}\right)^2 + \left(\frac{C_{min}}{\sigma_c}\right)^2 + \left(\frac{D_s - D_k}{h_s + h_k}\right)^2$$

The quality factor for anti-parallel segments is

$$Q = \left(\frac{\theta_s - \theta_k - 180°}{\sigma_\theta}\right)^2 + \left(\frac{C_{min}}{\sigma_c}\right)^2 + \left(\frac{D_s - D_k}{h_s + h_k}\right)^2$$

8.2.5 Class 4 - Complex Groupings (Groupings of Groupings)

Detecting lines and contours in the presence of multiple gaps is notoriously hard. If attacked in a naive way, this process has an exponential complexity. By performing the detection as a group of groups, such forms may be detected with a theoretical complexity which is $O(N^3)$ and which is practically much smaller.

Y Junctions

Y Junctions are built up as pairs of pair-wise junctions. All pair-wise junctions involve the proximity of a head (P_2) with a tail (P_1) of a second segment. Only two types of Y junctions are possible: Y–junctions with tow tails (Figure 8.11a) and Y–junctions with two heads (Figure 8.11b).

The parameters for a Y junction are:

Type:	Two heads or two tails.
$\psi_1, \sigma_{\psi 1}$:	The internal angle of the first junction.
$\psi_2, \sigma_{\psi 2}$:	The internal angle of the second junction.
θ, σ_θ:	The sum of the orientations of the junctions.
σ_d:	The tolerance of the distance between points of the two junctions.

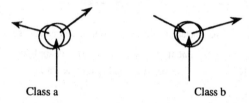

Class a   Class b

Fig. 8.11 The two classes of Y junctions. a) Junctions with one head and two tails. b) Junctions with two heads and one tail

A Y junction is detected based on coincidence of a segment in two junctions. That is, all two segment junctions within the region of interest are detected using the specified parameters for the first junction. All specified parameters for the second junction are then detected. Each operation returns a sorted list of junctions that includes the ID's of the segments. For each junction in the first list, the second list is scanned to search for the any junction that includes one of the two segments. For each case where the two junctions contain a common segment, each of the parameters is compared to its tolerance. If each parameter is within tolerance, then the quality factor is computed as:

$$Q = \left(\frac{d1}{\sigma_{d1}}\right)^2 + \left(\frac{d2}{\sigma_{d2}}\right)^2 + \left(\frac{\psi_1 - \psi_{s1}}{\sigma_{\psi 1}}\right)^2 + \left(\frac{\psi_2 - \psi_{s2}}{\sigma_{\psi 2}}\right)^2 + \left(\frac{\theta_p - \theta_s}{\sigma_\theta}\right)^2$$

The cost of the operation is

$$\text{Cost} = 2 (7A + 2M + 3C) N (N-1) + C N^2.$$

Complex edge lines

A common problem in edge segment descriptions is that segments break in unpredictable and often random places. Sequences of broken segments that are aligned can be grouped to form longer segments using a complex grouping operation. The parameters for the procedure for detecting a complex alignment are the same as those for detecting a simple alignment (section 8.4.3). These are:

σ_θ The tolerance for deviations of orientations within a segment.

σ_c The tolerance for perpendicular displacement

σ_d The maximum gap between end-points.

The resulting composite line segment has the same attributes as a simple line segment.

The algorithm begins by constructing a list of all pair-wise alignments of segments using the parameters σ_θ, σ_c, σ_d and the algorithm for detecting alignments described above. As before the cost of this step is $(6M+ 5A + 4C)N^2$. The result is a list of pairs of ID's of segments that are aligned, as shown in Figure 8.9 a.

Each pair is assigned a "pair-ID", to give a list composed of triples: (Pair-ID, (Seg-ID$_1$, Seg-ID$_2$)). This list of segment pairs is then scanned to detect alignments pairs that share a segment. This test is based uniquely on matching the Seg-IDs.

To test the alignment of two groups, a composite edge segment is computed by taking the end-point P$_1$ of seg-ID$_1$ and the point P$_2$ of seg-ID$_2$. The mid-point and orientation for this composition are computed. We then make exactly the same tests as for alignment of two segments. If the composite segments are aligned, with respect to the specified tolerances, a third level grouping is formed:

     (triple-ID, (pair-ID$_1$, pair-ID$_2$)).

The procedure is repeated for higher level groupings until no further groupings can be formed.

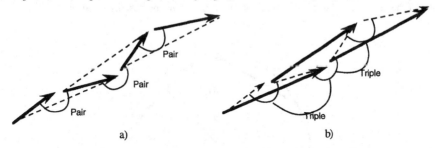

a)   b)

Fig. 8.12 a: Pair- wise groupings of segments, b: Groupings of groupings of segments

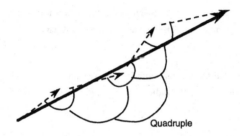

Fig. 8.12 c: A complex Fourth level grouping and the resulting line segment

Arcs

A procedure similar to that for complex edges can be used to detect complex arcs. This procedure is based on grouping junctions of two segments. The parameters for detecting an arc are:

σ_d: The maximum gap between end-points.

K: The desired Curvature

σ_K: The tolerance for deviations in curvature

The procedure returns

P_c: The center point of the composite arc

K: The average curvature of the arc.

P_1, P_2:   The end-points of the arc.

Our algorithm for detecting edge segments [Chehikian 91], approximates an arc with a segment whose end-points tend to lie on (or close to) the arc. The largest error tends to be at the center of the segment. Thus arcs are defined as passing through the end-points of segments.

The bisectrice of a segment defines the locus of centers of circles that through the end-points of a segment. The bisectrice has a line equation given by

$$B x + A y - (Bx_m + Ay_m) = 0$$

If two segments share a junction, and are not parallel, their bisectrices will have an intersection point. This intersection point defines a common center-point for the arcs for each of the segments, as illustrated in Figure 8.13. The radius of each arc can be computed as the distance of either end-point of the segment to this center-point. The two radii will not necessarily be equal.

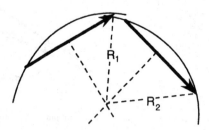

Fig. 8.13 A Junction of two segments can be used to define two arcs. The center point of the two arcs is the intersection of the segment normal lines. The radii are given by the distance of an end-point from this center-point

To arrive at a common radius, we can calculated a weighted average of the two radii, using the segments arc lengths, s, defined by the radius and the angle subtended by the arc, α.

$$s = \alpha R.$$

Thus, for segments 1 and 2, the joint radius is given by:

$$R = \frac{1}{s_1 + s_2}(R_1 s_1 + R_2 s_2)$$

Curvature, K, is defined by inverse of the radius. Two arcs form a complex arc if they share a segment. To decide if they are part of the same arc, we compare the difference in curvature to the tolerance σ_K.

The quality factor is based on the difference between the desired and observed curvature:

$$Q = \frac{K_s - K}{\sigma_K}$$

8.2.6 Compositions of Grouping Operations

The first two classes of operations described above each produce a list of ID's of edge segments that satisfy some relation. The third class produces a list of pairs of edge segments. The fourth class produces a list of arcs, curves or Y junctions, each of which contains a list segments. Each of these procedures contains a "mode" argument that will cause them to return the list of edge segments that were detected. The list of segments produced int his mode can be substituted for the edge segment model in subsequent model access procedures. This makes it possible to compose a chain of model access operations, using the list of segments from each operation as input to the next.

Chaining model access operations is a form of logical "AND" operation. For example, with such a chain it is possible to request all of the junctions of two segments that are near an ideal line segment, or all anti-parallel lines near a point. Chaining operations permits us to use the $O(N)$ operations from classes 1 and 2 to select the candidate segments considered by the more expensive $O(N^2)$ and $O(N^3)$ operations of classes 3 and 4.

The parameters for extracting segments near a line are the coefficients of the line equation, (A, B, C), and the tolerance distance, σ_d. Three values are compared to the tolerance distance: The perpendicular distance to each of the end-points, and the intersection point between the line and the segment. If the segment intersects the line, then its quality factor is zero.

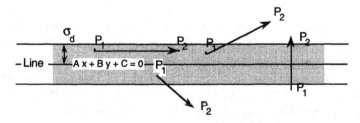

Fig. 8.14: Four cases for distance of a segment to a line

The distance from the end-points d_1 and d_2 of a segment to a line may be computed from the line coefficients, A, B, C by the formulae:

Case a: $d_1 = A\,X_1 + B\,Y_1 + C$

Case b: $d_2 = A\,X_2 + B\,Y_2 + C$

Each distance requires 2 multiplies and 2 adds. The overlap of the segment with the line equation requires comparing the sign of d_1 and d_2. If their signs are different, then the segment traverses the line, and the distance to the line is taken as zero.

Thus the cost of evaluating this function for each of N model segments is

$$Cost = (2M + 2A + 3C)\,N.$$

The quality factor is based on distance from the line compared to the tolerance distance d.

$$Q = \frac{d_{min}}{d}.$$

The ID's of the segments that pass the test are inserted in a list with a position based on their quality factor. The sorted list of segment ID's with their quality factors is returned.

8.3 Discussion and Conclusions

The problem of dynamically maintaining a description of the local environment is fundamental to perception. Dynamic world modeling can be organised as a cyclic process composed of the phases: predict, match and update. The Kalman Filter equations permit observations of a small number of properties to serve as constraints for a vector composed of a larger number of properties. This fact is counter-intuitive and easily mis-understood. In the 2D token tracking system, observations of the value of position, length and orientation permitted us to estimate the first temporal derivatives of these values.

Tracking preserves correspondence. This simple fact alone justifies the use of 2D tracking in a vision system. The 3D system exploits the estimates from the 2D tracking system as observations. This architecture shows that it is possible to compose perceptual systems with multiple independent predict-match-update loops operating in different perceptual coordinates and at different levels of abstraction.

The "observation" uncertainties are crucial to the function of the system. At the current time, the uncertainties for the segment tracking process are "tuned" by hand. That is, when the uncertainties are large, the number of tracked segments remains stable, but the precision for tracking is decreased. When the uncertainties are too small, the number of tracked segments decreases rapidly. This provides a method to "tune" the uncertainties for a particular situation. The process is made to run, and the number of segments with a confidence factor greater than 3 are displayed. The uncertainties are then systematically reduced until the number of segments in the model begins to decrease.

Interpretation requires geometric reasoning. Such reasoning requires frequent access to the geometric data. We have extracted the geometric portion of the interpretation into a vocabulary of procedures for associative model access and model grouping. We present a taxonomy for such procedures in terms of four classes, based on the nature of the data access and the complexity of the computation. The list of procedures which we describe is not exhaustive. We continue to construct additional grouping procedures as we write recognition procedures.

Some previous efforts at perceptual grouping had based the computations on the use of graph representation for the edge segments [Veillon et al. 90]. Our experience is that such representations are inappropriate for continuous operation. In any edge based description of an image there are always gaps along edge segments and at junctions. Closing such gaps requires specifying a tolerance. Such a tolerance will depend on the contents of the scene, the lighting, and the optical parameters of the cameras, and thus must be actively controlled. Our experiments have convinced us that it is more appropriate to treat the edge segments as a list of property vectors and to optimize the tests that are applied to this list.

Dynamic vision requires pre-attentive grouping for *event detection* as well as post-attentive grouping for *interpretation* [Triesman 88]. We have found that such pre-attentive event detection may be performed by arming a set of demon procedures that search the dynamically updated model for desired events. Some demons are armed by default, other are armed by request. We are currently trying to better understand the relative roles and computational costs of these different kinds of demons.

The grouping operations that we have presented have all concerned organization of 2D edge lines, a description which is inherently noisy. Another line of work in our group involves by-passing the extraction of edges and directly measuring interest points. We have experimented with receptive fields for such image events as end-stops, corners and junctions. Such operators are inspired by the complex and hyper-complex cells in the mammalian visual cortex. The corresponding simple cells are sets of Gabor filters of the sort described in chapters 10 through 14 below.

References

[Chehikian 91] Chehikian A., A One Pass Edge Exctraction Agorithm, 7th Scandanavian Conference on Image Analysis, Aalborg, 1991.

[Triesman 88] Triesman, A. Features and Objects, The Quarterly Journal of Experimental Psychology, 40A-2, 1988.

[Veillon et. al. 90] Horaud, P., F. Veillon, and T. Skordas, Finding Geometric and Relational Structures in an Image, Proceedings of the first ECCV, Springer Verlag, Antibes, April 1990.

9. Contextual Junction Finder

Jiri Matas and Josef Kittler

UOS

Abstract. A novel approach to junction detection using an explicit line finder model and contextual rules is presented. Contextual rules expressing properties of 3D-edges (surface orientation discontinuities) limit the number of line intersections interpreted as junctions. Probabilistic relaxation labelling scheme is used to combine the a priori world knowledge represented by contextual rules and the information contained in observed lines.

Junctions corresponding to a vertex (V-junctions) and an occlusion (T-junctions) of a 3D object are detected and stored in a *junction graph*. The information in the junction graph is used to extract higher level features. Results of the most promising method, *the polyhedral object face recovery*, are briefly discussed. The performance of the junction detection process is demonstrated on images from indoor, outdoor, and industrial environments.

9.1 Introduction

Perceptual groupings of image features have been widely used in computer vision systems to guide scene interpretation and 3D model matching [Bergevin and Levine, 1990; Coelho *et al.*, 1990; Horaud *et al.*, 1990; Lowe, 1987; Mohan and Nevatia, 1989]. Of all perceptual groupings studied by psychologists [Pomerantz, 1981; Haber and Hershenson, 1973] and computer vision researchers we focus our attention on *junctions of line segments* - points of co-termination of lines. As co-termination is a projection-invariant property, the task of junction detection would be relatively simple in an ideal noise-free world. A set of lines terminating at the same point could be interpreted as a projection of edges meeting at a vertex. However, due to the inherent inaccuracy of line (and edge) detection, endpoints of lines can be widely separated even if the lines emanate from a common vertex.

In a novel approach, an explicit error model for line detection in conjunction with *contextual rules* is used to recover junctions. The contextual rules express physical properties of 3D-edges (discontinuities in surface orientation): 1. projections of 3D-edges never cross and 2. visible parts of 3D-edges terminate at either a vertex or a point of occlusion. Both the use of context and an empirically tested explicit error model is a distinguishing feature of the work presented in the paper.

The problem we are facing can be stated as follows: For every line A, find line B which is most likely the line that occluded/had a common vertex in 3D with A. Of all possible assignments for A and B that don't violate rule 1. select the most probable one given the line detector error model. A probabilistic relaxation scheme developed in [Kittler and Hancock, 1989] is applied to the junction detection problem (Section 9.3). Section 9.2 specifies the line detection model. Implementation issues are addressed in Section 9.4. Section 9.5 presents intermediate level groupings built on top of the junction finder The results are summarised in Section 9.6.

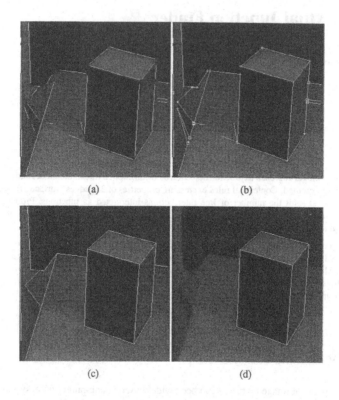

(a)   (b)

(c)   (d)

Fig. 9.1.: **Cube.** A simple scene with a test object used for stereo camera head calibration in the VAP project [project, 1989]. In Fig. (a), lines detected by Hough transform are superimposed over the original image. Note that errors in orientation and transversal position of line segments are negligible. Fig. (c) gives an example of an 'ideal line drawing' of the **Cube** scene. The image was prepared by editing the result of the junction finder shown in (b). Fig. (b) illustrates junction finder results. Endpoint errors are filtered out by 'stretching' line segments to junctions. The only significant structural error occurred at the top-left vertex of the cube. The left vertical edge of the cube is associated with the rear edge of the table. The result is explainable; the line corresponding to the vertical edge was terminated very close to the projected rear table edge because there is no gray level gradient between the front face of the cube and the background. Results of the *face recovery* postprocessing (Section 9.5) are shown in Fig. (d). Using geometric information enabled the recovery of the front face despite the top-left vertex problem described above. *Gap bridging* postprocessing (Fig. 9.6) joint the two lines on the rightmost vertical edge of the cube.

9.2   Line detector model

Conceptually, the first stage of the junction finder can be viewed as an attempt to recover *projected lines*. A *projected line* is, by definition, a projection of a 3D-edge and therefore must terminate at an intersection of 2 or more projected lines. A set of all projected lines would be very close to an 'ideal line drawing' (see Fig. 9.1(c)).

Any endpoint detected by a line finder can be treated as a noisy measurement of a projected endpoint. A statistical model of the noise affecting the line finder is a necessary prerequisite for any

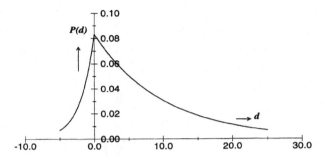

Fig. 9.2.: Assumed endpoint error distribution of the line finder. d denotes the distance from an enpoint along the line segment. Negative values refer to points inside the line segment.

attempt to recover projected lines. The junction detection becomes trivial if the filtering process is successful - any point where two lines touch is a junction. See Figs. 9.1(a)-(c) to compare the line finder output, junction finder output (filtered lines) and a set of projected lines.

The uncertainty in line parameters is a function of the particular edge and line finder used. After extensive (but subjectively evaluated) tests we selected Horaud's implementation [Horaud et al., 1990] of the Deriche filter [Deriche, 1987] for edge detection and a line detector based on Hough transform [Palmer et al., 1992; Princen, 1990] for line detection. In agreement with [Lowe, 1987, page 367] we observed:

- very precise estimation of the line angle and of the transversal position
- large uncertainty in the localisation of the endpoint

Simplifying the characteristics of the line finder the following model was adopted: 1. the projected endpoint lies on the straight line defined by the line segment. 2. if d denotes the (oriented) distance from detected endpoint E with positive values of d for point outside the line segment, then the endpoint error distribution is

$$P(d) = \begin{cases} \frac{k_{in}k_{out}}{k_{in}+k_{out}}e^{k_{in}d} & \text{if } d < 0 \\ \frac{k_{in}k_{out}}{k_{in}+k_{out}}e^{-k_{out}d} & \text{otherwise} \end{cases} \tag{1}$$

The two constants, k_{in} and k_{out} control the shape of the exponential. At present the values are set to 0.5 and 0.1 pixels respectively (Fig. 9.2). The choice of exponential is somewhat arbitrary, but Fig. 9.3 shows that it is in good agreement with empirical data. Test runs with different k_{in} (range 1-0.2) and k_{out} (0.2-0.03) produced similar results suggesting that the performance of the junction finder is not critically sensitive to the shape of the distribution.

9.3 Junction detection using probabilistic relaxation labelling

Visible parts of 3D edges terminate at either a vertex or a point of occlusion. A *junction* is a projection of such 3D point. The line finder model (section 9.2) guarantees that all junctions lie at an intersection of straight lines passing through a detected 2D line. Consider the example of Fig. 9.4. If line A is a projection of a 3D edge it must terminate at a junction that is identical to one of the intersection I_{ab}, I_{ac}, I_{ad}. We are concerned with the problem of computing the probabilities of all events $\{A_e = I_{ai}, \forall i\}$. To exploit the contextual information conveyed by these interacting events

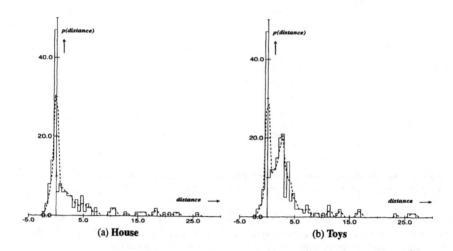

Fig. 9.3.: Histogram of enpoint distance errors. The output of the junction finder, the *junction graph* is assumed to represent the ground truth (see Fig. 9.8 to check the validity of this assumptions). Histogram of line end to corresponding junction distances (solid line) shows good agreement with the line detection model (Fig. 9.2). The dashed line graph shows the running average of two consecutive histogram bins (of 0.5 pixel size).

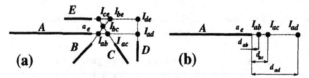

Fig. 9.4.: A set of interacting lines

we compute the probabilities using a dictionary based relaxation scheme whereby the probabilities at stage $n + 1$ are obtained from probabilities $P^n(A_e = I_{ai})$ at the previous iteration, ie.

$$P^{n+1}(A_e = I_{ai}) = \frac{P^n(A_e = I_{ai}) \sum_{\Lambda^k} P(\Lambda^k) \prod_l \frac{P^n(l_e = I_{em})}{P(l_e = I_{em})}}{\sum_j P^n(A_e = I_{aj}) \sum_{\Lambda^k | A_e = I_{aj}} P(\Lambda^k) \prod_l \frac{P^n(l_e = I_{em})}{P(l_e = I_{em})}} \tag{2}$$

The initial probabilities $P^0(A_e = I_{ai})$ are computed using the Bayes formula from the probability distribution of endpoint errors (Eq. 1, Fig. 9.2) and prior probabilities. All prior probabilities in Eq. 2 are assumed to be equal. The context is introduced using a dictionary Λ that contains all permissible configurations of junction assignments. The dictionary is constructed using to rules obeyed by projections of 3D edges: 1. projected lines must not cross and 2. every projected line must terminate at a single junction.

9.4   Implementation of the Junction finder

The junction detection process is performed in three stages - initialisation and preprocessing, relaxation, and postprocessing. In the first stage, all intersections of line pairs are considered as possible junctions. Any intersection with endpoint distance outside the margin of error of the line detector is immediately discarded. Each intersection is assigned an initial, non-contextual probability according to Bayes formula. The information associated with every intersection is stored in a node of an *intersection network*. Each node in the network is linked to four other nodes representing its predecessor and successor (with respect to distance) in the list of intersections of one line (Fig. 9.5(a)).

In the relaxation stage, repeated sweeps through the intersection network are made. At each endpoint, the probability distribution is updated according to context-conveying formula 2. Generally, the probabilities of an intersection being a junction gradually shift either towards 0 or 1. At the end of the sweep, intersections with 0 probabilities are deleted (it follows from formula 2 that the probability would remain 0). The originally dense network (Fig. 9.5(a)) is gradually transformed into a structure similar to Fig. 9.5(b). The iteration loop is exited when the average relative change of intersection probability falls under a preset threshold (default: 2%).

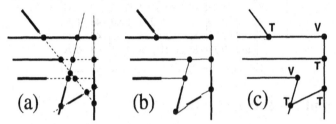

Fig. 9.5.: From *intersection network* to *junction graph*. The large margin of error of the line detector allows for multiple interpretations of an endpoint. An *intersection network* structure is created to facilitate the probabilistic updating of intersection probabilities. During the relaxation process, the network becomes sparser as probabilities of some intersections drop to 0 (see Tab. 9.1). Finally, all non-maximum junctions are discarded - a *junction graph* is created (c).

Finally, only intersections with maximum probability at an endpoint are retained and labelled as either V or T junctions. An intersection having a maximum probability with respect to both participating lines is assumed to be a projection of a vertex; the intersection is labelled a V-junction. If the intersection probability is a maximum with respect to one line only then the other line must 'pass through' (other possibilities are suppressed by the relaxation process). The situation indicates that the intersection is a projection of a point where a 3D-edge was occluded; the intersection is labelled as T-junction. The network of intersections is transformed into an attributed graph structure called a *junction graph*, (Fig. 9.5(c)). Every node of the junction graph represents a junction relation (either V or T) between a pair of lines. The junction graph is a semi-symbolic structure; numerical information is attached to both junctions (position, probability) and lines (position). The junction graph can be used for both symbolic and geometric reasoning about the scene (see section 9.5).

The efficiency-minded readers may express doubt about the speed of the process. The computational complexity of the implementation of preprocessing is $O(N^2)$ (where N is the number of input lines) as all line pairs are examined (an $O(N log N)$ algorithm can be found in [Sedgewick, 1983]). The theoretical worst-case complexity of the iterative relaxation is even worse. Fortunately, the worst-case complexity is not of practical importance (representing a situation where all lines terminate in a tight cluster); the *average* complexity is extremely hard to derive analytically, but empirical results suggest $O(N log N)$. Table 9.1 summarises the junction finder performance. For run-times in the order of seconds 1. optimising performance of the junction finder is not of particular importance and 2. the performance depends more on system specific constants (i/o speed etc.) than on the computational complexity. The good performance is a consequence of the *focus of attention* property of the relaxation labelling. After two or three iterations, app. $2N$ (not the theoretical N^2) intersection probabilities are updated. Most of the processing in later iterations revolves around a set of ambiguous intersections (Tab. 9.1, column 'inters.'). Empirical data (column 'relative change' in Tab. 9.1) suggest good stability and fast, although not monotonic, convergence.

Table 9.1.: Junction finder performance. The processing time was measured on a SPARC 2 machine. The 'inters.' column shows the number of intersections processed in the $n - th$ iteration. The 'change' column contains information about an average change (in %) of the intersection probabilities

image	House		Widget		Toys		Cube	
time	1.6s		0.7s		2.1s		0.4s	
lines	78		44		90		28	
Iteration	inters.	change	inters.	change	inters.	change	inters.	change
1	563	84.16	247	84.01	800	86.60	107	83.72
2	368	75.08	164	67.00	489	73.34	64	47.00
3	142	24.03	85	50.29	204	25.65	43	7.87
4	142	2.09	85	7.15	204	2.80	43	1.26
5	142	5.73	85	3.20	204	5.65	43	6.49
6	142	0.38	85	20.41	204	1.04	43	0.99
7			85	3.36				

9.5   Beyond the Junction Graph

The *junction graph* can be viewed as a final result of the junction finder. The richness of the information represented by the junction graph invites further exploitation. Three methods, *gap bridging, V3-junction detection and polyhedral object face recovery*, are presented in this section.

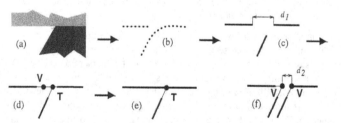

Fig. 9.6.: Gap bridging. Conventional edge detectors perform poorly (subfig. (b)) in areas where 3 regions (subfig. (a)) meet (see [Nobel, 1988; Li *et al.*, 1989]). This causes a projection of one 3D-edge to be broken into two line segments (c). Spatial arrangement (c) of line segments give rise to subgraph (d) of the junction graph. Situation (d) can be interpreted as either a view of a vertex from a highly improbable, accidental viewpoint or as a manifestation of the above mentioned edge detector problem. The latter interpretation is assumed (the former having a negligible probability) and the subgraph is transformed into the form depicted in subfig. (e). The gap bridging process is *context dependent*; gap of subfig. (f) of length d_2, although significantly smaller then d_1, is left intact. Examples of the gap bridging can be found in the **House** (see e.g. roof), **Widget** (upper vertical edge of the from face) and **Cube** images (Figs. 9.8, 9.1)

Gap bridging (see fig. 9.6 for full description) corrects edge detector failures at T-junctions. The V3-junction detector finds subgraphs of the junction graph that are likely to be projection of a *3-vertex*, a vertex where 3 visible 3D-edges meet. In Section 9.2 we made an assumption that transversal positions of lines are error-free. This implies that all 3 (or, more generally n) lines terminating at a common 3(or n)-vertex should intersect at a single point. In practise, the line intersections are clustered in a small region. The implementation is straightforward; for every V-junction: check for V-junctions in a small (default: 1 pixel) neighbourhood. The size of the neighbourhood makes false positives virtually impossible. Examples of V3-junctions can be found in figs. 9.8 and 9.1.

Recovery of polyhedral object faces is more complex. If 2D line segments are projections of 3D-edges and junctions projections of vertices then a completely visible face of a polyhedral object must project into a closed loop of V-junctions in the junction graph. Fig. 9.7 illustrates results of the *face recovery* procedure. Application of *face recovery* for fast 3D pose estimations is described in [Kittler *et al.*, 1992].

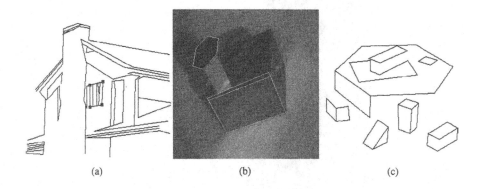

(a)   (b)   (c)

Fig. 9.7.: Results of *polyhedral face recovery*. Closed loops in the *junction graph* are assumed to correspond to 3D-edges and vertices of a polyhedral object face. Only one incorrect face is recovered in the **Toys** scene (the non-convex polygon inside the large hexagon of Fig. (c); a careful scrutiny of (c) and Fig. 9.8(f) reveals an accidental alignment of the block and pad leading to junction graph distortion). Fig. (a) shows *corrected lines* - lines stretched to terminate at a junction (compare with Fig. 9.8(b)). The single face recovered in the **House** scene is a projection of the window frame (corners marked by □).

9.6 Conclusion

We have presented an algorithm for junction detection with two new features: exploitation of spatial context and use of an explicit line detector model. The combination of contextual evidence is not based on an ad hoc method; a well established method, *probabilistic relaxation*, is employed to accomplish the task. The tests performed to validate the line detector model could prove valuable in its own right as an evaluation tool for line detectors.

The junction finder performance has been tested on hundreds of images, mostly running as a part of a continuously operating vision system [Kittler *et al.*, 1992]. Four scenes, **House, Toys, Widget,** and **Cube,** were selected as representative of different environments (indoor, outdoor, industrial). Results (Fig. 9.8) show that our main objective has been achieved - vast majority of V and T junctions indeed correspond to vertices and occlusions in the 3D world. Success of the polyhedral face recovery strongly supports this claim. The junction finder possesses two key features vital for continuous operation - it is fast and it doesn't require any user-defined thresholds even in changing conditions. Experiments have shown [Kittler *et al.*, 1992] that the junction finder output provides salient intermediate level features for model invocation and 3D pose estimation.

References

[Bergevin and Levine, 1990] R. Bergevin and M. D. Levine. Extraction of line drawing features for object recognition. In *Proceedings of IEEE International Conference Pattern Recognition*, pages 496–501, 1990.

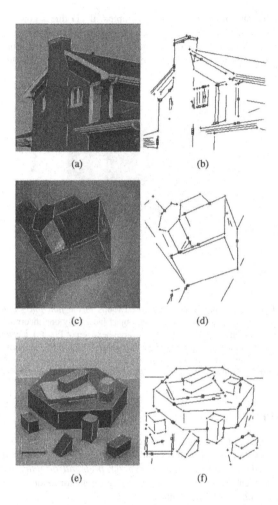

Fig. 9.8.: **House, Widget** and **Toys** scenes. Figures (a), (b) and (c) show lines detected by Hough transform superimposed over the original image . Figures (b), (c) and (d) depict lines corrected by contextual gap filling (described in section 9.6) and the results of the junction finder. *T-junctions* (indicating an occlusion) are marked by □; *V-junctions* (indicating a vertex of two 3D edges) by + and *V3-junctions* (indicating a vertex of three 3D-edges) by ×.

[Coelho *et al.*, 1990] C. Coelho, M. Straforini, and M. Campani. Using geometrical rules and a priori knowledge for the understanding of indoor scenes. In *Proc. British Machine Vision Conference*, pages 229–234, 1990.

[Deriche, 1987] R. Deriche. Using Canny's criteria to derive a recursively implemented optimal edge detector. *International Journal of Computer Vision*, 1(2):167, 1987.

[Haber and Hershenson, 1973] R.N. Haber and M. Hershenson. *The Psychology of Visual Perception*. Holt, Rinehart and Winston Inc., U.S.A, 1973.

[Horaud *et al.*, 1990] R. Horaud, F. Veillon, and T. Skordas. Finding geometric and relational structures in an image. In *European Conference on Computer Vision*, pages 373–384, 1990.

[Kittler and Hancock, 1989] J. Kittler and E.R. Hancock. Combining evidence in probabilistic relaxation. *International Journal of Pattern Recognition and Artificial Intelligence*, 3:29–51, 1989.

[Kittler et al., 1992] J. Kittler, J. Illingworth, J. Matas, P. Remagnino, K. C. Wong, H. Christensen, J-O. Eklundh, G. Olofsonn, and M. Li. Symbolic scene interpretation and control of perception. Technical report, ESPRIT BRA Project 3038, March 1992.

[Li et al., 1989] Du Li, G.D. Sullivan, and K.D. Baker. Edge detection at junctions. In *AVC*, pages 121–125, 1989.

[Lowe, 1987] D. Lowe. Three-dimensional object recognition from single two-dimensional images. *Artifical Intelligence*, 31:355–395, 1987.

[Mohan and Nevatia, 1989] R. Mohan and R. Nevatia. Using perceptual organisation to extract 3-d structures. *IEEE Transactions on Pattern Analysis and Machine Intelligence*, 11(11):1121–1139, 1989.

[Nobel, 1988] J. A. Nobel. Finding corners. In *AVC*, pages 267–274, 1988.

[Palmer et al., 1992] P. L. Palmer, J. Kittler, and M. Petrou. A Hough transform algorithm with a 2D hypothesis testing kernel. In *Proceedings of IEEE International Conference Pattern Recognition*, September 1992.

[Pomerantz, 1981] J.R Pomerantz. Perceptual organization and information processing. In *Perceptual Organization*, pages 141–180, 365 Broadway, Hillsdale, New Jersey, 1981. Lawrence ERLBRAUM associates.

[Princen, 1990] J. Princen. *Hough Transform Methods for Curve Detection and Parameter Estimation*. PhD thesis, University of Surrey, June 1990.

[project, 1989] VAP project. Vision as process. Technical annex, EEC, March 1989.

[Sedgewick, 1983] R. Sedgewick. *Algorithms*. Addison-Wesley, Reading, 1983.

Part III

Advanced Image Processing

In the VAP project, two different approaches to image processing and description have been pursued. The first approach, which was described in part II, is to a large extent based on use of existing techniques which have been refined to fit into the framework of the VAP skeleton system. The alternative approach, which is described in this part, was pursued to ensure a more long-term, but unproven and expensive research strategy.

A review of some of the methods in the first category of methods reveals that they often have two dominating characteristics: The spatial processing is carried out with no reference to the temporal phenomena in the scene being observed and the temporal processing (e.g., motion detection or estimation) is carried out in a local context where only information from a few frames are taken into account. This category, which includes most of the methods reported in the literature, contains the methods exploited in the integrated VAP skeleton system, as described in part II.

A major focal point for the VAP project is exploitation of temporal context and goal/event driven processing. To fully exploit contextual information it is thus necessary to integrate both spatial and temporal information into a coherent representation, and a related set of operators must be developed. Only a few researchers have pursued a line of research which fully integrates space and time. The most well known group of people being those at the Computer Vision Laboratory, Linköping University. Other notable efforts, in term of image processing, are [Fleet-Jepson 1990], [Mayhew-Friesby 1991] and [Burt 1988].

The integration of information in space and time may be complemented by an integration of features in a hierarchical representation which indirectly allow top-down projection of expectations (between feature levels) and bottom-up integration in the context of space, time and goal derived expectations, as described by [Granlund 1978]. The exploitation of a unified space-time representation has been part of the VAP effort, as it at an early stage was realised that they might provide increased robustness and simplified control. At the time when the effort was initiated, as part of VAP, the methods were computationally intractable, but based on predicted advances in computer hardware it was estimated that the methods, in due course, might be implemented as part of a vision system, which exploits traditional workstation hardware. During the coarse of the VAP project (BR-3038-VAP) the computational power of desktop workstations has been increased by a factor of 10 and consequently the methods may today (1993) be integrated into a continuously operating system. During the course of the project it has also been demonstrated how focus of attention methods may be used to control the amount of computations needed.

In this part the effort in image processing and description based on a unified spatio-temporal representation is described. Chapter 10 outlines the basic terminology and descriptors used in the remainder of the part, and outlines how a spatio-temporal signal may be transformed into a tensor based description.

In the common framework, described in chapter 10, it is possible to define new methods for most of the well known low-level vision problems. In chapter 11 it is, as an example, described how disparity may be estimated from phase information and it is described how the matching may exploit a recursive coarse to fine matching strategy, which enables efficient processing and handling of disparity over a large range of values. The developed disparity estimation method has been used both for full image reconstruction of depth and for driving the vergence mechanism on some of the camera heads developed as part of the VAP project (see part IV).

In active vision (as described in part I and IV) the selection of a gaze point for the vision system is performed in response to high level goals in the system, but for exploration type of problems it may be beneficial to direct the change of gaze point based directly on image information. One such approach to image based gaze control is described in chapter 12. Here local frequency and phase information is used for construction of a potential map that allow exploration of "interesting" features through use of a simple search in the gaze control map. Experimental results using both a simulation environment and real images are also presented.

The tensor based signal description introduced in chapter 10 is described in detail in chapter 13. It is here described how the tensor description may be used for description of both spatial and temporal phenomena (with a common framework). It is also described how the tensor description may be generated using banks of quadrature filters.

In the final chapter in this section, chapter 14, it is demonstrated how the tensor description may be used for extraction of line features. The chapters outlines how "perceptual grouping" operations in principle may be carried out. Through exploitation of a reparameterisation (using Möbius type of mapping) it is demonstrated that line extraction may be performed in the presence of a large amount of noise.

Bibliography

[Burt 1998] Burt, P.: Smart Sensing in Machine Vision, *IEEE Proceedings.* **76** (8), pp. 996-1006, August 1988.

[Fleet-Jepson 1990] Fleet, D.J. & A.D. Jepson: Computation of Component Image Velocity from Local Phase Information. *Intl. Jour. of Computer Vision*, **5**(1), pp. 77-104, 1990.

[Friesby-Mayhew 1991] Friesby, J. & J. Mayhew: 3D recognition from Stereoscopic Cues, MIT Press, 1991.

[Granlund 1978] Granlund, G.: In search of a general picture processing operator. *Computer Graphics and Image Processing*, **8**(2), pp. 155-178, 1978.

10. Introduction and Background

Carl-Fredrik Westin, Gösta Granlund and Hans Knutsson

CVL, LiTH

10.1 Introduction

This part contains a description of the advanced image description modules developed for the VAP project. These modules will, when requested, make the following information and/or functions available:

Depth	**Phase**
Disparity	**Motion discontinuity**
Vergence control	**Velocity**
Robot simulator	**Spatio-temporal frequency**
Camera control	**Möbius mapping**
Adaptive environment scanning	**Line extraction**
Orientation	

The presentation of the different techniques used to produce these features is divided into five sections. This section provides a background and an outline of some basic signal analysis tools and concepts which will be necessary for the following presentation.

In the second section, the problem of estimating depth information from two or more images is addressed. This problem has received considerable attention over the years and a wide variety of methods has been proposed to solve it [Barnard and Fichsler, 1982; Fleck, 1991]. The presented disparity algorithm is based on local phase estimations [Wilson and Knutsson, 1989]. Local phase is measured in the two images using quadrature filters in an iterative coarse to fine scheme.

The third section discusses low level gaze control. The proposed gaze control procedure includes focus of attention mechanisms and active vergence control. The active vergence control builds on the mentioned disparity algorithm but uses a fovea representation instead of a standard resolution pyramid. The algorithms are developed in a simulated robot environment. The active vergence control proved to be easily adapted and implemented on one of the robot heads in the consortium (LIFIA, Grenoble).

In the fourth section we present a theory for the analysis of three-dimensional signals. Spatio-temporal quadrature filters are combined into a tensor description of local structure [Knutsson, 1989]. This technique enables production of robust estimates of a number of important features, e.g. phase, orientation, spatio-temporal frequency, velocity and motion parameters. The use of time as an extra dimension provides more robust and less noise sensitive estimates compared to traditional 2-dimensional techniques.

In the last section we discuss line extraction using a new parameterization, the Möbius strip parameterization. One of the key features with this parameterization is that the parameters does not have any discontinuities for continuous line movements in the image. The continuity aspect becomes critical when sequences of images, rather than single images, are to be analyzed.

10.2   Background

Classically, the methodology of image analysis contains many appropriate procedures to perform various tasks [Duda and Hart, 1973; Ballard and Brown, 1982; Horn, 1986]. A common problem is that these procedures are often not suitable as components of a larger system where such procedures interact. One reason is that information is represented in different ways for different types of features. It is difficult to have such descriptors cooperate and to control each other in a parametrical way.

An important feature of current state of the art is the view that sufficiently efficient processing of complex scenes can only be implemented using an adaptive model structure. In the infancy of computer vision, it was believed that objects of interest could unequivocally be separated from the background using a few standard operations applied over the entire image. It turns out, however, that this simple methodology only works on simple images having a good separation between object and background. In the case of more difficult problems with noise, ambiguities and disturbances of different types, more sophisticated algorithms are required with provisions to adapt themselves to the image content.

It consequently turns out to be necessary to use what we may call different sub-algorithms on different parts of an image. The selection of a particular sub-algorithm is based upon a tentative analysis of the image content. The reason for using different sub-algorithms is the simple fact that all possible events can not be expected in a particular context. In order for this handling of sub-algorithms to be manageable, it has to be implemented as a parameterization of more general algorithms in such a way that the metric of the parameter reflects the contextual distance.

In our own work, as well as in some of the work cited, there has been taken a great deal of impression from what is known about biological visual systems [Hubel and Wiesel, 1962; Hubel, 1988; Linsker, 1988]. This is not to say that we assume that the structures presented are indeed models of phenomena used in biological visual systems. Too little is so far known to form any firm opinion on this. The ultimate criterion is simply performance from a technical point of view.

We will in this chapter discuss various aspects related to hierarchical vision systems, and review some results. Due to the format of such an overview, it is unfortunately not possible to give due references to all contributions within the area.

10.2.1   Hierarchical Computing Structures

Hierarchical structures is nothing new in information processing in general, or in computer vision in particular. A regular organization of algorithms has always been a desired goal for computer scientists.

Among the first structured approaches were those motivated by knowledge about biological visual systems. The perceptron approach by Rosenblatt [Rosenblatt, 1962], has attracted new attention as neural network theory has become a hot research topic [Hopfield, 1982]. The work on layered networks continued, where such networks would accept image data at their bottom level [Uhr, 1971; Tanimoto and Pavlidis, 1975; Hansen and Riseman, 1976].

Burt introduced an approach to hierarchical image decomposition using the Laplacian or DOLP (Difference Of Low Pass) pyramid [Burt and Adelson, 1985]. In this way an image is transformed into a set of descriptor elements. The image can then be reconstructed from its set of primitives.

The concept of scale or size as a dimension, was further extended in the so called *scale space* representation of images [Witkin, 1983; Koenderink and van Doorn, 1984; Lifshitz, 1987].

Most of the work so far has dealt with hierarchies relating to size or scale, although they have indirectly given structural properties. Granlund introduced an explicit abstraction hierarchy [Granlund, 1978], employing symmetry properties implemented by Gaussian wavelets in what today is commonly referred to as Gabor functions [Gabor, 1946].

It is apparent that there are two issues related to hierarchies and pyramid structures. One has to do with level of abstraction, and the other with size or scale. Although they are conceptually different, there are certain relations. With increased level of abstraction generally follows an increase of the scale over which we relate phenomena.

In order for a system modeling a high structural complexity to be manageable and extendable, it is necessary that it exhibits modularity in various respects. This implies for example standardized information representations for interaction between operator modules. Otherwise the complexity will be overwhelming and functional mechanisms will be completely obscure. One way to satisfy these requirements is to implement the model structure in a hierarchical fashion. In order for such a structure to work effectively, however, certain requirements have to be fulfilled for information representation and for operations. It is in this context important to remember that the purpose of an information processing system is to produce a response, be it immediate or delayed. This response can be the actuation of a mechanical arm to move an object from one place to another. Or it can be to point out an object of interest with reference to the input image, a procedure which we customarily denote classification. Another example is enhancement, where the system response acts upon the input image in order to modify it according to the results of the analysis. In this case, the input image is a part of the external world with respect to the system, upon which the system can act.

10.3   Local signal analysis

The Fourier transform has found considerable use in signal analysis. In image analysis, however, this type of representation gives problems due to the loss of spatial localization in the transform domain. Methods such as Gabor and wavelet transforms have proved to be useful alternatives.

The general reason for transforming or filtering a signal is to extract specific features from the signal. The Fourier transform is an example of a transform where the interesting feature to extract is the frequency. The assumption is that the signal is stationary, e.g. sine waves. If the signal is non-stationary, any abrupt change of the signal will be spread over the whole frequency range and the spatial position of the discontinuity will be obscured. The Fourier transform is apparently not well suited for analyzing such signals.

The Short Time Fourier Transform, or windowed Fourier transform, is one way to modify the Fourier transform for better performance on non-stationary signals. The widely chosen windowing function is the Gabor function due to its simultaneous concentrated in both domains [Gabor, 1946].

10.3.1   Phase and local frequency

Most people are familiar with the global Fourier phase. The shift theorem describing how the Fourier phase is affected by moving the signal, is common knowledge. The phase in signal representations based on local operations, e.g. Gabor filters or wavelets, is on the other hand not so well known, but the increasing interest in wavelets and their applications will probably change this in a near future.

The local phase has a number of interesting and robust properties. Phase estimates are independent of mean luminance and contrast. This feature makes phase estimates suitable for matching, since it reduces the need for camera exposure calibration and illumination control. Furthermore, it is a continuous variable which can measure changes much smaller than the spatial quantization, enabling subpixel accuracy without a subpixel representation of image features. Phase has also been proved to be stable against scaling up to 20 percent, [Fleet et al., 1991].

The output from a local phase filter, e.g. a Gabor filter, is an analytic function which can be represented as an even real part and an odd imaginary part, or an amplitude and an argument. The rate of change in phase depends on the frequency content of the input image. The phase derivative is therefore called the local, or instantaneous, frequency:

$$\omega = \frac{d\phi}{dx} \tag{1}$$

10.3.2 Gabor filters

In the literature, the Gabor filters are chosen since they minimize space–frequency uncertainty and have separable center frequency and bandwidth [Wilson and Granlund, 1984]. A Gabor filter tuned to a frequency ω_0 is created, spatially, by multiplying an envelope function with a complex exponential function of angular frequency ω_0, equation 4. Gabor showed that a Gaussian envelope minimizes a space–frequency uncertainty product [Gabor, 1946],

$$\int_\omega \|G(\omega)\|^2 d\omega \int_x \|g(x)\|^2 dx \tag{2}$$

a review of which can be found in [Maclennan, 1981]. This means that the Gabor filters are well localized in both domains simultaneously. Logarithmic partitioning of the frequency axis is equivalent to using filters with constant relative bandwidth.

$$\frac{\Delta\omega}{\omega_0} = \text{constant} \tag{3}$$

The bandwidth, $\Delta\omega$, will thus vary with the center frequency of the filter. It is now possible to get good time *or* space resolution at high frequencies [Wilson and Spann, 1988; Porat and Zeevi, 1988]. The hypothesis is that high frequency bursts are of short duration, while low frequency components have long duration. In images this hypothesis implies that "interesting" features have fixed shape but unknown size. This is a more realistic model than for example the global Fourier transform, where all frequencies are supposed to have infinite support in space.

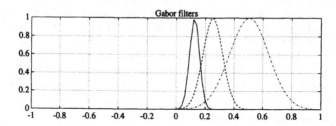

Fig. 10.1. The magnitude of three Gabor filters in the frequency domain. $\omega_0 = \{\pi/8, \pi/4, \pi/2\}$ and $\beta = 0.8$.

The parameters used when designing a Gabor filter are the standard deviation, σ, and the center frequency, ω_0. These also effect the size and bandwidth of the filter. The spatial domain definition of a Gabor filter is

$$g_{\sigma\omega_0}(x) = e^{ix\omega_0} \frac{1}{\sigma\sqrt{2\pi}} e^{-\frac{x^2}{2\sigma^2}} \tag{4}$$

and the frequency domain definition is

$$G_{\sigma_\omega\omega_0}(\omega) = e^{-\frac{(\omega-\omega_0)^2}{2(\sigma_\omega)^2}} \tag{5}$$

where $\sigma_\omega = 1/\sigma$.

10.3.3 The lognorm quadrature filter

A desirable feature of low level filters is that the output can be expressed independently in terms of magnitude and phase. This requires that the output magnitude is invariant to the phase of the input signal. This can be obtained by the use of quadrature filters.

Quadrature filters can be defined as filters being equal to zero for nonpositive frequencies, i.e. for $\omega <= 0$. Since Gabor functions have infinite support in the frequency domain it follows that these functions are not quadrature filters and thus incapable of producing a strictly phase invariant output. In practice, however, if the upper limit of the relative bandwidth is set to 0.8 octaves, a filter in relatively good quadrature is obtained [Fleet *et al.*, 1991], see figure 10.1.

A number of different types of quadrature filters having low uncertainty products exist. We will in this subsection introduce the lognorm quadrature filter, first suggested by Knutsson [Knutsson, 1982], and continue with other types in section 11.3.3 in relation to the discussion on disparity estimation. Lognorm filters are a class of quadrature filters used for orientation, phase and frequency estimation.

$$F(\omega) = \begin{cases} e^{-\frac{4}{\ln(2)B^2}\ln^2(\omega/\omega_i)} & if \ \omega > 0 \\ 0 & otherwise \end{cases} \qquad (6)$$

B is the relative bandwidth in octaves and ω_i is the center frequency. Three lognormal filters are shown in figure 10.2.

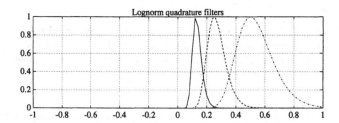

Fig. 10.2. Three lognorm filters in the frequency domain. $\omega_0 = \{\pi/8, \pi/4, \pi/2\}$ and $\beta = 0.8$.

Spatially these filters can be viewed as the combination of a line filter and an edge filter, see figure 10.3.

A real-valued image has a hermitian Fourier transform. This means that the imaginary components have opposite signs in the two halfplanes (e.g. a sinewave) while the real components have the same. Using a filter having equal support in the two halfplanes results in all the imaginary components being canceled. Filters having the main support in one halfplane ensures that all components can be detected although the signal strength will vary with the signal phase. The only way of achieving equal sensitivity for signals with arbitrary phase angle is to use a quadrature filtering approach. Since the quadrature filters only have support in one half of the domain, there will be no interference between the "negative" and "positive" frequency components.

10.3.4 Singular points in phase scale-space

The phase is generally very stable in scale-space. This is visualized in figure 10.4 where lognorm filters have been applied to a one-dimensional signal in scalespace. Note that the phase curves are almost vertical over scale.

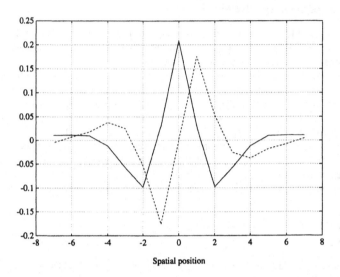

Fig. 10.3. The spatial lognormal filter realized as a 15 tap filter. The real part (solid) and the imaginary part (dashed) are plotted.

However, there are points around which the phase has a very unpleasant behavior. In these points the signal goes through the origin of the complex plane i.e. they are singular points of the analytic function. Figure 10.5 shows a stylized example of a phase resolution pyramid. Suppose we have a signal consisting of two positive frequencies,

$$F(\omega) = \delta(\omega_0) + 2\delta(3\omega_0)$$

Having a Dirac pulse in one halfplane of the Fourier domain corresponds to having a complex signal in the spatial domain:

$$f(x) = e^{i\omega_0 x} + 2e^{i3\omega_0 x}$$

Suppose now that we create the resolution pyramid using a filter that attenuates the high frequency part of the signal a factor $1/\sqrt{2}$ but leaves the low frequency part unaffected. The behavior of the signal can be studied using a polar plot of the signal vector, since the signal is periodic.

Figure 10.5a shows a polar plot of the original signal vector. The magnitude of the signal vector is fairly large and runs counter clockwise all the time. The phase is therefore monotonous and increasing, except for the wrap-around caused by the modulo 2π representation, see figure 10.5b. The local frequency, i.e. the slope, is positive and almost constant. In figure 10.5c the amplitude of the high frequency signal is reduced and the signal vector now circles the origin closely. The local frequency is much higher when the signal vector passes close to the origin, causing the phase curve to bend, though it still is monotonous and increasing. Further LP-filtering causes the signal vector to go through the origin, figure 10.5e. In these points the phase jumps discontinuously and the local frequency becomes impulsive. In figure 10.5g, the signal vector moves clockwise when going through the small loops. This means that the phase decreases and that the local frequency is negative, see figure 10.5h.

To avoid singular points, let us assume that we do not consider points with very low magnitude, remembering that the magnitude is zero in the singular points. Unfortunately, it is not that simple.

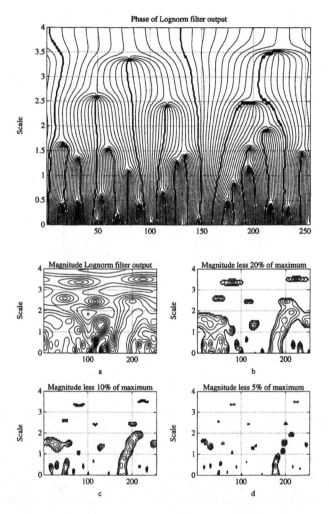

Fig. 10.4. Above: Isophase plot of Lognorm phase scalespace. The dark broad lines are due to phase wrap around. The positions where the isophase curves converges are singular points. Below: Isomagnitude plot of Lognorm phase scalespace. The dark areas indicates where the magnitude of the signal is below the threshold. $\omega_0 = \frac{\pi/4}{scale}$ and $\beta = 0.8$.

The impact of a singular point is spread in scale; negative frequencies on coarser resolution and very high frequencies on finer resolution. In these points the magnitude can not be neglected and a high enough threshold will also cut out many useful phase estimates. Fleet [Fleet and Jepson, 1991] describes how singular points can be detected and how their influence can be reduced.

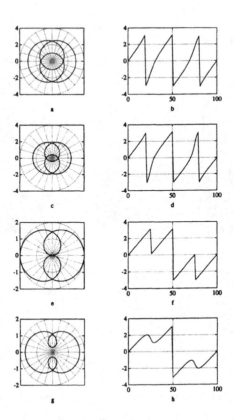

Fig. 10.5. The periodic signal, $f(x) = e^{i\omega_0 x} + 2e^{i3\omega_0 x}$, is LP filtered in four steps, attenuating the high frequency of the signal. To the left are polar plots of the signal vector and to the right is the phase for the same signal plotted. a) The signal circles the origin at a distance and always in the same direction. b) The slope of the phase plot, i.e. the local frequency, is positive and almost constant. c) The small loops round the origin closely. d) The phase curve bends i.e. the local frequency is locally very high, but continuous. e) The small loops go through the origin producing singular points. f) The local frequency is impulsive. g) The small loops do not round the origin. h) The phase curve bends downward i.e. the local frequency is locally negative, but continuous.

10.4 Spatio-temporal channels

The human visual system has difficulties handling high spatial frequencies simultaneously with high temporal frequencies [Arbib and Hanson, 1987; Davson, 1976]. This means that objects with high velocity cannot be seen sharply without tracking. A possible explanation to this is that the visual system performs an effective data reduction. The data reduction is made in such a way that high spatial frequencies can be handled if the temporal frequency is low, and vice versa. The same strategy can be used in a computer vision model for time sequences [Wiklund et al., 1989].

If an image sequence is subsampled spatially and/or temporally we term the output from this operation a *channel*. For convenience the channels is indexed in accordance to the level of subsampling:

$$\chi_{kl} \tag{7}$$

where k indicates level of spatial subsampling in octaves and l indicates the same for the level of temporal subsampling, relative to a reference sequence termed χ_{00}. The data content in channel χ_{kl} relative to the reference channel, χ_{00}, is consequently:

$$\mathcal{D}_{kl} = 2^{-(nk+l)} \tag{8}$$

where the weighting, n, is equal to the spatial dimension of the sequence. Examples of the frequency content in a set of spatio-temporal channels are shown in figure 10.6.

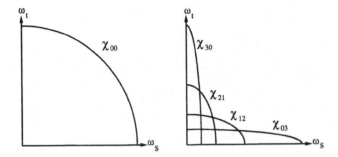

Fig. 10.6. Frequency contents in some spatio-temporal channels. For comparison, the frequency content in the reference channels is presented to the left.

Note that the reference channel, χ_{00}, is not used. The idea is that the content in the rest of channels contains most of the important information and are hence sufficient for the system. As mentioned above, the appealing reason for not using channel χ_{00} is the huge data reduction that is made when not allowing for high resolution temporally and spatially simultaneously.

10.5   Computer vision as signal analysis

As mentioned in section 10.2, hierarchies and modular structuring of a problem is necessary for solving complex problems. In the VAP project we have hierarchies of modular structures. A schematic overview of the top modular level VAP system structure is found in figure 10.7. This figure also approximately indicate the main area of contributions from the Computer Vision Laboratory in Linköping.

The modules in figure 10.7 is each a *collection* of vision modules providing various kind of information available when requested. The name of the used modules are presented in the beginning of each of the following chapters. The circular region roughly indicates the part of the system where LiTH has been active. Most of our modules was developed in a software environment called AVS [Wiklund *et al.*, 1993]. This environment allows for simple visual programming of module communication and handles all the data flow communication between them. The vergence control and the depth estimation modules have been adapted to the SAVA system described in Part I.

The concepts and algorithms used in the presented structure is based on signal theoretical approaches. It will be shown that vision as signal analysis has many interesting features and advantages. Another very important issue that will be addressed is the one of information representation. Finding a good or natural representation for an entity or an event often considerable reduces the complexity of subsequent operations. This aspect is not only of philosophical value but have practical implications as well. Having operators and data represented appropriately simplifies coding the operations into a computer. It normally reduces the requirement for conditional tests. One simple example is relaxation. A uniform representation ensures that relaxation can be achieved by ordinary convolution.

The VAP system structure

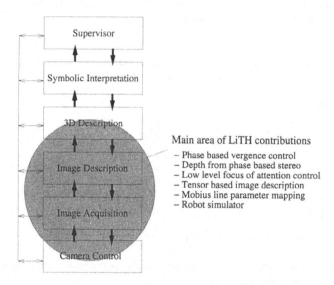

Main area of LiTH contributions

– Phase based vergence control
– Depth from phase based stereo
– Low level focus of attention control
– Tensor based image description
– Mobius line parameter mapping
– Robot simulator

Fig. 10.7. Outline of the VAP system structure. The arrows at the left indicates a message passing structure for passing messages between the modules e.g. requests for new information. Bold arrows indicate high bandwidth channels for sending large amounts of information, e.g. images. The circular region roughly indicates parts of the system where the Computer Vision Laboratory in Linköping has made contributions.

Another example is low level focus of attention control which will be discussed in detail in chapter 12.2. The simplification of the operators due to careful selection of data representations enables the solving of fairly complex tasks at a relatively low level in a hierarchical system.

The channel discussion in section 10.4 is related to conventional scale and multiresolution analysis, although these discussions in most cases only refer to spatial descriptions. We will not explicitly continue the discussion of these important issues in the following chapters. However, in the chapters describing disparity estimation and local focus of attention control scale is explicitly used and a coarse-to-fine strategy employed to solve the problem. When it comes to more general discussions such as image descriptions using tensors we have limited the presentation by only using one scale. Due to the format of this book we have chosen not to focus on combination of events over scale. The presentation gain clarity omitting this extra level of sophistication. The local tensor image description should be made in each channel separately. The estimates are then to be combined in different ways depending on the system goal. The combination of different kind of data is simplified due to a uniform data representation (tensors) and the fact that all statements carry a certainty measure defining the validity of the information represented.

References

[Arbib and Hanson, 1987] M. A. Arbib and A. Hanson, editors. *Vision, Brain and Cooperative Computation*, pages 187–207. MIT Press, 1987.

[Ballard and Brown, 1982] D. H. Ballard and C. M. Brown. *Computer Vision*. Prentice-Hall, 1982.

[Barnard and Fichsler, 1982] S. T. Barnard and M. A. Fichsler. Computational Stereo. *ACM Comput. Surv.*, 14:553–572, 1982.

[Burt and Adelson, 1985] P. J. Burt and E. H. Adelson. Merging images through pattern decomposition. In *Applications of digital image procesing VIII*. SPIE, 1985. vol. 575.

[Davson, 1976] Hugh Davson, editor. *The Eye*, volume 2A. Academic Press, New York, 2nd edition, 1976.

[Duda and Hart, 1973] R. O. Duda and P. E. Hart. *Pattern classification and scene analysis*. Wiley-Interscience, New York, 1973.

[Fleck, 1991] Margaret M. Fleck. A topological stereo matcher. *Int. Journal of Computer Vision*, 6(3):197–226, August 1991.

[Fleet and Jepson, 1991] D. J. Fleet and A. D. Jepson. Stability of phase information. In *Proceedings of IEEE Workshop on Visual Motion*, pages 52–60, Princeton, USA, October 1991. IEEE, IEEE Society Press.

[Fleet et al., 1991] David J. Fleet, Allan D. Jepson, and Michael R. M. Jenkin. Phase-based disparity measurement. *CVGIP Image Understanding*, 53(2):198–210, March 1991.

[Gabor, 1946] D. Gabor. Theory of communication. *Proc. Inst. Elec. Eng.*, 93(26):429–441, 1946.

[Granlund, 1978] G. H. Granlund. In search of a general picture processing operator. *Computer Graphics and Image Processing*, 8(2):155–178, 1978.

[Hansen and Riseman, 1976] A. R. Hansen and E. M. Riseman. Constructing semantic models in the visual analysis of scenes. In *Proceedings Milwaukee Symp. Auto. & Contr. 4*, pages 97–102, 1976.

[Hopfield, 1982] J. J. Hopfield. Neural networks and physical systems with emergent collective computational capabilities. *Proceedings of the National Academy of Sciences*, 79:2554–2558, 1982.

[Horn, 1986] B. K. P. Horn. *Robot vision*. The MIT Press, 1986.

[Hubel and Wiesel, 1962] D. H. Hubel and T. N. Wiesel. Receptive fields, binocular interaction and functional architecture in the cat's visual cortex. *J. Physiol.*, 160:106–154, 1962.

[Hubel, 1988] David H. Hubel. *Eye, Brain and Vision*, volume 22 of *Scientific American Library*. W. H. Freeman and Company, 1988.

[Knutsson, 1982] Hans Knutsson. *Filtering and Reconstruction in Image Processing*. PhD thesis, Linköping University, Sweden, 1982. Diss. No. 88.

[Knutsson, 1989] H. Knutsson. Representing local structure using tensors. In *The 6th Scandinavian Conference on Image Analysis*, pages 244–251, Oulu, Finland, June 1989. Report LiTH-ISY-I-1019, Computer Vision Laboratory, Linköping University, Sweden, 1989.

[Koenderink and van Doorn, 1984] J. J. Koenderink and A. J. van Doorn. The structure of images. *Biological Cybernetics*, 50:363–370, 1984.

[Lifshitz, 1987] L. M. Lifshitz. Image segmentation via multiresolution extrema following. Tech. Report 87-012, University of North Carolina, 1987.

[Linsker, 1988] Ralph Linsker. Development of feature-analyzing cells and their columnar organization in a layered self-adaptive network. In Rodney M. L. Cotteril, editor, *Computer Simulation in Brain Science*, chapter 27, pages 416–431. Cambridge University Press, 1988.

[Maclennan, 1981] B. Maclennan. Gabor representations of spatiotemporal visual images. Technical Report CS-91-144, Computer Science Department, University of Tennesse, September 1981.

[Porat and Zeevi, 1988] M. Porat and Y. Y. Zeevi. The Generalized Gabor Scheme of Image Representation in Biological and Machine Vision. *IEEE Trans. on PAMI*, 10(4):452–467, 1988.

[Rosenblatt, 1962] F. Rosenblatt. *Principles of Neurodynamics: Perceptrons and the theory of brain mechanisms*. Spartan Books, Washington, D.C., 1962.

[Tanimoto and Pavlidis, 1975] S. L. Tanimoto and T. Pavlidis. A hierarchical data structure for picture processing. *Computer Graphics and Image Processing*, 2:104–119, June 1975.

[Uhr, 1971] L. Uhr. Layered 'recognition cone' networks that preprocess, classify and describe. In *Proceedings Conf. on Two-Dimensional Image Processing*, 1971.

[Wiklund et al., 1989] J. Wiklund, L. Haglund, H. Knutsson, and G. H. Granlund. Time sequence analysis using multi-resolution spatio-temporal filters. In *The 3rd International Workshop on Time-Varying Image Processing and Moving Object Recognition*, pages 258–265, Florence, Italy, May 1989. Invited Paper. Report LiTH-ISY-I-1014, Computer Vision Laboratory, Linköping University, Sweden, 1989.

[Wiklund et al., 1993] J. Wiklund, C-F Westin, and C-J Westelius. Avs, application visualization system, software evaluation report. Report LiTH-ISY-R-1469, Computer Vision Laboratory, S–581 83 Linköping, Sweden, 1993.

[Wilson and Granlund, 1984] R. Wilson and G. H. Granlund. The uncertainty principle in image processing. *IEEE Transactions on Pattern Analysis and Machine Intelligence*, PAMI–6(6), November 1984. Report

LiTH–ISY–I–0576, Computer Vision Laboratory, Linköping University, Sweden, 1983.

[Wilson and Knutsson, 1989] R. Wilson and H. Knutsson. A multiresolution stereopsis algorithm based on the Gabor representation. In *3rd International Conference on Image Processing and Its Applications*, pages 19–22, Warwick, Great Britain, July 1989. IEE. ISBN 0 85296382 3 ISSN 0537–9989.

[Wilson and Spann, 1988] R. Wilson and M. Spann. *Image segmentation and uncertainty*. Research Studies Press, 1988.

[Witkin, 1983] A. Witkin. Scale-space filtering. In *8th Int. Joint Conf. Artificial Intelligence*, pages 1019–1022, Karlsruhe, 1983.

11. Phase-based Disparity Estimation

Carl-Johan Westelius, Hans Knutsson, Johan Wiklund and Carl-Fredrik Westin

CVL, LiTH

11.1 Modules

This chapter describes theories and algorithms used in the following modules: **Depth, Disparity** and **Vergence control**.

11.2 Introduction

The problem of estimating depth information from two or more images of a scene is one which has received considerable attention over the years and a wide variety of methods have been proposed to solve it [Barnard and Fichsler, 1982; Fleck, 1991]. Methods based on correlation and methods using some form of feature matching between the images have found most widespread use. Of these, the latter have attracted increasing attention since the work of Marr [Marr, 1982], in which the features are zero-crossings on varying scales. These methods share an underlying basis of spatial domain operations.

In recent years, however, increasing interest has been shown in computational models of vision based primarily on a localized frequency domain representation - the Gabor representation [Gabor, 1946; Adelson and Bergen, 1985], first suggested in the context of computer vision by Granlund [Granlund, 1978].

In [Sanger, 1988; Wilson and Knutsson, 1989; Jepson and Jenkin, 1989; Fleet *et al.*, 1991; Langley *et al.*, 1990] it is shown that such a representation also can be adapted to the solution of the stereopsis problem. The basis for the success of these methods is the robustness of the local Gabor-phase differences. The algorithm presented here is an extension of the work presented in [Wilson and Knutsson, 1989; Wiklund *et al.*, 1991; Wiklund *et al.*, 1992]. One of the key features is the consistency checks performed at a number of stages in the algorithm.

The consistency checks use vector, or complex number, representation of features where the magnitude is a confidence measure and the argument is the value. The vector average reflects the consistency in a neighborhood, since consistent vectors add up to a longer resultant than inconsistent vectors.

Fig. 11.1. Left: Vectors from an inconsistent neighborhood yields a short resultant vector, (dashed). Right: Vectors from a consistent neighborhood yields a long resultant vector, (dashed).

11.3   Choice of filters

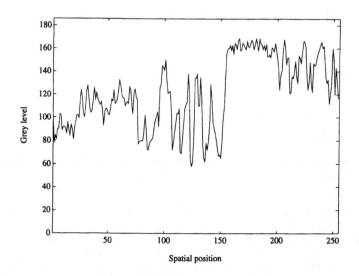

Fig. 11.2. The signal used to test the scalespace behavior of the filters.

When designing filters to be used as disparity estimators there are a number of requirements, some contradictory, to be considered. Different filter types have different characteristics and which one to use depends on the application.

There are a number of different filters that can be used when measuring phase disparities. Gabor filters are by far the most commonly used filter in phase-based disparity measurement. They have linear phase i.e. constant local frequency and are therefore intuitively appealing to use. Quadrature filters do not have any negative frequencies or any DC component. Difference of Gaussians approximating the first and second derivative of a Gaussian can also be used to estimate phase.

In subsections 11.3.2, 11.3.3 and 11.3.3 a number of filters are evaluated with regard to the following requirements:

No DC component The filters must not have a DC component. A DC component would make the signal vector wag back and forth instead of going round, see figure 11.3.

No wrap around It is desirable, but not absolutely necessary, that the phase of the impulse response runs from $-\pi$ to π without any wrap around. This maximizes the maximal measurable disparity for a given size of the filter.

Monotonous phase The phase must be monotonous, otherwise the phase difference between left and right image will not be a one to one function of the disparity. Below, the phase will be called monotonous even though it might wrap around, since the wrap around is caused by the modulo 2π representation.

Only positive frequencies It is also necessary that the filter only picks up positive frequencies, i.e. the phase must rotate in the same direction for all frequencies. If this is not the case the phase differences might change sign depending on the frequency content of the signal. This is a quadrature requirement which means that the spectrum has to be localized in the right halfplane of the frequency domain.

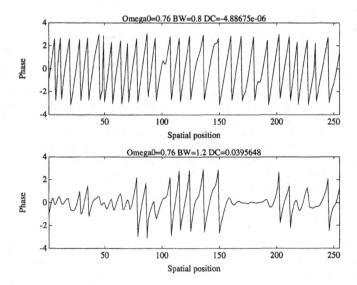

Fig. 11.3. Above, the phase from a Gabor filter with no DC-component. Below, the phase from a Gabor filter with broader bandwidth and thus a DC component on the same input image. Note that the phase is going back and forth instead of wrapping around when the signal fluctuation is small compared the DC level.

In-sensitive to singular points The area affected by the singular points has to be as small as possible, both spatially and in scale. A rule of thumb is that the sensitivity to singular points decreases with decreasing bandwidth. This requirement is contradictory to the requirement of small spatial support below.

Small spatial support The spatial support of the filter function, the size of the filter, determines the computational cost of the convolution.

11.3.1   Creating a phase scale–space

The phase behavior in scalespace has been tested using the signal in figure 11.2 as input. All filtering has been done in the fourier domain. The most high frequent versions of the filter DFTs have been generated, using their definitions. The frequency function has then been multiplied with LP Gaussian function:

$$LP(\omega) = e^{-\frac{\omega^2}{2\sigma_\omega^2}} \tag{1}$$

where $\sigma_\omega = \pi/(2\sqrt{2})$. This emulates the total filtering in a subsampled resolution pyramid or the use of spread kernels in a non–subsampled resolution pyramid, where the LP filtering reduces aliasing. It could be argued that the LP filtering should not be used at the highest resolution level, but it can be motivated by taking the smoothing effects of the imaging system into account. The filter function for each level is calculated by scaling the frequency functions appropriately.

$$F_{\omega_1}(\omega) = F_{\omega_0}(\omega\frac{\omega_0}{\omega_1}) \tag{2}$$

Using linear interpolation between nearest neighbours enables non-integer scaling.

11.3.2 Gabor filters

The equations describing Gabor filters were presented in the introductory section in this chapter. Here follows a description of these filters in relation to phase estimation. The Gabor filters have linear, and thus monotonous, phase by definition. Keeping the filter in the right halfplane, i.e. no negative frequencies and no DC component, is impossible since the Gaussian has infinite support. The center frequency ω_0 is connected to the number of pixels per cycle of the phase and the frequency standard deviation σ_ω is connected to the spatial and frequency support. By adjusting them it is possible to get any number of phase cycles over the size of the spatial support. But all combinations of ω_0 and σ_ω do not yield a useful filter. To see this, suppose that a certain center frequency, ω_0 is wanted. The radius of the frequency support must then be smaller than ω_0 so that the frequency function is sufficiently low at $\omega = 0$, i.e. no DC component. This gives an upper limit on the bandwidth, or rather frequency standard deviation, of the filter. Say, we allow the ratio between the DC component and the top value to be maximally P_{DC} i.e.

$$\frac{G_{\sigma_\omega \omega_0}(0)}{G_{\sigma_\omega \omega_0}(\omega_0)} \le P_{DC} \Rightarrow \sigma_\omega \le \frac{\omega_0}{\sqrt{-2\ln(P_{DC})}} \tag{3}$$

See for instance figure 11.3 where the DC component were a few percent. Using the dual relationship between the frequency and spatial domains, it is possible to use inequality 3 as a lower limit of the spatial standard deviation.

$$\sigma_\omega \le \frac{\omega_0}{\sqrt{-2\ln(P_{DC})}} \Rightarrow \frac{1}{\sigma_\omega} \ge \frac{\sqrt{-2\ln(P_{DC})}}{\omega_0} \Rightarrow \sigma \ge \frac{\sqrt{-2\ln(P_{DC})}}{\omega_0} \tag{4}$$

The spatial support of a Gabor filter is infinite just as the frequency support. A threshold, P_{cut}, must therefore be set in order to get a finite size. The spatial radius, R, of the filter is then

$$\frac{\|g_{\sigma_\omega \omega_0}(R)\|}{\|g_{\sigma_\omega \omega_0}(0)\|} \le P_{cut} \Rightarrow R \ge \sigma\sqrt{-2\ln(P_{cut})} \tag{5}$$

Using the lower limit of the standard deviation i.e.

$$R = \sqrt{-2\ln(P_{cut})}\sigma = \frac{2\sqrt{\ln(P_{cut})\ln(P_{DC})}}{\omega_0} \tag{6}$$

allows the phase difference between the filter ends to be calculated.

$$\Delta\phi = \omega_0 R - \omega_0(-R) = 2\omega_0 R = 4\sqrt{\ln(P_{cut})\ln(P_{DC})} \tag{7}$$

It should be pointed out that the truncation threshold, P_{cut}, affects the DC component of the filter. The DC component should therefore be checked after truncation of the filter to see if it is still less than P_{DC}.

The upper limit of the relative bandwidth, β of the Gabor filters used has heuristically been set to approximately 0.8 octaves, see figure 10.1. This is also the bandwidth used by Fleet et al [Fleet et al., 1991; Jepson and Fleet, 1990].

The behavior of the Gabor filters around the singular points has been thoroughly investigated by Fleet et al,[Fleet et al., 1991]. They used a Gabor scale-space function defined as

$$g(x, \lambda) = g_{\sigma(\lambda)\omega_0(\lambda)}(x) \tag{8}$$

where λ is the scale parameter. The center frequency decreases when the scale parameter increases i.e.

$$\omega_0(\lambda) = \frac{2\pi}{\lambda} \tag{9}$$

In theory it would be possible to keep the absolute bandwidth constant i.e. fix σ_w at the standard deviation used at the lowest ω_0 and then vary ω_0. But doing so the number of phase cycles over the filter would vary with the scale. If the relative bandwidth is kept constant, increasing λ can be seen as stretching out the same filter to cover larger areas, [Granlund, 1978].

Approximating the upper and lower half-height cutoff frequencies as one standard deviation over and one under the center frequency, i.e.

$$\beta = \log_2 \frac{\omega_0(\lambda) + \sigma_w}{\omega_0(\lambda) - \sigma_w} \tag{10}$$

gives the expression for the spatial standard deviation of the filter.

$$\sigma(\lambda) = \frac{1}{\omega_0(\lambda)} \left(\frac{2^\beta + 1}{2^\beta - 1} \right) \tag{11}$$

It is easy to verify that this procedure is equivalent with the scaling procedure described in section 10.3.4,

The isophase curves in figure 11.4 show the phase on a number of scales. The dark broad lines are due to phase wrap around. A feature that is stable in scale-space keeps its spatial position in all scales. If the phase was completely stable in scale the isophase pattern would then only consist of vertical lines. In section 10.3.1 the existence of singular points was shown, and they can easily be observed in phase diagram. The positions where the isophase curves converges are singular points. Just above them the isophase curves turn downwards indicating areas with decreasing phase i.e. negative local frequency, compare with figure 10.5g. The high density of isophase curves just below singular points shows that the local frequency is very high, see figure 10.5d. Langley suggests that the mean DC level should be subtracted from the input images in order to enhance the results,[Langley et al., 1990]. The reason is probably that the DC component of the filter then is less critical. The best would be to calculate the weighted average in every image point using the Gaussian envelope of the the Gabor filter and subtract it from the original image, which is the same as constructing a new filter with no DC component.

11.3.3   Quadrature filters

In this section we proceed the discussion from section about quadrature filters.

Lognorm filter For a certain relative bandwidth the phase will go through a certain number of cycles independent of the size of the filter support. Recalling, from the Gabor case, that the center frequency is related to the number of pixels per phase cycle and the bandwidth is related to the size of the spatial support, it is evident that ensuring no wrap around then is the same as using a wide relative bandwidth.The long tail of the lognorm frequency function makes this possible only for relatively low center frequencies, see figure 10.2. If too much of the tail is cut it will no longer be a lognorm filter.

The isophase curves in figure 10.4 show the phase on a number of scales. It is generated using the same parameters as in the Gabor case above i.e. $\omega_0 = \pi/4$ and $\beta = 0.8$. The similarity makes it easy to identify the singular points and compare the behavior of the phase around them.

Studying the behavior of the phase around the singular points indicates that the disturbance region is smaller than for Gabor filters i.e. the areas with negative frequencies are smaller. On the other hand, the size of a lognorm filter is approximately 50 percent larger than a Gabor filter with the same center frequency and bandwidth when truncating on one percent of the maximal value.

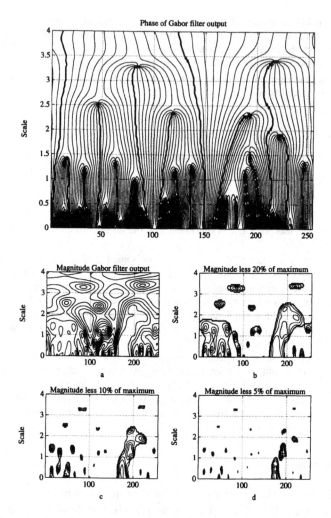

Fig. 11.4. Above: Isophase plot of Gabor phase scale-space. The dark broad lines are due to phase wrap around. The positions where the isophase curves converges are singular points. Below: Isomagnitude plot of Gabor phase scale-space. The dark areas indicates where the magnitude of the signal is below the threshold. $\omega_0 = \frac{\pi/4}{scale}$ and $\beta = 0.8$

Powexp filter There is a type of quadrature filters where the number of cycles of the phase is directly controllable. The frequency domain definition is:

$$F(\omega) = \begin{cases} \left(\frac{\omega}{\omega_0}\right)^{\alpha} e^{\left(-\alpha\left(\frac{\omega}{\omega_0}-1\right)\right)} & \text{if } \omega > 0 \\ 0 & \text{otherwise} \end{cases} \qquad (12)$$

and the spatial definition is

$$f(x) = \frac{C}{(1+i\frac{\omega_0 x}{\alpha})^{(\alpha+1)}}$$
where
$$C = \frac{1}{\alpha^\alpha e^{-\alpha}\|\frac{\alpha}{\omega_0}\|}$$

(13)

Both the bandwidth and the number of phase cycles of these filters are a function of α. For $\alpha = 1$ the relative bandwidth is approximately 2 octaves and the phase cycles once. However, both the frequency and spatial support of the filter are very large, which reduces the usefulness of the filter type e.g. the spatial support is approximately 60 percent larger than a lognorm filter with the same center frequency and bandwidth when truncating at one percent of the maximum value.

Other even–odd pairs The filters described above all consist of an even real part and an odd imaginary part and the phase is calculated from the ratio between these parts. There are a few other types of filters that are neither Gabor nor quadrature filters, but which can be interpreted as odd–even pars.

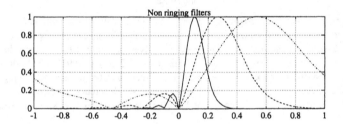

Fig. 11.5. Three non-ringing filters. $\omega \approx \pi/8, \pi/4, \pi/2$ and $\beta \approx 2.2$. The filters where generated in the spatial domain using equation 16. The radii of spatial support where $R = 14, 5, 2$.

Non-ringing filters A filter type that has only one phase cycle over the spatial support and that can be designed to have almost quadrature features has been suggested by Knutsson,[Knutsson, 1982]. Using a monotonous antisymmetric phase function a filter having a phase span of $n2\pi$ and no DC component can be defined as:

$$f(x) = g'(x)\exp(iC_0 g(x))$$
where
$$g'(x) = \begin{cases} > 0 \text{ if } -R \le x \le R \\ 0 \quad \text{otherwise} \end{cases}$$
and
$$C_0 = \frac{n\pi}{g(R)} \qquad n = 0, 1, 2, ..$$

(14)

The function $g(x)$ can be any monotonous antisymmetric function, but since the derivative controls the envelope it is advisable to use a function with a smooth and unimodal derivative. How well such a filter approximates a quadrature filter depends on the size of the filter and how smooth the filter function is. It is easily shown that the DC component is zero,

$$F(0) = \int_\infty^\infty f(x)dx =$$

$$\int_{-R}^R f(x)dx =$$

$$\int_{-R}^R g'(x)\exp(iC_0 g(x))dx =$$

$$\int_{-R}^R g'(x)\exp(i\frac{n\pi g(x)}{g(R)})dx = \tag{15}$$

$$\exp(i\frac{n\pi g(R)}{g(R)}) - \exp(i\frac{n\pi g(-R)}{g(R)})$$

$$\exp(in\pi) - \exp(-in\pi) = 0$$

Choosing $n = 1 \Rightarrow C_0 = \pi/g(R)$ yields a filter with no DC component, and no wrap around.

The isophase curves have been calculated using the primitive function to a squared cosine as argument function and, thus, the squared cosine as envelope.

$$f(x) = \begin{cases} \cos^2(\frac{x\pi}{2R})\exp(-i\pi(\frac{x}{R} + \sin(\frac{x\pi}{R}))) & \text{if } -R < x < R \\ 0 & \text{otherwise} \end{cases} \tag{16}$$

The center frequency is approximately $3 * \pi/2R$ and the relative bandwidth is 2.2 octaves. The phase is monotonous in the filter but the filter has a considerable support for negative frequencies.

Gaussian differences Gaussian filters and their derivatives, or rather differences, can be efficiently implemented using binomial filter kernels, [Chehikian and Crowley, 1991]. The basic kernels are a LP kernel and a difference kernel, see figure 11.7. These can be implemented using shifts and summations. Crowley has suggested that the first and second difference of the binomial Gaussians can be used as a phase estimator, see figure 11.7. The spatial support is extremely small, only 5 pixels.

From the design it is evident that there is no DC component and that the phase does not wrap around. The first difference filter, dx, is the odd kernel, and changing sign on the second difference, d^2x, gives the even kernel. Instead of just using the kernels as they are, it is possible to give them different relative weights, producing a range of complex filters.

$$f(x) = -\alpha d^2 x + i(1 - \alpha)dx$$
where
$$0 \le \alpha \le 1 \tag{17}$$

The energy in the left halfplane of the frequency domain is minimized by setting $\alpha = 0.3660$, see figure 11.8. This design method is a special case of a method for producing quadrature filters called prolate spheroidals. In the general case there are an arbitrary number of basis filters that are weighted together, using a multivariable optimizing technique. The method produces the best possible quadrature filter of a given size, in the sense that it has minimum energy in the left half plane.

If it is essential to the implementation to use only summations and shifts the weights can be chosen to 1 for d^2x and 2 for dx, corresponding to $\alpha = 0.3333$. The relative bandwidth is approximately 2 octaves and independent of α.

11.3.4   Discussion on filter choice

The choice of filter is not evident from these investigations. Different characteristics might have different priorities in different applications. The size of the kernel might be less important if special

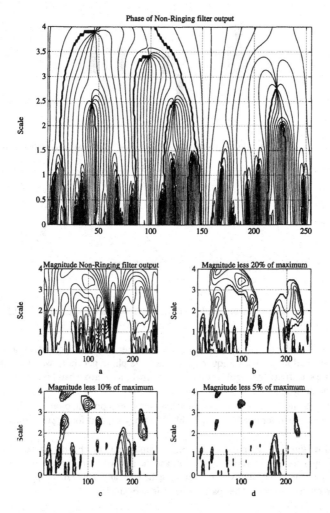

Fig. 11.6. Above: Isophaseplot of non-ringing phase scale-space. The dark broad lines are due to phase wrap around. The positions where the isophase curves converges are singular points. Below: Isomagnitude plot of non-ringing phase scale-space. The dark areas indicates where the magnitude of the signal is below the threshold. Note the difference from the other threshold plots. $\omega_0 = \frac{\pi/4}{scale}$

purpose hardware is used, making scale-space behavior a critical issue. On the other hand, if the convolution time depends directly on the kernel size, a less robust, but smaller, kernel might be accepted. The most relevant test is to use the filters in the intended application and measure the overall performance. For convenience the filter characteristics are summed below.

Gabor filters The Gabor filters might have a DC component if not designed carefully. The phase is monotonous but wraps around. The frequency support is localized in the right halfplane, and the sensitivity to singular points is moderate.

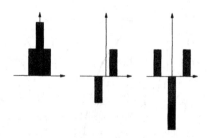

Fig. 11.7. Left, the LP kernel. Middle, the first difference kernel. Right, the second difference kernel.

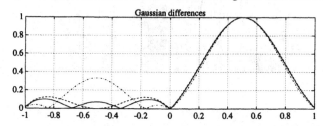

Fig. 11.8. The DFT magnitude of the binomial phase filter for $\alpha = 0.5$ (dot-dashed), $\alpha = 0.3333$ (dashed) and $\alpha = 0.3660$ (solid).

Lognorm filters The lognorm quadrature filters have neither a DC component nor any frequency support in the left halfplane of the frequency domain. The phase generally wraps around but it is monotonous. The sensitivity to singular points is very small, for narrow band filters. The sensitivity increases with bandwidth.

Non-ring filters The non-ring filter investigated has no DC component, monotonous phase and no phase wrap around. The filter has a not negligible frequency support for negative frequencies. The spatial support is small.

Difference of Gaussians Gaussian derivatives filters implemented with binomial coefficients do not have any DC component. The phase is monotonous and there is no phase wrap around. The sensitivity for negative frequencies can be adjusted by weighing the even and odd kernels appropriately. It can, however, not be reduced to zero. The spatial support is very small.

11.4 Algorithm description

A hierarchical stereo algorithm that uses a phase based disparity estimator has been developed. To optimize the computational performance, a multiresolution representation of the left and right image is used. The algorithm has been implemented using non-subsampled, subsampled and fovea resolution pyramids [Westelius, 1992]. An edge detector, tuned to vertical structures, is used to produce a pair of images containing edge information. The edge images reduces the influence of singular points since the singular points in the original images and the edge images generally do not coincide, see figure 11.11. The impact of any DC component in the disparity filter is also reduced with the edge images. The edge images together with the corresponding original image pair are used to build the resolution pyramids. It is one octave between the levels. The number of levels needed depends on the maximum disparity in the stereo pair.

The algorithm starts at the lowest resolution. The disparity accumulator holds and updates

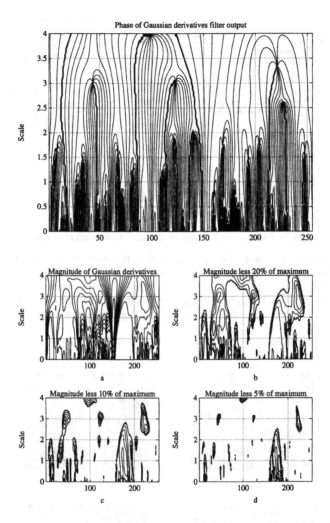

Fig. 11.9. Above: Isophase plot of Gaussian differences phase scale-space. The dark broad lines are due to phase wrap around. The positions where the isophase curves converges are singular points. Below: Isomagnitude plot of Gaussian derivatives phase scale-space. The dark areas indicates where the magnitude of the signal is below the threshold. Note the difference from the other threshold plots. $\omega_0 = \frac{\pi/2}{scale}$

disparity estimates and confidence measures for each pixel. The four input images are shifted according to the current disparity estimates. After the shift, the new disparity estimate is calculated using the phase differences, the local frequency and their confidence values. The disparity estimate from the edge image pair has high confidence close to edges, while the confidence is low in between them. The estimates from the original image pair resolve possible problems of matching incompatible edges, that is, only edges with the same sign of the gradient should be matched. Both these disparity estimates are weighted together by a consistency function to form the disparity measure between the shifted images. The new disparity measure updates the current estimate in the disparity accumulator.

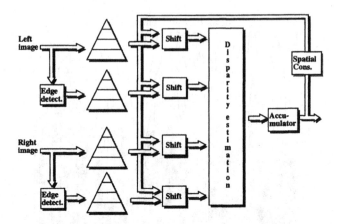

Fig. 11.10. Computation structure for the hierarchical stereo algorithm.

For each resolution level a refinement of the disparity estimate can be done by iterating these steps. It should get closer and closer to zero during the iterations.

Between each level the disparity image is resampled to the new resolution, if applicable, and a local spatial consistency check is performed. The steps above are repeated until the finest resolution is reached. The accumulator image then contains the final disparity estimates and certainties.

11.4.1   Edge extraction

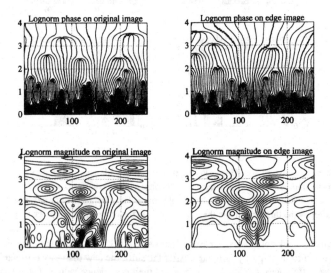

Fig. 11.11. Isophase and isomagnitude plots for the original and edge images. Note that the singular points of the grey level image and the edge image do generally not coincide.

Creating edge images can be done using any edge extraction algorithm. Here the edge extraction is performed using the same filter as for the disparity estimation. The magnitude of the filter response is stored in the edge image, creating a sort of line drawing. The disparity filters are sensitive only to more or less vertically oriented structures, but this is no limitation since horizontal lines does not contain any disparity information. The produced edge image is used as input to create a resolution pyramid in the same way as described above. There are a total of four pyramids that are generated before starting the disparity estimation.

11.4.2 Local image shifts

The images from the current level in the resolution pyramid are shifted according to the disparity accumulator, which is initialized to zero. The shift procedure decreases the disparity since the left and right images are shifted towards each other. This means that if a disparity is estimated fairly well at a low resolution, the reduction of the disparity will enable the next level to further refine the result.

The shift is implemented as a "picking at a distance" procedure, which means that a value is picked from a the old image to the new image at a distance determined by the disparity. This ensures that there will be no areas without a value. Linear interpolation between neighbors allow non-integer shifts.

11.4.3 Disparity estimation

The disparity is measured on both the grey level images and the edge images. The phase can be estimated using any of the filters described in section 11.3. The result will of course vary with the filter characteristics, but a number of consistency checks reduces the variation between filter types, see separate discussion.

The disparity is estimated in the grey level images and the edge images separately and the result is weighted together. The filter responses in a point can be represented with a complex number. The real and imaginary parts of the complex number represents the even and odd filter responses respectively. The magnitude is a measure of how strong the signal is and how well it fits the filter. The magnitude will therefore be used as a confidence measure of the filter response. The argument of the complex number is the phase in the signal.

Let the responses from the phase estimating filter be represented with the complex numbers Z_L and Z_R for the left and right image respectively. The filters are normalized so that $0 \leq \|Z_{L,R}\| \leq 1$. Calculating $D = Z_L Z_R^*$, where $*$ denotes the complex conjugate, yields a phase difference measure and a confidence value,

$$
\begin{aligned}
\|D\| &= \|Z_L\|\|Z_R\|, & 0 &\leq \|D\| \leq 1 \\
\arg(D) &= \arg(Z_L) - \arg(Z_R), & -\pi &\leq \arg(D) \leq \pi
\end{aligned}
\tag{18}
$$

The magnitude, $\|D\|$, is large only if both filter magnitudes are large. It consequently indicates how reliable the phase difference is. If a filter sees a completely homogeneous neighborhood its magnitude will be zero and its argument will be undefined. Calculating the phase difference without any confidence values will then produce an arbitrary result.

If the images are captured under similar conditions and they are covering approximately the same area, it is reasonable that the magnitudes of the filter responses are approximately the same for both images. This can be used to check the validity of the disparity estimate. A substantial difference in magnitude can be due to noise or too large disparity, i.e. the image neighborhoods do not depict the same part of reality. It can also be due to a singular point in one of the signals, since the magnitude is reduced considerably in such neighborhoods. In any of these cases the confidence value of the estimate should be reduced, so the consistency checks later on can weigh the estimate accordingly. Sanger used the ratio between the smaller and the larger of the magnitudes as a confidence value, [Sanger, 1988]. Such a confidence value does not differentiate between strong and weak signals.

The confidence function below depends both on the relation between the filter magnitudes and the absolute value. The confidence value will therefore reflect both the similarity and the signal strength.

$$C_1 = \sqrt{\|Z_L Z_R\|} \left(\frac{2\|Z_L Z_R\|}{\|Z_L\|^2 + \|Z_R\|^2} \right)^\gamma \tag{19}$$

The square root of $\|Z_L Z_R\|$ is geometrical average between the filter magnitudes i.e. a measure on the combined signal strength. The exponent γ controls how much a magnitude difference should be punished. The expression within the parenthesis is equal to one if $\|Z_L\| = \|Z_R\|$ and decays with increasing magnitude difference. Setting $M^2 = \|Z_L Z_R\|$ and $\alpha = \|Z_R\|/\|Z_L\|$ transforms equation 19 into a more intuitively understandable form:

$$C_1 = M \left(\frac{2\alpha}{1 + \alpha^2} \right)^\gamma \tag{20}$$

If $\|Z_L\| = \|Z_R\| = M$, i.e. $\alpha = 1$, then $C_1 = M$. This means that if the magnitudes are almost the same the confidence value will also be the same. If the magnitudes differ the confidence will go down with a rate which is controlled by γ. Figure 11.12 shows how the confidence depends on the filter magnitude ratio, α and the exponent γ. Throughout the testing of the algorithm the exponent γ has heuristically been set to 4.

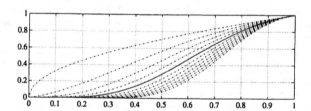

Fig. 11.12. The magnitude difference penalty function. The plots show the function for $0 \leq \gamma \leq 10$ from left to right. The abscissa (α) is the ratio between the smaller and the larger magnitude.

If the phase difference is very large it might wrap around and indicate a disparity with the opposite sign. Very large phase differences should therefore be given a lower confidence value, [Wilson and Knutsson, 1989].

$$C_2 = C_1 \cos^2 \left(\frac{\arg(D)}{2} \right) \tag{21}$$

In section 10.3.1 it was shown that the phase derivative varies with the frequency content of the signal. In order to correctly interpret the phase difference, $\Delta\phi = \arg(D)$, as disparity it is necessary to estimate the phase derivative i.e. local frequency, [Langley et al., 1990; Fleet et al., 1991].

Let $Z(x)$ be a phase estimate in position x. The phase differences between position x and its two neighbors is a measure of how fast the phase is varying, i.e. the local frequency, in the neighborhood. The local frequency can be approximated using the the phase difference to the left and right of the current position:

$$
\begin{aligned}
d_{L-} &= Z_L(x-1)Z_L^*(x) \\
d_{L+} &= Z_L(x)Z_L^*(x+1) \\
d_{R-} &= Z_R(x-1)Z_R^*(x) \\
d_{R+} &= Z_R(x)Z_R^*(x+1)
\end{aligned}
\tag{22}
$$

The arguments of $d_{L-,L+,R-,R+}$ are estimates of the local frequency, that are combined using

$$\phi' = \arg\left(\frac{d_{L-} + d_{L+} + d_{R-} + d_{R+}}{4}\right) \tag{23}$$

The confidence value is updated with a factor depending only on the similarity between the local frequency estimates and not on their magnitudes. If the local frequency is zero or negative the confidence value is set to zero since the phase difference then is completely unreliable.

$$C_3 = \begin{cases} C_2 \left\| \frac{1}{4} \left(e^{i\arg(d_{L-})} + e^{i\arg(d_{L+})} + e^{i\arg(d_{R-})} + e^{i\arg(d_{R+})} \right) \right\|^4 & \text{if } \phi' > 0 \\ 0 & \text{if } \phi' \leq 0 \end{cases} \tag{24}$$

Let subscript g and e denote grey level and edge image values respectively. The estimated disparities are then

$$\Delta x_g = \frac{\arg(D_g)}{\phi'_g} \tag{25}$$

$$\Delta x_e = \frac{\arg(D_e)}{\phi'_e}$$

The total disparity estimate and its confidence value are calculated from the disparity estimates, and the confidence values in both the grey level images and the edge images.

$$\Delta x = \frac{C_{g3}\Delta x_g + C_{e3}\Delta x_e}{C_{g3} + C_{e3}} \tag{26}$$

The confidence value for the disparity estimate depends on C_{g3}, C_{e3} and the similarity between the phase differences $\arg(D_g)$ and $\arg(D_e)$. This is accomplished by adding the confidence values as vectors with the phase differences as arguments.

$$C_{tot} = \left\| C_{g3} \exp(i\frac{\arg(D_g)}{2}) + C_{e3}\exp(i\frac{\arg(D_e)}{2}) \right\| \tag{27}$$

The phase differences, $\arg(D_{e,g})$ are divided by two in order to ensure that C_{tot} is large only for $\arg(D_g) \approx \arg(D_e)$ and not for $\arg(D_g) \approx \arg(D_e) \pm 2\pi$ as well.

11.4.4 Disparity accumulation

The disparity accumulator is updated using the disparity estimate and its confidence value. The accumulator holds the cumulative sum of disparity estimates. Since the images are shifted according to the current accumulator value, the value to be added is just a correction towards the true disparity. Thus, the disparity value is simply added to the accumulator.

$$\Delta x_{new} = \Delta x_{old} + \Delta x \tag{28}$$

When updating the confidence value of the accumulator, high confidence values are emphasized and low values are attenuated.

$$C_{new} = \left(\frac{\sqrt{C_{old}} + \sqrt{C_{total}}}{2}\right)^2 \tag{29}$$

11.4.5   Spatial consistency

In most images there are areas where the phase estimates are very weak or contradictory. In these areas the disparity estimates are not reliable. This results in tearing the image apart when making the shift before disparity refinement, creating unnecessary distortion of the image. It is then desirable to spread the estimates from nearby areas with higher confidence values. On the other hand, it is not desirable to average between areas with different disparity and high confidence. A filter function fulfilling these requirements has a spatial function with a large peak in the middle and then decays rapidly towards the periphery, such as a Gaussian with a small σ:

$$h(x) = \frac{1}{\sigma\sqrt{2\pi}} e^{-\frac{x^2}{2\sigma^2}} \quad -R \le x \le R \tag{30}$$

A kernel with $R = 7$ and $\sigma = 1.0$ has been used when testing the algorithm. The filter is used in the x and y direction separately.

The filter is convolved with both the confidence values alone and the disparity estimates weighted with the confidence values.

$$m = h * C \tag{31}$$

$$v = h * C \Delta x \tag{32}$$

If the filter is positioned on a point with a high confidence value the disparity estimate will be left virtually untouched, but if the confidence value is weak it changes towards the average of the neighborhood. The new disparity estimate and its confidence value are

$$C_{new} = m \tag{33}$$

$$\Delta x_{new} = \frac{v}{m} \tag{34}$$

After the spatial consistency operation the accumulator is used to shift the input images either on the same level once more or on the next finer level, depending on how many iterations are used on each level.

11.5   Experimental results

The algorithm has been tested both on synthetic test images and on real images, using a wide variety of filters, see table 11.1. All types of filters discussed in section 11.3 are represented. Also a number of design parameter combinations have been tested. The filter based on differences of binomial Gaussians was designed with $\alpha = 1/3$, see function 17. The non-ringing filters were designed using function 16. The spatial size of these two types of filters are given by their definitions. Strictly speaking, the Gabor and Lognorm filters have infinite support and must be truncated to get a finite size. The size could be set large enough to make the truncation negligible, but this often gives very large filters. The criterions for setting the size of the Gabor and lognorm filters have been that the DC level must not be more than one percent and the envelope must have decreased to less than ten percent of the peak value.

11.5.1   Generating stereo image pairs

The quantitative results have been obtained by using synthetically generated images and comparing the estimated disparities with truth values. The often used method of taking an image and simply shift it a few pixels in order to create a known disparity will not show the advantage of the local phase estimation. For such image pairs the difference in *global* fourier phase would do just as well.

Table 11.1. The filters used testing the phase based stereo algorithm.

Filter	Peak frequency	Bandwidth	Size
Gaussian diff.	$\pi/2$	2.0	5
Non-ringing 7	$\pi/2$	2.2	7
Non-ringing 11	$3\pi/10$	2.2	11
Non-ringing 15	$3\pi/14$	2.2	15
Gabor 1	$\pi/2$	0.8	15
Gabor 2	$\pi/(2\sqrt{2})$	0.8	17
Gabor 3	$\pi/4$	0.8	19
Lognorm 1	$\pi/2$	1.0	13
Lognorm 2a	$\pi/(2\sqrt{2})$	1.0	15
Lognorm 2b	$\pi/(2\sqrt{2})$	2.0	15
Lognorm 3a	$\pi/4$	1.0	19
Lognorm 3b	$\pi/4$	2.0	19
Lognorm 3c	$\pi/4$	4.0	19

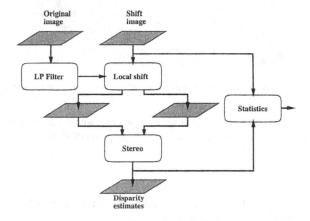

Fig. 11.13. The synthetic stereo pairs are generated using a shift image describing the local shifts. The image to be shifted is LP filtered in order to avoid aliasing. The shift image is also used as truth values for the stereo estimates.

A method for evaluation of a disparity estimator must be based on locally varying shifts in order to resemble real life situations. A scheme for generation of locally varying disparity in a controlled fashion can be found in figure 11.13. The test image is shifted locally according to the values in a syntheically generated disparity image, which also is used as truth values when evaluating the results.

Global shifting creates an image with the same properties as the original images at a certain distance. Local shifts deform the original image, stretching it in some areas and compressing it in other areas. The stretching do not create any problems but the compressing might do so. The spatial frequency increases when the image is compressed and it might exceed the Nyquist frequency. The image to be shifted is therefore LP filtered using a Gaussian with $\sigma = 1.273$.

In real life images, the structures in the image belong to real objects and so do the disparities. In synthetic test images generated as shown in figure 11.13 the disparities are not necessarily related to the image structure. A random image, e.g. white noise, random dots etc, is therefore best suited

Fig. 11.14. A noise image with an energy function inversely proportional to the radius in the frequency domain.

for testing purposes, since all parts of the image then gets the same structure. In the tests below a noise image with a spatial frequency spectrum inversely proportional to the radius of the frequency domain

$$\|F(\omega)\| \sim \frac{1}{\|\omega\|}$$

has been used, see figure 11.14. This is justified by the fact that it resembles a natural image more than e.g. a white noise image, [Knutsson, 1982].

Two different shift images have been used in the tests, see figure 11.15. "Twin Peaks" consists one positive and one negative peak. The disparity is zero along the image edges and the magnitude changes linearly in both the vertical and horizontal direction. The other shift image, "Flip Flop" consists of alternating positive and negative peaks with exponentially increasing frequency downwards. It is also zero along the image edges. The magnitude is a sinus function vertically and a exponential function horizontally. The shift values in the shift images are normalized to the interval $[-1, 1]$, and the maximum disparity is then controlled by a parameter to the shift module, see figure 11.13.

11.5.2 Statistics

The result of the stereo algorithm is evaluated by measuring the mean and the standard deviation of the error between the shift image used to create the stereo pair and the disparity image. Since the algorithm also provides a confidence value, the mean and standard deviation weighted with the confidence values are also calculated. Let Δx_i denote the true disparity and $\Delta \tilde{x}_i$ denote the estimated disparity. The statistics are then calculated as:

$$m = \frac{1}{n} \sum_{i=1}^{n} (\Delta x_i - \Delta \tilde{x}_i) \tag{35}$$

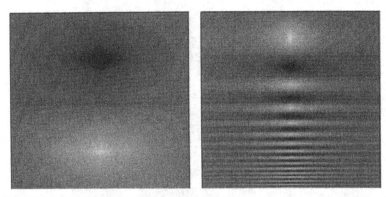

Fig. 11.15. Left: "Twin Peaks", one positive peak, one negative peak and zero around the edges. The magnitude changes linearly in both the horizontal and the vertical direction. Right: "Flipflop", alternating positive and negative peaks, zero around the edges. The magnitude is a sinus function vertically and a exponential function horizontally.

$$s^2 = \frac{1}{n-1} \sum_{i=1}^{n} (\Delta x_i - \Delta \tilde{x}_i - m)^2 \tag{36}$$

$$m_w = \frac{\sum_{i=1}^{n} C_i (\Delta x_i - \Delta \tilde{x}_i)}{\sum_{i=1}^{n} C_i} \tag{37}$$

$$s_w^2 = \frac{\sum_{i=1}^{n} C_i (\Delta x_i - \Delta \tilde{x}_i - m_w)^2}{\sum_{i=1}^{n} C_i} \tag{38}$$

The unweighted values furnish a measure of the how well the algorithm has done over the whole image. The weighted values, on the other hand, indicates how well the confidence value reflects the reliability of the measurements. If the confidence value always is low when the disparity estimate is wrong, the weighted statistics will show better values than the unweighted.

If the disparity is captured at the coarsest level the finer levels will refine the estimate to very high precision, while if the algorithm fails on a coarse resolution there is no way to recover. In the areas with too large disparity, the estimates will be arbitrary. The algorithm is also likely to mismatch structures that accidently coincides due to insufficient shifting of the images. A consequence of this is that it is hard to compare the statistics when using different shift images since they are dependent on the ratio between the image area with measurable disparity and the area with too large disparity. The statistics diagrams should therefore be compared quantitatively only if they belong to the same shift image. The qualitative behavior is comparable for different shift images, though.

The zero estimator is used for comparison. It always estimates zero disparity with full confidence i.e. the statistics for the zero estimator measures the mean and standard deviation of the shift image.

11.5.3   Increasing number of resolution levels

A test where the maximum disparity was fixed at 40 pixels and the number of iterations on each level were fixed at 2 has also been carried out. The number of levels of resolution was varied between one and six. The test showed [Westelius, 1992] that it is possible for all filters to correctly estimate the disparity for the whole image if the number of levels are more than four. In "Twin Peaks" there are two well separated areas, which do not interfere significantly with each other as the image is LP

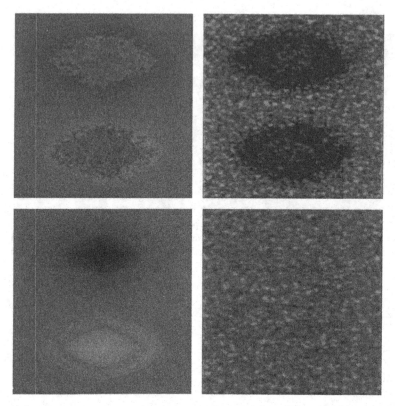

Fig. 11.16. Disparity estimates, left, and their confidence values, right, for one resolution level, above, and three resolution levels, below. The maximum disparity was fixed to 40 pixels, the number of iterations were fixed to two, and the filter used was lognorm2b. Note how the confidence decreases in neighborhoods were the disparity estimation fails.

filtered. In "Flip Flop" the areas with different disparity are smaller and closer to each other. A large enough filter will always collect information from areas with opposite disparities averaging to zero. Consequently, the interference within the filter cancels out the advantage of reaching over a greater distance.

11.6   Conclusion

The test results show that the overall performance of the stereo algorithm is not critically dependent on the type of disparity filter used. The consistency checks and edge image filtering applied in order to enhance the performance of the algorithm, do indeed reduce the impact of the actual filter shape.

There are, however, some conclusions to be made. The results of the phase based stereo algorithm in section 11.5, are somewhat contradictory to the results of the investigation of the singular points in section 11.3. The stereo tests indicate that filters with a wide bandwidth are preferable, while a narrow bandwidth is more advantageous from a phase scale space point of view. The reason is that broad band filters have less phase cycles and therefore can handle greater disparities for a given filter size. The computational efficiency that follows from this implies that the best filter choice is the filters without wrap around; Difference of Gaussians are very small and thus requires increasing levels of resolution, while the non-ringing are larger and manages with a smaller number of levels. As pointed out in subsection 11.5.3 it is not always possible to compensate small size with increasing number of levels.

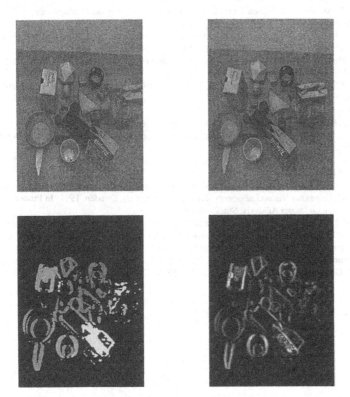

Fig. 11.17. Above: Left and right image, (courtesy to to CVAP, Royal Institute of Techology, Stockholm). Lower left: Disparity estiamtes threshold using the confidence values. Lower right: confidence values. Note that the confidence values are strong on image structures and weak on flat surfaces. Also note how the very large disparities make the image look very different from each other. The result is obtained with six resolution levels and two iterations on each level. The filter used is nonring7.

References

[Adelson and Bergen, 1985] E. H. Adelson and J. R. Bergen. Spatiotemporal energy models for the perception of motion. *Jour. of the Opt. Soc. of America*, 2:284–299, 1985.

[Barnard and Fichsler, 1982] S. T. Barnard and M. A. Fichsler. Computational Stereo. *ACM Comput. Surv.*, 14:553–572, 1982.

[Chehikian and Crowley, 1991] A. Chehikian and J. L. Crowley. Fast computation of optimal semi–octave pyramids. In *Proceedings of the 7th scandinavian Conf. on image analysis*, pages 18–27, Aalborg, Denmark, 1991. Pattern recognition of Denmark and Aalborg University.

[Fleck, 1991] Margaret M. Fleck. A topological stereo matcher. *Int. Journal of Computer Vision*, 6(3):197–226, August 1991.

[Fleet et al., 1991] David J. Fleet, Allan D. Jepson, and Michael R. M. Jenkin. Phase-based disparity measurement. *CVGIP Image Understanding*, 53(2):198–210, March 1991.

[Gabor, 1946] D. Gabor. Theory of communication. *Proc. Inst. Elec. Eng.*, 93(26):429–441, 1946.

[Granlund, 1978] G. H. Granlund. In search of a general picture processing operator. *Computer Graphics and Image Processing*, 8(2):155–178, 1978.

[Jepson and Fleet, 1990] Allan D. Jepson and David J. Fleet. Scale-space singularities. In O. Faugeras, editor, *Computer Vision-ECCV90*, pages 50–55. Springer-Verlag, 1990.

[Jepson and Jenkin, 1989] A. D. Jepson and M. Jenkin. The fast computation of disparity from phase differences. In *Proceedings CVPR*, pages 386–398, San Diego, California, USA, 1989.

[Knutsson, 1982] Hans Knutsson. *Filtering and Reconstruction in Image Processing*. PhD thesis, Linköping University, Sweden, 1982. Diss. No. 88.

[Langley et al., 1990] K. Langley, T. J. Atherton, R.G. Wilson, and M. H. E. Larcombe. Vertical and horizontal disparities from phase. In O. Faugeras, editor, *Computer Vision-ECCV90*, pages 315–325. Springer-Verlag, April 1990.

[Marr, 1982] David Marr. *Vision*. W. H. Freeman and Company, New York, 1982.

[Sanger, 1988] T. D. Sanger. Stereo disparity computation using gabor filters. *Biological cybernetics*, 59:405–418, 1988.

[Westelius, 1992] C-J Westelius. Preattentive gaze control for robot vision, June 1992. Thesis No. 322, ISBN 91-7870-961-X.

[Wiklund et al., 1991] J. Wiklund, H. Knutsson, and R. Wilson. A hierarchical stereo algorithm. Report LiTH–ISY–I–1167, Computer Vision Laboratory, Linköping University, Sweden, 1991. In Proceedings of the SSAB Symposium on Image Analysis, Stockholm, March 1991.

[Wiklund et al., 1992] J. Wiklund, C-J Westelius, and H. Knutsson. Hierarchical phase based disparity estimation. Report LiTH–ISY–I–1327, Computer Vision Laboratory, Linköping University, Sweden, 1992.

[Wilson and Knutsson, 1989] R. Wilson and H. Knutsson. A multiresolution stereopsis algorithm based on the Gabor representation. In *3rd International Conference on Image Processing and Its Applications*, pages 19–22, Warwick, Great Britain, July 1989. IEE. ISBN 0 85296382 3 ISSN 0537-9989.

12. Low Level Focus of Attention Mechanisms

Carl-Johan Westelius, Hans Knutsson and Gösta Granlund

CVL, LiTH

12.1 Modules

This chapter describes theories and algorithms used in the following modules: **Robot simulator, Camera control** and **Adaptive environment scanning**

12.2 Introduction

12.2.1 Human focus of attention

Humans can shift the attention either by moving the fixation point or by concentrate on a part of the field of view. The two types are called *overt* and *covert* attention respectively. The covert attention shifts are about four times as fast as the overt shifts. This speed difference can be used to check a potential fixation point to see if it is worthwhile moving the gaze to that position. High speed reading techniques depend on the covert attention shifts. Such techniques are based on positioning the fixation point on every third or fourth word instead of smoothly scanning the text. Since the number of fixations per line of text is reduced the speed of reading is increased. Note that these techniques are not the same as browsing a book searching for a special item.

A number of paradigms describing human focus of attention has been developed over the years, [Milanese, 1990]. In the *zoom-lens* metaphor computational resources can either be spread on the whole field of view, 'wide angle lens', or concentrated on a portion, 'telephoto lens'. This metaphor is founded on the assumption that it is the computational resources that are the main cause for having to focus the attention on one thing at the time. Thus, the problem is to allocate the resources properly.

The work presented below relates to the *search light* metaphor, [Julesz, 1991]. A basic assumption in this metaphor is the division between preattentive and attentive perception. The idea is that the preattentive part of the system makes a crude analysis of the field of view. The attentive part then analyzes areas with indications on being of particular interest more closely. The two systems should not be seen as taking turns in a time multiplex manner, but rather as a pipeline where the attentive part uses the continuous stream of results from the preattentive part as clues. The term 'search light' reflects how the attentive system analyzes parts of the information available by illuminating it with a attentional search light. The reason for having to focus the attention in this metaphor is that some tasks are inherently of sequential nature.

What features or properties are then important for positioning the fixation point? For attentional shifts the criterion is closely connected to the task at hand. Yarbus pioneered the work on studying how human move the fixation point in images depending on the wanted information,[Yarbus, 1969]. For preattentional shifts gradients in space and time, i.e high contrast areas or motion, are considered to be the important features. Abbott and Ahuja present a list of criterions for the choice of the next fixation point, [Abbott and Ahuja, 1989]. Many of the items in the list relate to computational considerations concerning the surface reconstruction algorithm presented. However, a few clues from human visual behavior was also included:

Absolute distance and direction If multiple candidates for fixation points are present, the ones closer to the center of the viewing field are more likely to be chosen. Upward movement is generally preferred over downward movement.

2D images characteristics If polygonal objects are presented, points close to corners are likely to be chosen as fixation points. When symmetries are present the fixation point tends to be chosen along symmetry lines.

Temporal changes When peripheral stimuli suddenly appear, a strong temporal cue often leads to a movement of fixation point towards the stimuli.

Since fixation point control is a highly task dependent action, it is probably easy to construct situations that contradict the list above. The reader is urged to go back for the appropriate references in order to get a full description of how the results where obtained.

12.2.2 Machine focus of attention

A number of research groups are currently working on incorporating focus of attention mechanisms in computer vision algorithms. This section is by no means a comprehensive overview, but rather a few interesting examples.

The Vision as Process consortium was united by the scientific hypothesis that vision should be studied as a continuous process. The project is aimed at bringing together knowhow from a wide variety of research fields ranging low level feature extraction and ocular reflexes through object recognition and task planning, [VAP consortium, 1992].

Ballard and Brown have produced a series of experiments with ocular reflexes and visual skills, [Ballard, 1990; Brown, 1988; Brown, 1990b; Brown, 1990a; Ballard and Ozcandarli, 1988]. The basic idea is to use simple and fast image processing algorithms in combination with a flexible, active percepting system.

A focus of attention system based on salient features has been developed by Milanese, [Milanese, 1991]. A number of features are extracted from the input image and are represented in a set of feature maps. Features differing from their surroundings are moved to a corresponding set of conspicuity maps. These maps consists of interesting regions of each feature. The conspicuity maps are then merged into a central saliency map where the attention system generates a sequence of attention shifts based on the activity in the map.

Brunnström, Eklund and Lindeberg has presented an active vision approach to classifying corner points in order to examine the structure of the scene. Interesting areas are detected and potential corner points scrutinized by zooming in on them, [Brunnström *et al.*, 1990]. The possibility of actively choosing the imaging parameters, e.g. point of view and focal length, allows the classification algorithm to be much simpler than for static images or pre–recorded sequences.

A variation of the search light metaphor, called the attentional beam has been developed by Tsotsos and Culhane, [Culhane and Tsotsos, 1992; Tsotsos, 1991; Tsotsos, 1992]. It is based on a hierarchical information representation where a search light on the top is passed downwards in the hierarchy to all processing units that contributes to the attended unit. Neighboring units are inhibited. The information in the 'beamed' part of the hierarchy is reprocessed, without the interference from the neighbors, the beam is then used to inhibit the processing elements and a new beam is chosen.

12.3 Foveate Vision

The human eye has the highest resolution in the center and it decays towards the surround. There are a number of advantages in such an arrangement. To mention a few:

- Data reduction compared to having the whole field of view in full resolution.
- High resolution is combined with a broad field of view.
- The fovea marks the area of interest, and disturbing details in the surround are blurred.

These advantages can be utilized in a robot vision system as well. It is actually only the central part of the human retina that is called fovea, but in robot vision the name is often used on all systems that make a difference between the central and the peripheral parts of the field of view.

Fig. 12.1. Left, the levels of the fovea representation marked in the original image. Right, the field of view represented with varying resolution, see also the left column of figure 12.7.

In the simulation the fovea is generated by repeated LP-filtering, subsampling and cropping in octaves. The result is a number of levels with the same number of pixels covering different fields of view. In a real system it would be preferable to have an optic system that generates the fovea as the input images are digitized. There are a number of research projects developing both hardware and algorithms for heterogeneous sampled image arrays, implementing the fovea concept in one form or another, e.g. [Tistarelli and Sandini, 1991].

Starting with a 512x512 image and using 5 levels the LP-filtering and subsampling reduce the size of the final level to 32x32 pixels. Cutting out the center 32x32 pixels on each level reduces the data with a factor $\frac{512 \cdot 512}{5 \cdot 32 \cdot 32} = 51.2$. The total field of view is represented as five 32x32 images, see left column of figure 12.7, but can be visualized by interpolating and combining the images, see figure 12.1.

The image operations used are applied on all levels of resolution. The images have the same size but since they cover different visual fields they can be seen as a subsampled resolution pyramid. Unlike ordinary scale-spaces it is only the center part of the visual field that is represented in all scales. This is compensated for by moving the fixation point around.

12.4 Sequentially defined, data modified focus of attention

12.4.1 Control mechanism components

Having full resolution only in the center part of the visual field makes it obvious that a good algorithm for positioning the fixation point is necessary. A number of focus-of-attention control mechanisms must be active simultaneously to be able to both handle unexpected events and perform an effective search. The different components can roughly be divided into the following groups:

1. Preattentive, data driven control. Non-predicted structured image information and events attracts the focus-of-attention in order to get the information analyzed.

2. Attentive, model driven control. The focus-of-attention is directed toward an interesting region according to predictions using already acquired image information or a priori knowledge.
3. Habituation. As image structures are analyzed and modeled their impact on preattentive gaze control is reduced.

The distinction between the preattentive and attentive parts is floating. It is more of a spectrum from pure reflexes to pure attentional movements of the fixation point.

12.4.2   The concept of nested regions of interest

Let's say we have the following four major levels in the information processing hierarchy on the left hand side of the pyramid (the names are borrowed from Esprit project BRA 3038, Vision as Process):

1. System Supervisor
2. Symbolic Scene Interpreter
3. 3D Geometric Interpreter
4. Low Level Feature Extractor

Assume further that we have a table with a few objects, see top left of figure 12.2, and the task is simply "watch the table". Assuming that the system supervisor knows where the table is, it determines a region of interest, see top right of figure 12.2. Note that the circles only refer to the positioning of the fixation point, not any other limitations in the viewing field. The fixation point is in the middle of the table. The symbolic scene interpretator have found indications on an object or group of objects in the upper left corner of the table and sets its region of interest accordingly, see lower left of figure 12.2. The 3D Geometric Interpreter starts its task by a further refinement of the region of interest by marking the light object, see lower right of figure 12.2. Finally, on the lowest level, the Low level feature extractor selects an area to start model the structures in the image. The fixation point is now moved towards the interesting object by the nested regions of interest. The general response command "watch the table" has been transformed into "focus on the set of objects in upper left corner of the table".

The border of the regions of interest are not to be absolute, rigid boundaries. They are more like recommendations that can be neglected if there are very good reasons. How good a reason must be is controlled with an importance value. This can be illustrated by viewing the region of interest as a basin, or potential well, within which the lower levels can move around freely. The slope and height of the wall is proportional to how important it is to keep the fixation point there.

There are, however, situations when the lower levels are supposed to violate the directives from superior levels. One such occasion is event detection. An event is any unpredicted gradient in space and/or time. When an interesting or important stimuli is detected the fixation point should be moved in that direction on a reflex basis. Suppose the system is watching the table and an object enters the scene and is detected in the 'corner of the eye'. The resolution in periphery will probably not be high enough to see what it is, only where it is. The low level feature extractor will be the first level to detect the event and will react by pulling the fixation point towards the event. By the time the higher levels react on the event there will be extracted low level features with high resolution available. The higher levels will now either move their regions of interest to analyze the event further or force the fixation point back to the original position. Thus, the different regions of interest do not have to be determined in the order indicated in figure 12.2. It all depends on whether it is an attentive or preattentive movement of the fixation point.

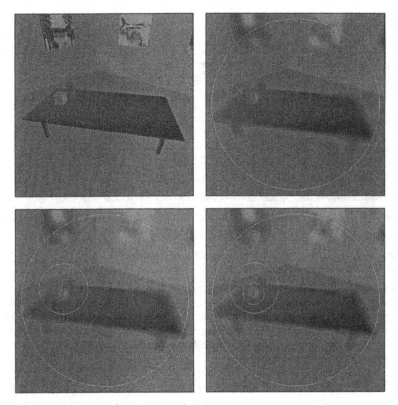

Fig. 12.2. Regions of interest for the different subsystems in the hierarchy.

12.5   Gaze control

In the experiments below the robot consists of an arm with a head, see figure 12.3. The head has two movable cameras. The purpose of this system might be automatic identification, inspection or even surveillance of the objects in the scene in front of it. This type of robotic vision system is very wide spread, see for instance [Abbott and Ahuja, 1989; Brown, 1988].

The outward response of this particular system is designed to enable information gathering. The robot does not move objects. It is only permitted to change the point of view.

The total field of view is 90° and the fovea consists of 5 levels. The fields of view are measured from diametrically opposite corners and are in the experiments 7°, 14°, 28°, 53° and 90° respectively.

12.5.1   Control hierarchy

A control system with three levels has been developed:

Camera Vergence. The cameras are verged towards the same fixation point using the disparity estimates from the stereo algorithm described in chapter .

Edge tracker. The magnitude and phase from quadrature filters form a vector field drawing the attention towards and along lines and edges in the image [Knutsson, 1982; Westelius *et al.*, 1991].

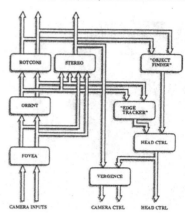

Fig. 12.3. Left: The robot configuration and the environment. Right: The preattentive focus of attention system.

Object finder. Rotation symmetries in the orientation estimates is used to indicate potential objects, [Bigün, 1988; Hansen and Bigun, 1992].

The refinement of the positioning of the fixation point is handled with potential fields in the robots parameter space. It can be visualized as an 'energy landscape' where the trajectory is the path a little ball freely rolling around would take, see figure 12.7. The fixation point can be moved to a certain position by forming a potential well around the position in the parameter space corresponding to the robot looking in that direction. The potential fields from the different controlling modules is weighted together to get the total behavior.

12.5.2   Disparity estimation and camera vergence

The disparity is measured using the multiscale method based on the phase in quadrature filters described in chapter . The fovea version have two major differences from the computation structure in figure 11.10. The first difference is that only the center of the field of view is represented in all resolutions. The accuracy of disparity estimates will therefore decay towards the periphery of the field of view. The second difference is that the edge extractor is used on every level of the input pyramid, instead of creating a pyramid from the edge representation of the finest resolution. This difference is motivated with a possible future fovea sensor array. In such a system it would not be possible to have the computational structure in figure 11.10, since a high resolution image of the total field of view does not exist.

The cameras are verged to get zero disparity in the center of the image. If the head is moving the vergence is calculated using the disparity from the part of the field of view that is predicted to be centered the next time step.

12.5.3   Edge tracker

The phase from quadrature filters is also used to generate a potential field drawing the attention towards and along lines and edges in the image [Knutsson, 1982; Westelius et al., 1991]. Both operations are complex valued where the magnitude is a measure of certainty and the argument is the estimated feature, see figure 12.4. In this application the operations are combined to ensure averagability of both orientation and phase estimates as described in [Haglund, 1989].

The magnitude of the phase operation forms a line sketch of the image where lines and edges look the same regardless if they are bright lines on a dark background, dark lines on bright background,

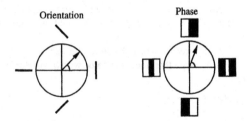

Fig. 12.4. GOP vector representation of orientation and phase.

bright-to-dark edges or dark-to-bright edges. This line sketch forms the energy landscape in the middle column of figure 12.7.

Rotation symmetries

Fig. 12.5. GOP vector representation of rotation symmetries.

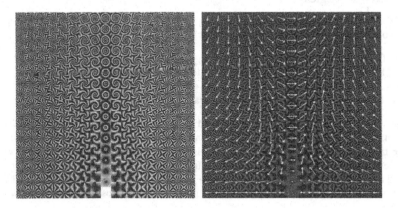

Fig. 12.6. Rotation symmetry test pattern and the results from the symmetry detector. This image pair is borrowed from [Westin, 1991]. The test pattern originally appeared in [Bigün, 1988].

12.5.4 Object finder

Rotation symmetries are detected using convolution with a complex valued kernel estimating the rotation of the orientation estimates within a neighborhood [Bigün, 1988]. Figure 12.5 shows the vector representation of the symmetries. A consistency algorithm is applied to enforce the neighborhoods that fits the symmetries well [Knutsson *et al.*, 1986a; Knutsson and Granlund, 1986; Knutsson *et al.*, 1986b]. Figure 12.6 shows a test pattern with rotation symmetries and the results of the symmetry detector.

Objects small enough to be covered with one glance can be seen as an imperfect instantiation of the concentric circles. The estimates are therefore attenuated with the vector angle, ϕ, i.e. attenuating the crossed lines estimates:

$$closed_a rea = \begin{cases} mag \cdot \cos^2(\phi) & \text{if } \frac{-\pi}{2} \leq \phi \leq \frac{\pi}{2} \\ 0 & \text{otherwise} \end{cases}$$

The result is a 'closed area detector'. It marks areas with evidence for being closed and the magnitude is a measure of how much evidence there is.

The symmetry operation can also be turned into a corner detector by attenuation of the concentric circles estimates instead:

$$corner = \begin{cases} mag \cdot \cos^2(\phi) & \text{if } \phi \leq \frac{-\pi}{2} \text{ and } \phi \geq \frac{\pi}{2} \\ 0 & \text{otherwise} \end{cases}$$

A vector field pointing toward the local mass center is produced by convolving the closed area or corner magnitude with two separable filters estimating the distance to the local mass center:

$$h_x(x, y) = -\cos^2(\tfrac{x \cdot \pi}{8}) \cdot \cos^2(\tfrac{y \cdot \pi}{8}) \cdot x \;\; -7 \leq x \leq 7 \text{ and } -7 \leq y \leq 7$$

$$h_y(x, y) = -\cos^2(\tfrac{x \cdot \pi}{8}) \cdot \cos^2(\tfrac{y \cdot \pi}{8}) \cdot y \;\; -7 \leq x \leq 7 \text{ and } -7 \leq y \leq 7$$

Interpreting the result as gradient vectors in an energy landscape gives the potential fields in the right column of figure 12.7.

12.5.5   Search strategies

The potential fields are weighted together differently depending on what the robot is currently doing. It has four modes of operation:

Search line   The phase estimate is used to locate an edge or line of an object.
Track line   The orientation and phase estimate is used to track an edge or line.
Avoid object   The rotation symmetries are used to move away from the object.
Locate object   The rotation symmetries are used to locate a possible object.

The transitions between the modes are determined by the type and quality of the data in the fixation point. When the system is locating and fixating a possible object it is using the rotation symmetry estimates on the coarser levels. When the distance is small enough the system starts to search for the lines and edges of the object. A search from coarse to fine gives the location of the structure and the tracking procedure starts. If the a line or edge is lost, the search starts again. The system moves away from an object when the fixation point returns to a position where the it has tracked before.

The decision scheme can be described in pseudo-code:

```
switch(mode){
      case AVOID:
            if ( NO_ROTATION_SYMMETRY){
                  mode=SEARCH;
            }else{
                  if( MOVING_TOWARD_NEW_OBJECT){
```

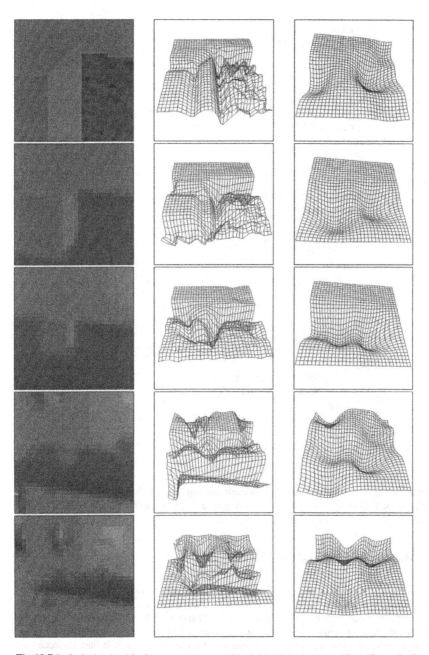

Fig. 12.7. Left, the levels of the fovea representation. The fields of view measured from diametrically opposite corners are 7°, 14°, 28°, 53°, 90°. Middle, potential fields from the edge tracker. Right, potential fields from the object finder.

```
                    mode=LOCOBJ;
            }else{
                    mode=AVOID;
            }
     }
     break;
case LOCOBJ:
     if((NO_ROTATION_SYMMETRY) or (CLOSE_TO_OBJECT)){
            mode=SEARCH;
     }else{
            mode=LOCOBJ;
     }
case TRACK:
     if ( BEEN_HERE_BEFORE ){
            mode=AVOID;
     }else{
            if( NOT_ON_EDGE ){
                    mode=SEARCH;
            }else{
                    mode=TRACK;
            }
     }
case SEARCH:
     if(NO_ENERGY){
            mode=SEARCH; /* continue in the same direction */
     }else{
            if( NOT_ON_EDGE ){
                    mode=SEARCH;
            }else{
                    mode=TRACK;
            }
     }
}
```

The mode of operation is determined for the left and right view independently and if they are not consistent the following ranking order is used from high to low: LOCOBJ, AVOID, TRACK, SEARCH.

12.5.6   Model acquisition and memory

The system marks the states in its parameter space that corresponds to the direction in which it has been tracking edges. This is the first step toward a rudimentary memory of where it has looked before. Since the robot only moves two joints the parameter space is 2D and there is only one way of looking at an object. In a general system where many points in the parameter space might correspond to looking at the same thing, an extended approach to memory is needed. It is then important to remember not only WHERE but also WHAT the system has seen. For non-static scenes WHEN becomes important. This leads to a procedure for model acquisition which is an ultimate goal for this process.

12.6   Results

A preattentive control structure for focus of attention in a robot vision system has been developed. It has been tested in a simulated environment. The system has a heterogeneous sampled imaging system, a fovea, reducing the amount data and computational effort.

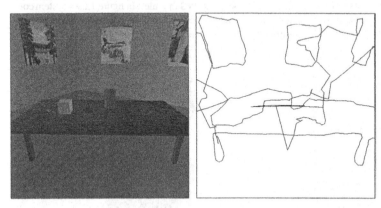

Fig. 12.8. Trajectory of the fixation point. The fixation point has followed the structures in the image and moved from object to object.

A three level control structure has been tested where the coordination between the control levels is handled with potential fields in the robot parameter space. A number of different scan modes can be produced by giving the potential fields from the control levels different weights. A higher weight on the potential field created from circular symmetries make the robot look for possible objects while a higher weight on the field from edges and lines make the system follow image structures. A trajectory of how the robot moves the fixation point between the objects in front of it one at a time can be found figure 12.8.

In these experiments only features extracted from gray scale structures have been used. A natural extension is to incorporate color edges, [Westin and Westelius, v1988; Westelius and Westin, 1989b; Westelius and Westin, 1989a], texture gradients etc. in order to get a better segmentation of the image. The potential fields give the possibility to do this fairly easy. If for instance a color and gray scale edge coincide, the potential well will be much deeper than if it is a gray scale edge on a constant color.

It has been shown that preattentive mechanisms are an important part of gaze control. The results also show that a set of individually simple processes can together produce a complex and purposive behavior.

12.7   Robot vision simulator

A robot vision simulator has been developed to facilitate testing of robot vision and control algorithms. The simulator reduces the need for expensive special purpose hardware since it can be run in 'slow motion'. A real-time process can be investigated although only limited computational resources are available. It also allows testing of different types of robots and robot configurations without any extra cost. The scene can be varied from a few very simple polyhedral objects to complex, realistic, texture mapped environments. The simulated reality, such as true 3D-structure, true distance etc, can easily be compared to the results obtained by the robot in order to evaluate the performance of the algorithms.

The simulator is implemented using the Application Visualization System (AVS). AVS is a tool for easy visualization of scientific data. The program modules are separate Unix processes communicating via pipes and shared memory using user defined data formats. The configuration of the flow networks is done graphically by the user.

The geometric model of the robot system consists of a set of subsystems that can be configured in various combinations by connecting the corresponding AVS modules. In figure 12.9 a system consisting of a car, a robot arm, and a stereo camera head has been assembled. A general transformation module has also been attached for positioning purposes.

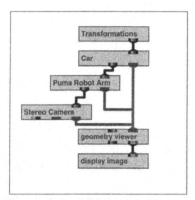

Fig. 12.9. The AVS network and the corresponding robot configuration. The 'Transformations' module supplies a general transformation matrix. The 'Car', 'Puma Robot Arm' and 'Stereo Camera' modules generates geometric descriptions and send them to the 'geometry viewer' module. They also furnishes a transformation matrix to the the next modules in line. Stereo image pairs require an additional 'geometry viewer' module.

The 'Transformations' module supplies a general transformation matrix to the 'Car' module, which generates a geometric model of a four wheeled vehicle. All modules generating geometric description pass the descriptions to the 'Render Geometry' modules, one for each camera. The 'Car' module adds its transform to the transformation matrix and passes it to the 'Puma Robot Arm' module, which generates a geometric description and updates the transformation matrix, and so on.

In figure 12.9 the robot is controlled manually by the user. It can also be controlled from another module by connecting it via a parameter input port. This other module might for instance be a robot path planner. A filter can be attached between the control module and the robot in order to emulate the dynamics of the system.

References

[Abbott and Ahuja, 1989] A. L. Abbott and N. Ahuja. Surface reconstruction by dynamic integration of focus, camera vergence and stereo. In *Proceedings IEEE Conf. on Computer Vision*, pages 523–543, 1989.

[Ballard and Ozcandarli, 1988] D.H. Ballard and Altan Ozcandarli. Eye fixation and early vision: kinetic depyh. In *Proceedings 2nd IEEE Int. Conf. on computer vision*, pages 524–531, december 1988.

[Ballard, 1990] Dana H. Ballard. Animate vision. Technical Report 329, Computer Science Department, University of Rochester, Feb. 1990.

[Bigün, 1988] J. Bigün. *Local Symmetry Features in Image Processing*. PhD thesis, Linköping University, Sweden, 1988. Dissertation No 179, ISBN 91–7870–334–4.

[Brown, 1988] C. M. Brown. The Rochester robot. Technical Report 257, Computer Science Department, University of Rochester, Aug. 1988.

[Brown, 1990a] C. M. Brown. Gaze control with interactions and delays. *IEEE systems, man and cybernetics*, 20(1):518–527, march 1990.

[Brown, 1990b] C. M. Brown. Prediction and cooperation in gaze control. *Biological cybernetics*, 63:61–70, 1990.

[Brunnström et al., 1990] K. Brunnström, J. O. Eklundh, and T. Lindeberg. Active detection and classification of junctions by foveating with a head–eye system guided by the scale–space primal sketch. Technical Report TRITA-NA-P9131, CVAP, NADA, Royal Institute of Technology, Stockholm, Sweden, 1990.

[Culhane and Tsotsos, 1992] S. Culhane and J. Tsotsos. An attentinal prototype for early vision. In *Proceedings of the 2:nd European Conf. on computer vision*, Santa Margharita Ligure, Italy, May 1992.

[Haglund, 1989] L. Haglund. Hierarchical scale analysis of images using phase description. Thesis 168, LIU–TEK–LIC–1989:08, Computer Vision Laboratory, Linköping University, Sweden, 1989.

[Hansen and Bigun, 1992] O. Hansen and J. Bigun. Local symmetry modeling in multidimensional images. In *Pattern Recognition Letters, Volume 13, Nr 4*, 1992.

[Julesz, 1991] B. Julesz. Early vision and focal attention. *Review of Modern physics*, 63(3):735–772, 1991.

[Knutsson and Granlund, 1986] H. Knutsson and G. H. Granlund. Apparatus for determining the degree of variation of a feature in a region of an image that is divided into descrete picture elements. Swedish patent 8502569-0 (US-Patent 4.747.151, 1988), 1986.

[Knutsson et al., 1986a] H. Knutsson, G. H. Granlund, and J. Bigun. Apparatus for detecting sudden changes of a feature in a region of an image that is divided into descrete picture elements. Swedish patent 8502571-6 (US-Patent 4.747.150, 1988, 1986.

[Knutsson et al., 1986b] H. Knutsson, M. Hedlund, and G. H. Granlund. Apparatus for determining the degree of consistency of a feature in a region of an image that is divided into descrete picture elements. Swedish patent 8502570-8 (US-Patent 4.747.152, 1988), 1986.

[Knutsson, 1982] Hans Knutsson. *Filtering and Reconstruction in Image Processing*. PhD thesis, Linköping University, Sweden, 1982. Diss. No. 88.

[Milanese, 1990] R. Milanese. Focus of attention in human vision: a survey. Technical Report 90.03, Computing Science Center, University of Geneva, Geneva, August 1990.

[Milanese, 1991] R. Milanese. Detection of salient features for focus of attention. In *Proc. of the 3rd Meeting of the Swiss Group for Artificial Intelligence and Cognitive Science*, Biel-Bienne, October 1991. World Scientific Publishing.

[Tistarelli and Sandini, 1991] M. Tistarelli and G. Sandini. Direct estimation of time–to–impact from optical flow. In *Proceedings of IEEE Workshop on Visual Motion*, pages 52–60, Princeton, USA, October 1991. IEEE, IEEE Society Press.

[Tsotsos, 1991] J. K. Tsotsos. Localizing stimuli in a sensory field using an inhibitory attentinal beam. Technical Report RBCV–TR–91–37, Department of Computer Science, University of Toronto, October 1991.

[Tsotsos, 1992] J. K. Tsotsos. On the relative complexity of active vs. passive visual search. *Int. Journal of Computer Vision*, 7(2):127–142, Januari 1992.

[VAP consortium, 1992] Esprit basic research action 3038, vision as process, final report. Project document, April 1992.

[Westelius and Westin, 1989a] C-J Westelius and C-F Westin. A colour representation for scale-spaces. In *The 6th Scandinavian Conference on Image Analysis*, pages 890–893, Oulu, Finland, June 1989.

[Westelius and Westin, 1989b] C-J Westelius and C-F Westin. Representation of colour in image processing. In *Proceedings of the SSAB Conference on Image Analysis*, Gothenburg, Sweden, March 1989. SSAB.

[Westelius et al., 1990] C-J Westelius, H. Knutsson, and G. H. Granlund. Focus of attention control. Report LiTH–ISY–I–1140, Computer Vision Laboratory, Linköping University, Sweden, 1990.

[Westelius et al., 1991] C-J Westelius, H. Knutsson, and G. H. Granlund. Focus of attention control. In *Proceedings of the 7th Scandinavian Conference on Image Analysis*, pages 667–674, Aalborg, Denmark, August 1991. Pattern Recognition Society of Denmark.

[Westelius et al., 1992a] C-J Westelius, H. Knutsson, and G. H. Granlund. Preattentive gaze control for robot vision. In *Proceedings of Third International Conference on Visual Search*. Taylor and Francis, 1992.

[Westelius et al., 1992b] C-J Westelius, H. Knutsson, and J. Wiklund. Robust vergence control using scale-space phase information. Report LiTH-ISY-I-1363, Computer Vision Laboratory, Linköping University, Sweden, 1992.

[Westin and Westelius, v1988] C-F Westin and C-J Westelius. A colour model for hierarchical image processing. Master's thesis, Linköping University, Sweden, August v1988. LiTH–ISY–EX–0857.

[Westin, 1991] C-F Westin. Feature extraction based on a tensor image description, September 1991. Thesis No. 288, ISBN 91–7870–815–X.

[Yarbus, 1969] A. L. Yarbus. *Eye movements and vision.* Plenum, New York, 1969.

13. Tensor Based Spatio-temporal Signal Analysis

Hans Knutsson

CVL, LiTH

13.1 Modules

This chapter describes theories and algorithms used in the following modules: **Orientation, Phase, Motion discontinuity, Velocity** and **Spatio-temporal frequency**

13.2 Features

In this section, techniques for mul-dimensional signal analysis will be presented. The theory is based on tensor representations and spatio-temporal filtering. The following features can be obtained, (all listed features describe spatio-temporal structure and are local in space-time):

Phase	**Velocity**
Orientation	**Motion discontinuity**
Plane-ness	**Spatio-temporal frequency**
Line-ness	

For each channel six spatio-temporal quadrature filters are employed. The filters are sensitive to signals in different orientations in the 3-dimensional (x,y,t) signal space. Each quadrature filter produces a magnitude and a phase. Phase can be used to describe local symmetry properties of the spatio-temporal signal.

The magnitudes of the different filters can combined to give a tensor description of local structure and orientation. The orientation tensor also contains information about how plane-like or line-like the spatio-temporal neighbourhood is.

Velocity estimates can be directly obtained from the orientation tensor. In line-like, i.e. 'moving point' type, areas a true velocity estimate is produced. In plane-like, i.e. 'moving line' type, areas the velocity component perpendicular to the local structure is produced. In addition, discontinuous motion can be detected by operating on the tensor field produced by the 3-D orientation algorithm.

A spatio-temporal frequency estimate can be obtained by combining outputs from different channels.

13.3 Background

The work presented here is intended to serve as a basis for the design of efficient multi-dimensional signal analysis algorithms. Other work relevant to the ideas presented in this chapter, but not explicitly referred to, can be found in [Granlund, 1978], [Wilson and Knutsson, 1988], citemera.

New techniques to produce and process multi-dimensional data are constantly being developed in a number of different fields. Higher dimensional signals can be seen as functions of more than 2 variables and there are in principle no restrictions on what these variables may represent. However, signals as functions of space or space-time are by far the most frequently used. A few examples of commonly available higher dimensional data sets are given below.

- **Image sequences (2D + time)**
    - Video
    - Ultrasound
    - Satellite
- **Volumes (3D)**
    - Magnetic Resonance (MR)
    - Computer Tomography (CT))
    - Positron Emission Tomography (PET)
- **Volume sequences (3D + time)**
    - Magnetic Resonance (MR)

It is obvious that processing of higher dimensional data sets puts high demands on computer power and storage capacity. Perhaps less obvious is that increasing the dimensionality of the data also has profound implications for the analysis of the data. In this section it will be described how higher dimensional data can be analyzed.

In the analysis of higher dimensional data it turns out that using only scalars and vectors is no longer convenient [Knutsson, 1989]. An important extension in this chapter is the introduction of the concept of a *tensor*.

13.4   Tensors - A short introduction

Tensor analysis is a generalization of the notions from vector analysis. The need for such a theory is motivated by the fact that there are many physical quantities of complicated nature that cannot naturally be described or represented by scalars or vectors. Examples are the stress at a point in a solid body due to internal forces, the deformation of an arbitrary infinitesimal element of volume of an elastic body, and the moments of inertia. These quantities can be described and represented adequately only by the more sophisticated mathematical entities called tensors. Scalars and vectors are actually also belonging to this family of elements. Thus, scalars and vectors are special cases of tensors.

Associated with a tensor is the order of that tensor. The order of a tensor can be thought of as the complexity of the entity it represents, e.g. a 0th order tensor is a scalar, a 1st order tensor is a vector and a 2nd order tensor can be thought of as a matrix. Tensors will be denoted in bold capitol letters e.g. $\mathbf{T}$.

The norm of a tensor As for vectors each tensor can be associated with a norm. The norm used in this chapter is the *Frobenius norm* which, again just as for vectors, is defined as the square root of the sum of the squares of the components, i.e. the *norm* of a 2nd order tensor $\mathbf{T}$ is defined by:

$$\|\mathbf{T}\|^2 \equiv \sum_{ij} t_{ij}^2 = \sum_n \lambda_n^2 \tag{1}$$

where:
t_{ij} are the components of $\mathbf{T}$ and
λ_n are the eigenvalues of $\mathbf{T}$.

A lot more could be said about tensors but that is beyond the scope of this book and using tensors of orders 0, 1 and 2 will suffice for the following presentation.

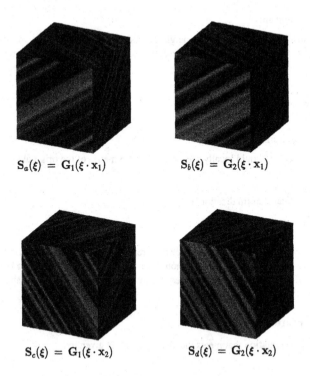

$$S_a(\xi) \;=\; G_1(\xi \cdot x_1) \qquad\qquad S_b(\xi) \;=\; G_2(\xi \cdot x_1)$$

$$S_c(\xi) \;=\; G_1(\xi \cdot x_2) \qquad\qquad S_d(\xi) \;=\; G_2(\xi \cdot x_2)$$

Fig. 13.1. Four simple neighborhoods in 3D. The neighborhoods are constructed using two different signal functions (G_1 and G_2) and two different signal orienting vectors (x_1 and x_2).

13.5 Local orientation

To begin, it should be noted that, in the mathematical analysis of local orientation, time and space can be treated equally. Correct interpretation of local orientation, however, requires knowledge of which dimension corresponds to time.

The analysis in this section is general in the sense that it is valid for signals of any dimension but will deal only with signals that locally can vary in one direction only. These signals will be denoted *simple* signals.

A *simple* neighborhood can be expressed as:

$$S(\xi) \;=\; G(\xi \cdot x) \tag{2}$$

where:
S and G are non-constant tensors of any order.
ξ is the spatial coordinate and
x is a constant vector oriented along the axis of maximal signal variation.

For this type of neighborhood it is possible to define the local orientation as a tensor **T** of order two.

13.5.1 Definition of orientation

The entity representing orientation should meet two basic requirements:

The *Invariance* requirement:

It is evident that the *orientation* of the neighborhood will be the same for all possible **G** in equation (2). In other words the entity representing orientation must be invariant to **G**.

$$\frac{\delta T}{\delta G} = 0 \tag{3}$$

where **T** represents orientation.

The *Equivariance* requirement:

The orientation tensor should locally preserve the angle metric of the original space, i.e

$$\|\delta\hat{T}\| \propto \|\delta\hat{x}\| \tag{4}$$

where the "^" indicates normalization, i.e. :
$\hat{T} = \frac{T}{\|T\|}$ and
$\hat{x} = \frac{x}{\|x\|}$.

The implications of the invariance and equivariance requirements can easily be explained by studying figure 13.4. Invariance to **G** implies that **T** should be the same for neighborhoods S_a and S_b even though **G** is different. Equivariance implies that moving from S_a to S_c, **T** should change proportionally to the change in local orientation.

13.5.2   The orientation tensor

A representation **T** for orientation that meets the above criteria is given by:

$$T \equiv A\hat{x}\hat{x}^T \tag{5}$$

where $A > 0$ can be any constant.

The norm   Calculating the norm of **T** gives:

$$\|T\|^2 = A^2(\sum_i x_i^2)^{-1} \sum_{ij} x_i^2 x_j^2 = A^2 \tag{6}$$

where $x_{i(j)}$ are the components of **x**.

Thus the norm of **T** is equal to A. Since the orientation represented by **T** is independent of the norm of **T** A can be used to represent another property. In section 13.6 it will be shown that A can be made to represent the local amplitude of the signal in a natural way.

Invariance   By its construction **T** is trivially invariant to **G**. However, as can be expected, making the actual orientation estimates invariant to **G** is by no means trivial. A discussion of this topic will be found in section 13.6.

Equivariance   To show that the *Equivariance* requirement is met by the mapping is fairly straightforward. Rotate $\hat{x}$ by adding a small perpendicular vector $\epsilon\hat{v}$ to $\hat{x}$ and calculate the difference in the norm of **T**. To start define **B** to be:

$$B = \lim_{\epsilon\to 0} \frac{T(\hat{x} + \epsilon\hat{v}) - T(\hat{x})}{\epsilon} \tag{7}$$

where: $\hat{v} \cdot \hat{x} = 0$, then

$$\|\delta T\| = \|B\|\|\delta\hat{x}\| \tag{8}$$

Carrying out the limit calculation is simple and yields:

$$B = A(\hat{x}\hat{v}^T + \hat{v}\hat{x}^T) \tag{9}$$

The eigenvectors and eigenvalues of $\mathbf{B}$ is easily found by letting $\mathbf{B}$ operate on the combined vectors $(\hat{\mathbf{x}} + \hat{\mathbf{v}})$ and $(\hat{\mathbf{x}} - \hat{\mathbf{v}})$.

$$\begin{cases} \mathbf{B}(\hat{\mathbf{x}} + \hat{\mathbf{v}}) = A(\hat{\mathbf{x}} + \hat{\mathbf{v}}) \\ \mathbf{B}(\hat{\mathbf{x}} - \hat{\mathbf{v}}) = A(\hat{\mathbf{v}} - \hat{\mathbf{x}}) \end{cases} \tag{10}$$

This shows that $(\hat{\mathbf{x}} + \hat{\mathbf{v}})$ and $(\hat{\mathbf{x}} - \hat{\mathbf{v}})$ are eigenvectors of $\mathbf{B}$ the eigenvalues being A and $-A$ respectively. Since all other eigenvectors of $\mathbf{B}$ are orthogonal to $\hat{\mathbf{x}}$ and $\hat{\mathbf{v}}$ it follows from equation (9) that all other eigenvalues are zero. Then, according to equation (1), the norm of $\mathbf{B}$ is given by:

$$\|\mathbf{B}\| = \sqrt{\lambda_1^2 + \lambda_2^2} = \sqrt{2}\,A \tag{11}$$

and since A is constant not depending on the orientation it is shown that the *Equivariance* requirement is met.

13.6   Orientation estimation

Having found a suitable orientation representation the question arises: *Can the representation be realized using measurements on actual image data, where lines (or other structures) are represented as local gray scale correlations?* It will be shown that by combining the outputs from polar separable quadrature filters, it is possible to produce a representation corresponding exactly to equation (5). The exactness relies on the image data being locally *simple*, equation (2), i.e. on the existence of a locally well defined orientation. The case where the simplicity assumption does not hold is discussed in section 13.6.3.

   In the following the above procedure will be discussed in detail. The analysis will deal only with *real* valued *simple* neighborhoods, ie neighborhoods that can be expressed as:

$$s(\xi) = g(\xi \cdot \hat{\mathbf{x}}) \tag{12}$$

where f and g are real functions.

The quadrature filter concept   As a part of trying to realize the *invariance* requirement the estimation procedure is designed using quadrature filters [Knutsson, 1982]. The quadrature filter concept forms a basis for minimizing the sensitivity to phase changes in the signal

   A quadrature filter can, independently of the dimensionality of the signal space, be defined as a filter being zero over one half of the Fourier space, or more precisely defined by:

$$F_k(\mathbf{u}) = 0 \quad if \quad \mathbf{u} \cdot \hat{\mathbf{n}}_k \leq 0 \tag{13}$$

   where:
   $\hat{\mathbf{n}}_k$ is the filter directing vector.
   $\mathbf{u}$ is the frequency.

   The output q_k of the corresponding quadrature filter will be a complex number. As an example of the phase insensitive property of quadrature filters it can be mentioned that, if g is a sinusoidal function, the magnitude $q_k = \|\mathbf{q}_k\|$ of $\mathbf{q}_k$ will be completely phase invariant. The argument $\arg(\mathbf{q}_k)$ will represent the local phase.

Spherically separable filters   For *simple* signals the Fourier transform is non-zero only on the line defined by:

$$\mathbf{u} \propto \mathbf{x} \tag{14}$$

More precisely the Fourier transform of $s(\xi)$ in equation 12 can be expressed as:

$$S(\mathbf{u}) = G(\mathbf{u} \cdot \hat{\mathbf{x}})\delta_{\hat{\mathbf{x}}}^{n-1}(\mathbf{u}) \tag{15}$$

Let the filter functions be real and spherically separable [Knutsson, 1982], ie separable into one function of radius, R, and one function of direction D.

$$F(\mathbf{u}\|) = R(u)\,D(\hat{\mathbf{u}}) \tag{16}$$

where: $u = \|\mathbf{u}\|$

The result $\mathbf{q}$ of filtering S by F then becomes:

$$\mathbf{q} = \int_{-\infty}^{\infty} F(\mathbf{u})S(\mathbf{u})\,d\mathbf{u} = \int_{-\infty}^{\infty} R(u)\,D(\hat{\mathbf{u}})G(\mathbf{u}\cdot\hat{\mathbf{x}})\delta_{\hat{\mathbf{x}}}^{n-1}(\mathbf{u})\,d\mathbf{u} \tag{17}$$

Setting $\mathbf{u} = u\hat{\mathbf{x}}$ yields:

$$\mathbf{q} = D(\hat{\mathbf{x}})\int_0^{\infty} R(u)G(u)du + D(-\hat{\mathbf{x}})\int_0^{\infty} R(u)G(-u)du \tag{18}$$

Note that the value of the integrals does not depend on the signal orientation. Let $\mathbf{d}$ be the value of the first integral in eq 18.

$$\mathbf{d} = \int_0^{\infty} R(u)G(u)du \tag{19}$$

Then, since the signal $f(\boldsymbol{\xi})$ is real, $G(u)$ will be Hermitian and the value of the second integral becomes $\mathbf{d}^*$. Thus eq 18 simplifies to:

$$\mathbf{q} = \mathbf{d}\,D(\hat{\mathbf{x}}) + \mathbf{d}^*\,D(-\hat{\mathbf{x}}) \tag{20}$$

Taking the quadrature requirement into account it is clear that either $D(\hat{\mathbf{x}})$ or $D(-\hat{\mathbf{x}})$ will be zero, implying that the two components will not interfere and the magnitude of the quadrature filter output can be written:

$$q = d\,[D(\hat{\mathbf{x}}) + D(-\hat{\mathbf{x}})] \tag{21}$$

showing that the *invariance* requirement is met and that the magnitude of the quadrature filter output is separable into two components: One orientation invariant component, $d = \|\mathbf{d}\|$, which can be thought of as the local signal amplitude, and one component (in brackets) invariant to the signal function g.

The directional function To meet the *equivariance* requirement, equation (13), it is required that the frequency response of the filters have particular interpolation properties. Directional functions having the necessary properties was first suggested by Knutsson in [Knutsson, 1982] for the 2D case and in [Knutsson, 1985] for the 3D case.

Regardless of dimension these functions can be written:

$$\begin{cases} D_k(\hat{\mathbf{u}}) = (\hat{\mathbf{u}}\cdot\hat{\mathbf{n}}_k)^2 & if \ \mathbf{u}\cdot\hat{\mathbf{n}}_k > 0 \\ D_k(\hat{\mathbf{u}}) = 0 & otherwise \end{cases} \tag{22}$$

where $\hat{\mathbf{n}}_k$ is the filter directing vector. ie $D(\hat{\mathbf{u}})$ varies as $\cos^2(\varphi)$, where φ is the difference in angle between $\mathbf{u}$ and the filter direction $\hat{\mathbf{n}}_k$. See figure 13.2 for a visualization of the directional function.

The filter outputs. Finally, combining equations (21) and (22), the output magnitude from a quadrature filter in direction k is found to be:

$$q_k = d(\hat{\mathbf{x}}\cdot\hat{\mathbf{n}}_k)^2 \tag{23}$$

where d is independent of the filter orientation and depends only on radial distribution of the signal spectrum $G(u)$ and the radial filter function $R(u)$. See figure 13.3 for a visualization of the output magnitude q as a function of $\hat{\mathbf{x}}$.

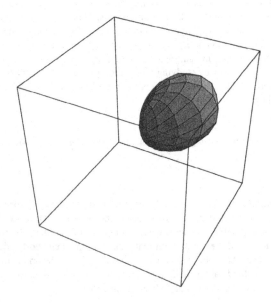

Fig. 13.2. Angular plot, ie $radius = D(\hat{\mathbf{u}})$, of the directional function in 3 - dimensional signal space.

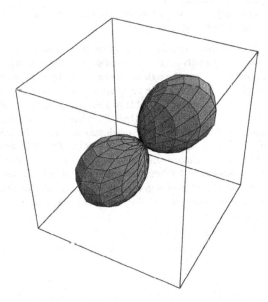

Fig. 13.3. Angular plot of the quadrature output magnitude, q_k, as a function of the signal orienting vector $\hat{\mathbf{x}}$ for 3-dimensional simple signals.

The radial function It is clear from the preceding analysis that the radial function $R(u)$ can be chosen arbitrarily without violating the basic requirements. This makes the choice of $R(u)$ subject to considerations similar to those traditionally found in *one − dimensional* filter design. Typically $R(u)$ is a band-pass function having design parameters such as *center frequency* and *bandwidth*. Perhaps even more important than in traditional $1D$ filter design are the concepts of locality and scale. Good $R(u)$:s are therefore found by studying the resulting filter simultaneously in the space-time and the Fourier domain. A radial filter function with useful properties, first suggested in [Knutsson *et al.*, 1980], is given by:

$$R(u) = e^{-\frac{4\ln 2}{B^2} \ln^2(u/u_i)} \tag{24}$$

This class of functions are termed *lognormal* functions. B is the relative bandwidth and u_i is the center frequency.

13.6.1 Filter sets

In the following the minimum number of quadrature filters, K, necessary for orientation estimation will be discussed. The result is that the minimum number is 3 for 2D, 6 for 3D and 12 for 4D [Knutsson, 1982; Knutsson, 1985; Knutsson, 1989; Knutsson and Bårman, 1992].

As before the analysis will deal only with real valued simple neighborhoods. In addition it is assumed that the filter axes should be symmetrically distributed over the orientation space. This is a reasonable assumption as the final result, **T**, by definition is invariant to rotation of the filters.

It is helpful in the following discussion to bear in mind that:

1. The Fourier transform is invariant to rotation of the coordinate system.
2. The Fourier transform of a simple signal is a line through the origin parallel to the signal orienting vector, equation (2) and (14).
3. The quadrature filter output is invariant to rotation around its axis (given by $\hat{n}_k$) and also diametrically symmetric so that $q(\mathbf{x}) = q(-\mathbf{x})$, equation (23).

Consider the case of 2^{N-1} quadrature filters, having symmetry axes passing through the corners of a cube in N dimensions, giving a fully symmetric distribution of filters. Consider the contribution to the filters from frequencies on a line through the center of two opposing cube faces. Since the angle between the line and any filter axis will be the same it is clear that all the filters will give the same output. Consequently the filter set is incapable of giving information sufficient to determine which pair of cube faces the line passes through and thus, the orientation of the signal is undecidable. It can be concluded, therefore, that more than 2^{N-1} quadrature filters must be used.

The minimum number of filters and their directions in 2D, 3D and 4D are given below. (For $N > 4$ no regular polyhedron having more vertices than a cube exists, it is, however, still possible to estimate orientation tensors for $N > 4$ although calculations will be more complex.)

2D filter set It has been shown in chapter 6 that the minimum number of filters required when $N = 2$ is 3, the orientation of the filters are given by vectors pointing to three adjacent vertices of a regular hexagon, see figure 13.4

The filter orienting vectors are given by:

$$\hat{n}_1 = (\ 1, \ \ 0\ )^T$$
$$\hat{n}_2 = (\ a, \ \ b\ )^T \tag{25}$$
$$\hat{n}_3 = (-a, \ \ b\ )^T$$

  where:
  $a = 0.5$
  $b = \sqrt{3}/2$

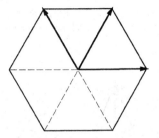

Fig. 13.4. A hexagon and the corresponding filter orienting vectors.

3D filter set For N = 3 the number of filters must be greater than 4 but, since there does not exist a way of distributing 5 filters in 3-D in a fully symmetrical fashion, the next possible number is 6. (In fact the only possible numbers are those given by half the number of vertices (or faces) of a diametrically symmetric regular polyhedron, leaving only the numbers 3, 4, 6 and 10. Note that this is in contrast to the 2-D case where the only symmetry restriction is K > 2.) It turns out that the minimum required number of quadrature filters K *is* 6.

The orientations of the filters are given by vectors pointing to the vertices of a *hemi-icosahedron*, see figure 13.5. The 6 normal vectors are given in cartesian coordinates by:

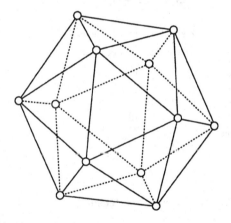

Fig. 13.5. An icosahedron (one of the 5 Platonic polyhedra).

$$\hat{n}_1 = c(\ a,\ 0,\ b\ )^T \quad \hat{n}_2 = c(-a,\ 0,\ b\ )^T$$
$$\hat{n}_3 = c(\ b,\ a,\ 0\ )^T \quad \hat{n}_4 = c(\ b,-a,\ 0\ )^T \tag{26}$$
$$\hat{n}_5 = c(\ 0,\ b,\ a\ )^T \quad \hat{n}_6 = c(\ 0,\ b,-a\ )^T$$

where:
$$a = 2$$
$$b = (1 + \sqrt{5})$$
$$c = (10 + 2\sqrt{5})^{-1/2}$$

4D filter set As before, the filters should be distributed symmetrically over half of the Fourier space. This implies that the filters should be distributed in accordance with the vertices of a regular polytope. The choice is further limited by the restriction that K, the number of filters, should be greater than 8. This leaves the 24–cell (Coxeter [Coxeter, 1961]) as the only alternative, see figure 13.6. (Computational complexity makes the 120–cell and the 600–cell unrealistic alternatives.)

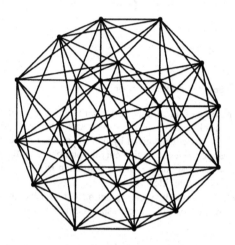

Fig. 13.6. A Projection of the 24–cell

The 12 filter directions are given in cartesian coordinates by:

$$
\begin{aligned}
\hat{n}_1 &= c\,(1,1,0,0)^T & \hat{n}_2 &= c\,(1,-1,\ 0,\ 0)^T \\
\hat{n}_3 &= c\,(1,0,1,0)^T & \hat{n}_4 &= c\,(1,\ 0,-1,\ 0)^T \\
\hat{n}_5 &= c\,(1,0,0,1)^T & \hat{n}_6 &= c\,(1,\ 0,\ 0,-1)^T \\
\hat{n}_7 &= c\,(0,1,1,0)^T & \hat{n}_8 &= c\,(0,\ 1,-1,\ 0)^T \\
\hat{n}_9 &= c\,(0,1,0,1)^T & \hat{n}_{10} &= c\,(0,\ 1,\ 0,-1)^T \\
\hat{n}_{11} &= c\,(0,0,1,1)^T & \hat{n}_{12} &= c\,(0,\ 0,\ 1,-1)^T
\end{aligned}
\tag{27}
$$

where $c = \frac{1}{\sqrt{2}}$.

13.6.2   Construction of the orientation tensor

Independent of signal dimensionality the final result $\mathbf{T}$ can be obtained by linear summation of the quadrature filter output magnitudes as indicated by equations (28) and (29).

$$
\mathbf{T} = \sum_k q_k(\,\hat{\mathbf{N}}_k - \alpha\,\mathbf{I}\,)
\tag{28}
$$

$$
\hat{\mathbf{N}}_k = \hat{\mathbf{n}}_k\hat{\mathbf{n}}_k^T
\tag{29}
$$

where:
q_k is the output from quadrature filter k,
$\hat{\mathbf{N}}_k$ is the tensor associated with quadrature filter k,
$\hat{\mathbf{n}}_k$ is the orientation of quadrature filter k,
$\mathbf{I}$ is the identity tensor and
α is 1/4 for 2D, 1/5 for 3D and 1/6 for 4D as will be shown below.

2D tensor construction Combining equations (25) and (29) gives the elements of the filter associated tensors $\hat{\mathbf{N}}_k$:

$$\hat{\mathbf{N}}_1 = \begin{pmatrix} 1 & 0 \\ 0 & 0 \end{pmatrix}$$

$$\hat{\mathbf{N}}_2 = \begin{pmatrix} a^2 & ab \\ ab & b^2 \end{pmatrix} \tag{30}$$

$$\hat{\mathbf{N}}_3 = \begin{pmatrix} a^2 & -ab \\ -ab & b^2 \end{pmatrix}$$

where:
$$a = 0.5$$
$$b = \sqrt{3}/2$$

Let the signal orienting vector be given by:

$$\mathbf{x} = (x_1, x_2)^T \tag{31}$$

then the magnitude of the outputs from the three quadrature filters are, according to equation (23), given by:

$$\begin{aligned}
q_1 &= dx^{-2}x_1^2 \\
q_2 &= dx^{-2}(a^2x_1^2 + 2abx_1x_2 + b^2x_2^2) \\
q_3 &= dx^{-2}(a^2x_1^2 - 2abx_1x_2 + b^2x_2^2)
\end{aligned} \tag{32}$$

Next, calculating the sum

$$\mathbf{T}' = \sum_k q_k \hat{\mathbf{N}}_k \tag{33}$$

yields the components of $\mathbf{T}'$:

$$\begin{aligned}
t_{11} &= d'(x_1^2 x^{-2} + \tfrac{1}{2}) \\
t_{22} &= d'(x_2^2 x^{-2} + \tfrac{1}{2}) \\
t_{12} &= t_{21} = d'x_1 x_2 x^{-2}
\end{aligned} \tag{34}$$

where $d' = \tfrac{3}{4}d$.

It is evident that if the quantity $\tfrac{1}{2}d' = \tfrac{3}{8}d$ is subtracted from the diagonal elements of $\mathbf{T}'$ the result will be of the desired form.

$$\mathbf{T} = \mathbf{T}' - \tfrac{1}{2}d'\mathbf{I} = d'\hat{\mathbf{x}}\hat{\mathbf{x}}^T \tag{35}$$

Finally calculate the sum of all quadrature filter output magnitudes.

$$\sum_k q_k = \frac{3}{2}d \tag{36}$$

Combining equation (33), (35) and (36) yields the desired result:

$$\mathbf{T} = \sum_k q_k(\hat{\mathbf{N}}_k - \frac{1}{4}\mathbf{I}) \tag{37}$$

Note that $(\hat{\mathbf{N}}_k - \frac{1}{4}\mathbf{I})$ are constant tensors and can be precalculated. Thus the orientation tensor can be estimated by weighted summation of fixed tensors, the weights being the quadrature filter output magnitudes q_k. In the following it will be shown that this statement is also true for 3D and 4D signal spaces. Although the numbers and number of numbers differ in higher dimensions, the method for arriving at the final result is identical to the one used in 2D.

3D tensor construction Combining equations (26) and (29) gives the elements of the filter associated tensors $\hat{N}_k$:

$$\hat{N}_1 = c^2 \begin{pmatrix} a^2 & 0 & ab \\ 0 & 0 & 0 \\ ab & 0 & b^2 \end{pmatrix} \quad \hat{N}_2 = c^2 \begin{pmatrix} a^2 & 0 & -ab \\ 0 & 0 & 0 \\ -ab & 0 & b^2 \end{pmatrix}$$

$$\hat{N}_3 = c^2 \begin{pmatrix} b^2 & ab & 0 \\ ab & a^2 & 0 \\ 0 & 0 & 0 \end{pmatrix} \quad \hat{N}_4 = c^2 \begin{pmatrix} b^2 & -ab & 0 \\ -ab & a^2 & 0 \\ 0 & 0 & 0 \end{pmatrix} \tag{38}$$

$$\hat{N}_5 = c^2 \begin{pmatrix} 0 & 0 & 0 \\ 0 & b^2 & ab \\ 0 & ab & a^2 \end{pmatrix} \quad \hat{N}_6 = c^2 \begin{pmatrix} 0 & 0 & 0 \\ 0 & b^2 & -ab \\ 0 & -ab & a^2 \end{pmatrix}$$

where:
$$a = 2$$
$$b = (1 + \sqrt{5})$$
$$c = (10 + 2\sqrt{5})^{-1/2}$$

Let the signal orienting vector be given by:

$$\mathbf{x} = (x_1, x_2, x_3)^T \tag{39}$$

then the magnitude of the outputs from the 6 quadrature filters are, according to equation 23, given by:

$$q_1 = dc^2 x^{-2}(a^2 x_1^2 + 2abx_1x_3 + b^2 x_3^2) \quad q_2 = dc^2 x^{-2}(a^2 x_1^2 - 2abx_1x_3 + b^2 x_3^2)$$

$$q_3 = dc^2 x^{-2}(b^2 x_1^2 + 2abx_1x_2 + a^2 x_2^2) \quad q_4 = dc^2 x^{-2}(b^2 x_1^2 - 2abx_1x_2 + a^2 x_2^2) \tag{40}$$

$$q_5 = dc^2 x^{-2}(b^2 x_2^2 + 2abx_2x_3 + a^2 x_3^2) \quad q_6 = dc^2 x^{-2}(b^2 x_2^2 - 2abx_2x_3 + a^2 x_3^2)$$

Next, calculating the sum

$$T' = \sum_k q_k \hat{N}_k \tag{41}$$

yields the components of T':

$$t_{11} = d'(x_1^2 x^{-2} + \tfrac{1}{2}) \quad t_{12} = t_{21} = d'x_1x_2x^{-2}$$

$$t_{22} = d'(x_2^2 x^{-2} + \tfrac{1}{2}) \quad t_{13} = t_{31} = d'x_1x_3x^{-2} \tag{42}$$

$$t_{33} = d'(x_3^2 x^{-2} + \tfrac{1}{2}) \quad t_{23} = t_{32} = d'x_2x_3x^{-2}$$

where $d' = \tfrac{4}{5}d$.

It is evident that if the quantity $\tfrac{1}{2}d' = \tfrac{2}{5}d$ is subtracted from the diagonal elements of T' the result will be of the desired form.

$$T = T' - \tfrac{1}{2}d' = d'\hat{x}\hat{x}^T \tag{43}$$

Finally calculate the sum of all quadrature filter output magnitudes.

$$\sum_k q_k = 2d \tag{44}$$

Combining eqns.(35), (37) and (38) yields the desired result:

$$T = \sum_k q_k(\hat{N}_k - \tfrac{1}{5}I) \tag{45}$$

4D tensor construction Combining equations (27) and (29) gives the elements of the filter associated tensors $\hat{\mathbf{N}}_k$:

$$
\hat{\mathbf{N}}_1 = c^2 \begin{pmatrix} 1&1&0&0 \\ 1&1&0&0 \\ 0&0&0&0 \\ 0&0&0&0 \end{pmatrix} \quad
\hat{\mathbf{N}}_2 = c^2 \begin{pmatrix} 1&-1&0&0 \\ -1&1&0&0 \\ 0&0&0&0 \\ 0&0&0&0 \end{pmatrix}
$$

$$
\hat{\mathbf{N}}_3 = c^2 \begin{pmatrix} 1&0&1&0 \\ 0&0&0&0 \\ 1&0&1&0 \\ 0&0&0&0 \end{pmatrix} \quad
\hat{\mathbf{N}}_4 = c^2 \begin{pmatrix} 1&0&-1&0 \\ 0&0&0&0 \\ -1&0&1&0 \\ 0&0&0&0 \end{pmatrix}
$$

$$
\hat{\mathbf{N}}_5 = c^2 \begin{pmatrix} 1&0&0&1 \\ 0&0&0&0 \\ 0&0&0&0 \\ 1&0&0&1 \end{pmatrix} \quad
\hat{\mathbf{N}}_6 = c^2 \begin{pmatrix} 1&0&0&-1 \\ 0&0&0&0 \\ 0&0&0&0 \\ -1&0&0&1 \end{pmatrix}
$$

$$
\hat{\mathbf{N}}_7 = c^2 \begin{pmatrix} 0&0&0&0 \\ 0&1&1&0 \\ 0&1&1&0 \\ 0&0&0&0 \end{pmatrix} \quad
\hat{\mathbf{N}}_8 = c^2 \begin{pmatrix} 0&0&0&0 \\ 0&1&-1&0 \\ 0&-1&1&0 \\ 0&0&0&0 \end{pmatrix}
$$

$$
\hat{\mathbf{N}}_9 = c^2 \begin{pmatrix} 0&0&0&0 \\ 0&1&0&1 \\ 0&0&0&0 \\ 0&1&0&1 \end{pmatrix} \quad
\hat{\mathbf{N}}_{10} = c^2 \begin{pmatrix} 0&0&0&0 \\ 0&1&0&-1 \\ 0&0&0&0 \\ 0&-1&0&1 \end{pmatrix}
$$

$$
\hat{\mathbf{N}}_{11} = c^2 \begin{pmatrix} 0&0&0&0 \\ 0&0&0&0 \\ 0&0&1&1 \\ 0&0&1&1 \end{pmatrix} \quad
\hat{\mathbf{N}}_{12} = c^2 \begin{pmatrix} 0&0&0&0 \\ 0&0&0&0 \\ 0&0&1&-1 \\ 0&0&-1&1 \end{pmatrix}
$$

(46)

where $c = \frac{1}{\sqrt{2}}$.

Let the signal orienting vector be given by:

$$
\mathbf{x} = (x_1, x_2, x_3, x_4)^T \tag{47}
$$

Then the magnitude of the outputs from the 12 quadrature filters are given by

$$
\begin{aligned}
q_1 &= dc^2 x^{-2}(x_1^2 + 2x_1x_2 + x_2^2) & q_2 &= dc^2 x^{-2}(x_1^2 - 2x_1x_2 + x_2^2) \\
q_3 &= dc^2 x^{-2}(x_1^2 + 2x_1x_3 + x_3^2) & q_4 &= dc^2 x^{-2}(x_1^2 - 2x_1x_3 + x_3^2) \\
q_5 &= dc^2 x^{-2}(x_1^2 + 2x_1x_4 + x_4^2) & q_6 &= dc^2 x^{-2}(x_1^2 - 2x_1x_4 + x_4^2) \\
q_7 &= dc^2 x^{-2}(x_2^2 + 2x_2x_3 + x_3^2) & q_8 &= dc^2 x^{-2}(x_2^2 - 2x_2x_3 + x_3^2) \\
q_9 &= dc^2 x^{-2}(x_2^2 + 2x_2x_4 + x_4^2) & q_{10} &= dc^2 x^{-2}(x_2^2 - 2x_2x_4 + x_4^2) \\
q_{11} &= dc^2 x^{-2}(x_3^2 + 2x_3x_4 + x_4^2) & q_{12} &= dc^2 x^{-2}(x_3^2 - 2x_3x_4 + x_4^2)
\end{aligned}
\tag{48}
$$

Next, calculating the sum

$$
\mathbf{T}' = \sum_k q_k \hat{\mathbf{N}}_k \tag{49}
$$

yields the components of $\mathbf{T}'$

$$t_{11} = d(x_1^2 x^{-2} + \tfrac{1}{2}) \qquad t_{22} = d(x_2^2 x^{-2} + \tfrac{1}{2})$$

$$t_{33} = d(x_3^2 x^{-2} + \tfrac{1}{2}) \qquad t_{44} = d(x_4^2 x^{-2} + \tfrac{1}{2})$$

$$t_{12} = t_{21} = d x_1 x_2 x^{-2} \quad t_{13} = t_{31} = d x_1 x_3 x^{-2}$$

$$t_{14} = t_{41} = d x_1 x_4 x^{-2} \quad t_{23} = t_{32} = d x_2 x_3 x^{-2} \tag{50}$$

$$t_{24} = t_{42} = d x_2 x_4 x^{-2} \quad t_{34} = t_{43} = d x_3 x_4 x^{-2}$$

It is evident that if the quantity $\tfrac{1}{2}d$ is subtracted from the diagonal elements of $\mathbf{T}'$ the result will be of the desired form.

$$\mathbf{T} = \mathbf{T}' - \tfrac{1}{2}d = d\hat{\mathbf{x}}\hat{\mathbf{x}}^T \tag{51}$$

Finally calculate the sum of all quadrature filter output magnitudes.

$$\sum_k q_k = 3d \tag{52}$$

Combining equations (49), (51) and (52) yields the desired result:

$$\mathbf{T} = \sum_k q_k (\hat{\mathbf{N}}_k - \frac{1}{6}\mathbf{I}) \tag{53}$$

13.6.3 Interpretation of the orientation tensor

In real life acquired data are seldom exactly *simple*. It is, however, still possible to find a best approximation to $\mathbf{T}$ corresponding to a *simple* neighborhood, equation 2. This is done by finding the $\mathbf{T}_s$ that minimizes:

$$\Delta = \|\mathbf{T} - \mathbf{T}_s\| \tag{54}$$

where:

$$\mathbf{T}_s = A\hat{\mathbf{x}}\hat{\mathbf{x}}^T \tag{55}$$

The norm of a tensor is invariant under rotation of the coordinate system and equation (54) can be rewritten as:

$$\Delta = \|\mathbf{C}^{-1}(\mathbf{T} - A\hat{\mathbf{x}}\hat{\mathbf{x}}^T)\mathbf{C}\| \tag{56}$$

giving:

$$\Delta = \|\mathbf{C}^{-1}\mathbf{T}\mathbf{C} - \mathbf{C}^{-1}A\hat{\mathbf{x}}\hat{\mathbf{x}}^T\mathbf{C}\| \tag{57}$$

where $\mathbf{C}$ is an orthogonal matrix.

Let $\mathbf{C}$ be such that $\mathbf{C}^{-1}\mathbf{T}\mathbf{C}$ is diagonal and note that only one eigenvalue of $\hat{\mathbf{x}}\hat{\mathbf{x}}^T$ is nonzero. Then, since the norm of $\mathbf{C}^{-1}\mathbf{T}\mathbf{C}$ is the sum of the squares of its elements, it is clear that Δ is minimized if $\mathbf{C}^{-1}A\hat{\mathbf{x}}\hat{\mathbf{x}}^T\mathbf{C}$ removes the largest of these values, i.e. the largest eigenvalue of $\mathbf{C}^{-1}\mathbf{T}\mathbf{C}$. Thus, if the eigenvalues are numbered in decreasing order, Δ is given by:

$$\Delta = \sqrt{\sum_{n\neq 1} \lambda_n^2} \tag{58}$$

Then, since $\mathbf{T}$ and $A\hat{\mathbf{x}}\hat{\mathbf{x}}^T$ are subject to identical rotation, it is clear that the A and $\hat{\mathbf{x}}$ which minimizes Δ are given by:

$$\hat{\mathbf{x}} = \hat{\mathbf{e}}_1$$
$$A = \lambda_1 \tag{59}$$

where e_1 is the eigenvector corresponding to the largest eigenvalue of **T**.

Thus, the tensor most similar to **T** corresponding to a *simple* neighborhood is given by:

$$\mathbf{T_s} = \lambda_1 \hat{e}_1 \hat{e}_1^T \qquad (60)$$

where:
λ_1 is the largest eigenvalue of **T** and
$\hat{e}_1$ is the corresponding eigenvector.

The value of Δ indicates how well the 1-dimensionality hypothesis fits the neighborhood, the smaller the value the better the fit.

Higher rank neighborhoods Simple neighborhoods are represented by tensors, $\mathbf{T_s}$, having rank 1. In higher dimensional data there exists highly structured neighborhoods which are not simple. The rank of the representing tensor will then reflect the complexity of the neighborhood. Below the eigenvalue distributions and the corresponding tensor representations are given for three particular cases of **T** in 3D. $\lambda_1 \geq \lambda_2 \geq \lambda_3 \geq 0$ are the eigenvalues of **T** in decreasing order, and $\hat{e}_i$ is the eigenvector corresponding to λ_i.

1. **Plane case** (*simple neighborhood:* $\lambda_1 = \lambda_{plane}$; $\lambda_2, \lambda_3 = 0$)

$$\mathbf{T} = \lambda_{plane}\mathbf{T_1} = \lambda_{plane}\hat{e}_1\hat{e}_1^T \qquad (61)$$

This case corresponds to a neighborhood that is perfectly *planar*, i.e. is constant on planes in a given orientation. The orientation of the *normal vectors* to the planes is given by $\hat{e}_1$.

2. **Line case** (*rank 2 neighborhood:* $\lambda_1, \lambda_2 = \lambda_{line}$; $\lambda_3 = 0$)

$$\mathbf{T} = \lambda_{line}\mathbf{T_2} = \lambda_{line}\left(\hat{e}_1\hat{e}_1^T + \hat{e}_2\hat{e}_2^T\right) \qquad (62)$$

This case corresponds to a neighborhood that is constant on *lines*. The orientation of the lines is given by the eigenvector corresponding to the smallest eigenvalue, $\hat{e}_3$.

3. **Isotropic case** (*rank 3 neighborhood:* $\lambda_1, \lambda_2, \lambda_3 = \lambda_{iso}$)

$$\mathbf{T} = \lambda_{iso}\mathbf{T_3} = \lambda_{iso}\left(\hat{e}_1\hat{e}_1^T + \hat{e}_2\hat{e}_2^T + \hat{e}_3\hat{e}_3^T\right) \qquad (63)$$

This case corresponds to an *isotropic* neighborhood, meaning that there exists energy in the neighborhood but no orientation, e.g. in the case of noise.

The eigenvalues and eigenvectors are easily computed with standard methods such as the Jacobi method, e.g. [Press *et al.*, 1986]. In general **T** will be some where in between these cases but, note that the spectrum theorem states that all 3D tensors can be expressed as a linear combination of these three cases, i.e. **T** can always be expressed as:

$$\mathbf{T} = (\lambda_1 - \lambda_2)\,\mathbf{T_1} + (\lambda_2 - \lambda_3)\,\mathbf{T_2} + \lambda_3\,\mathbf{T_3} \qquad (64)$$

13.7 Accuracy of the orientation estimate

The performance of the algorithm was evaluated by Haglund [Haglund, 1992] using a synthetic test pattern. The test pattern is generated to be locally planar with all directions equally represented. The test patterns reflect case 1 and case 2 in section 13.6.3. The quadrature filters for the estimation were optimized using the principals described [Knutsson, 1982].

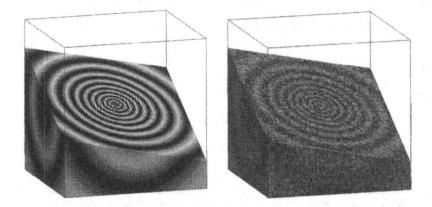

Fig. 13.7. Test volumes for 3D orientation estimation. The volumes have been "cut" along a slanted plane to show the interior signal. Left - without noise, Right SNR = 0 dB.

The test pattern is spherically symmetric containing all 3D plane orientations over a wide frequency range. The pattern is realized as a 64 x 64 x 64 data volume, see figure 13.7.

Four instances of the test pattern were used for the evaluation, one without noise and three with added Gaussian distributed noise, see Figure 13.7. The volumes with noise have a SNR of 20 dB, 10 dB and 0 dB respectively, with SNR defined as:

$$SNR = 20\log[\frac{\mathbf{sdev}(\text{pattern})}{\mathbf{sdev}(\text{noise})}] \tag{65}$$

Since this is a synthetic test pattern it is possible to generate the correct normalized tensor field, $\hat{\mathbf{T}}_1$. The pattern is a one variable function of the radius, $r = \|\boldsymbol{\xi}\|$. Thus, the signal orienting vector $\mathbf{x}$ for the test pattern is given by the spatial coordinate $\boldsymbol{\xi}$ and the correct tensor field is given by:

$$\hat{\mathbf{T}}_1 = \hat{\boldsymbol{\xi}}\hat{\boldsymbol{\xi}}^T \tag{66}$$

The filters used for estimation of the orientation tensor had a size of 7x7x7. The estimated orientation tensor $\mathbf{T}$ was compared with the ideal, rank one, tensor $\mathbf{T}_1$ for all points in the volume. The comparison was done using the error estimate:

$$\mathbf{err} = \sqrt{\frac{1}{P}\sum_{i=1}^{P}\|\hat{\mathbf{T}}_i - \hat{\mathbf{T}}_{1i}\|^2} \tag{67}$$

where P is the number of points involved in the error calculation. The results are given in Table 13.1 demonstrating that the orientation algorithm produces robust and accurate estimates of the 3D orientation.

Note that it is possible to average the orientation tensor to get even more accurate estimates. This is a virtue of the representation and does not depend upon the filters used to achieve the initial estimates.

13.8 Time sequences - velocity

A natural way of estimating velocity in an image sequence is to estimate 3D-orientation in the sequence, as described above. The orientation estimate in three dimensions (two spatial and one

Table 13.1. Performance figures for the orientation estimation algorithm.

SNR	ERR
∞ dB	0.04
20 dB	0.06
10 dB	0.11
0 dB	0.32

time-dimension) contains information of both the local spatial orientation and the local velocity. Note that for time sequences, a 3D plane means a line in the 2D image plane and a 3D line means a point in the image plane.

The velocity can be obtained by an eigenvalue analysis of the estimated representation tensor. The projection of the eigenvector corresponding to the largest eigenvalue onto the image plane will give the line velocity field. For moving lines or linear structures only the velocity component perpendicular to the structure can be determined. The velocity component along the line is indeterminable since motion in this direction induces no change in the local signal. This fact is commonly, but somewhat misleading, referred to as the "aperture problem". It is a fundamental problem for all velocity algorithms if the entire neighborhood is constant on parallel lines. The "aperture problem" does not exist for moving non-linear structures, e.g. points, and in this case the correct velocity can be estimated.

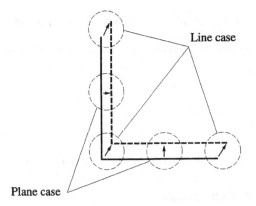

Fig. 13.8. Illustration of the "aperture problem".

By examining the relations between the eigenvalues in the orientation tensor it is possible to estimate which of the above categories the neighborhood belongs to. Depending on the category different strategies can be chosen, see Section 13.6.3. Case number one in Section 13.6.3, the plane case, corresponds to a moving line and gives a correct estimation only for the velocity component perpendicular to the line. Case number two in Section 13.6.3, the line case, corresponds to a moving point and gives a correct estimation of the velocity in the image plane.

Haglund [Haglund, 1992] tested the algorithm using a synthetic test sequence consisting of a rotating and translating star together with a fair amount of Gaussian noise, figure 13.9 (top left). The star is rotated 1.8° counter-clockwise around its center, and translates 0.5 pixel up and 1 pixel to the right between each frame. In Figure 13.9 (top right) the correct velocity field is given with arrows, white arrows correspond to the moving line case and black arrows to the moving point case.

To categorize the tensor the following functions were chosen, (see also equation 64).

$$p_{plane} = \frac{\lambda_1 - \lambda_2}{\lambda_1} \tag{68}$$

$$p_{line} = \frac{\lambda_2 - \lambda_3}{\lambda_1} \tag{69}$$

$$p_{iso} = \frac{\lambda_3}{\lambda_1} \tag{70}$$

These expressions can be seen as the probability for each case. the division is made by selecting the case having the highest probability. In Figure 13.9(bottom left) the probability for the moving line case is shown.

The calculation of the velocity is done using Eq. 71 for the moving line case and Eq. 72 for the moving point case. In neighborhoods classified as 'isotropic' no velocity is computed. In neighborhoods classified as belonging to the moving line case the velocity is computed by:

$$\begin{cases} \mathbf{v}_{line} = -x_3 \left(x_1 \hat{\xi}_1 + x_2 \hat{\xi}_2 \right) / (x_1^2 + x_2^2) \\ x_1 \quad = \hat{\mathbf{e}}_1 \cdot \hat{\xi}_1 \\ x_2 \quad = \hat{\mathbf{e}}_1 \cdot \hat{\xi}_2 \\ x_3 \quad = \hat{\mathbf{e}}_1 \cdot \hat{\mathbf{t}} \end{cases} \tag{71}$$

where $\hat{\xi}_1$ and $\hat{\xi}_2$ are the orthogonal unit vectors defining the image plane and $\hat{\mathbf{t}}$ is a unit vector in the time direction.

In neighborhoods classified as belonging to the moving point case the velocity is computed by:

$$\begin{cases} \mathbf{v}_{point} = \left(x_1 \hat{\xi}_1 + x_2 \hat{\xi}_2 \right) / x_3 \\ x_1 \quad = \hat{\mathbf{e}}_3 \cdot \hat{\xi}_1 \\ x_2 \quad = \hat{\mathbf{e}}_3 \cdot \hat{\xi}_2 \\ x_3 \quad = \hat{\mathbf{e}}_3 \cdot \hat{\mathbf{t}} \end{cases} \tag{72}$$

In Figure 13.9 the results from Eq. 71 (white arrows) and Eq. 72 (black arrows) are given.

13.9   Discontinuous motion

Objects suddenly appearing, disappearing or abruptly changing direction or speed often signifies important events. Such events will be reflected by corresponding changes in the orientation tensor and can be detected robustly using a tensor field filtering technique termed 'differential convolution' [Knutsson and Westin, 1993]. Gradual changes such as curvature and acceleration can also be estimated using local operators [Bårman, 1991].

13.10   Spatio-temporal frequency

The fact that the norm of the orientation tensor is rotation invariant allows for simple estimation of local spatio-temporal frequency. This estimation can be done in several ways dependent on system needs. One simple way is to produce one spatial frequency estimate and one temporal frequency estimate. Let the norm of the orientation tensor from channel χ_{kl} be denoted $\|\mathbf{T}\|_{kl}$, where k is the

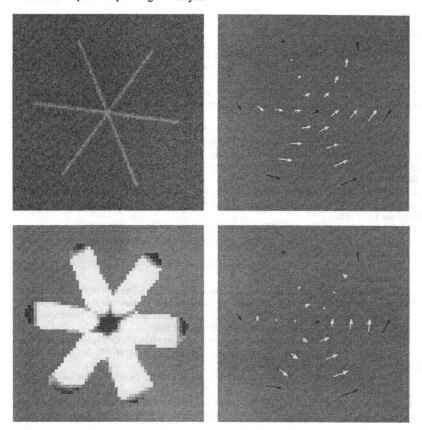

Fig. 13.9. Top left: One frame from the original sequence of the translating and rotating star, with white Gaussian noise added. **Top right:** The correct velocity vectors from the star sequence. Black vectors correspond to the moving point case and white ones to the moving line case. **Bottom left:** The probability for neighborhood belonging to the moving line case according to equation 68 for the test sequence. **Bottom right:** The result of the velocity algorithm. Black vectors correspond to the moving point case and white ones to the moving line case.

spatial channel index and l the temporal channel index. A spatial frequency estimate can then be obtained by:

$$\omega_s = \sum_{kl} \|\mathbf{T}\|_{kl}\, e^{i\alpha k} \tag{73}$$

where α is a constant.

The argument of ω_s will represent the local spatial frequency. Analogously a temporal frequency estimate is obtained by:

$$\omega_t = \sum_{kl} \|\mathbf{T}\|_{kl}\, e^{i\alpha l} \tag{74}$$

This is a straight forward extension of the technique presented in [Haglund, 1992].

References

[Bårman, 1991] H. Bårman. *Hierachical Curvature Estimation in Computer Vision*. PhD thesis, Linköping University, Sweden, S–581 83 Linköping, Sweden, September 1991. Dissertation No 253, ISBN 91–7870–797–8.

[Coxeter, 1961] H. S. M. Coxeter. *Introduction to Geometry*. John Wiley & Sons, Inc., 1961.

[Granlund, 1978] G. H. Granlund. In search of a general picture processing operator. *Computer Graphics and Image Processing*, 8(2):155–178, 1978.

[Haglund, 1992] L. Haglund. *Adaptive Multidimensional Filtering*. PhD thesis, Linköping University, Sweden, S–581 83 Linköping, Sweden, October 1992. Dissertation No 284, ISBN 91–7870–988–1.

[Knutsson and Bårman, 1992] H. Knutsson and H. Bårman. Robust orientation estimation in 2D, 3D and 4D using tensors. In *Proceedings of International Conference on Automation, Robotics and Computer Vision*, September 1992.

[Knutsson and Westin, 1993] H. Knutsson and C-F Westin. Normalized and differential convolution: Methods for interpolation and filtering of incomplete and uncertain data. In *Proceedings of CVPR*, New York City, USA, June 1993. IEEE.

[Knutsson et al., 1980] H. Knutsson, B. von Post, and G. H. Granlund. Optimization of arithmetic neighborhood operations for image processing. In *Proceedings of the First Scandinavian Conference on Image Analysis*, Linköping, Sweden, January 1980.

[Knutsson, 1982] Hans Knutsson. *Filtering and Reconstruction in Image Processing*. PhD thesis, Linköping University, Sweden, 1982. Diss. No. 88.

[Knutsson, 1985] Hans Knutsson. Producing a continuous and distance preserving 5-D vector representation of 3-D orientation. In *IEEE Computer Society Workshop on Computer Architecture for Pattern Analysis and Image Database Management - CAPAIDM*, pages 175–182, Miami Beach, Florida, November 1985. IEEE. Report LiTH–ISY–I–0843, Linköping University, Sweden, 1986.

[Knutsson, 1989] H. Knutsson. Representing local structure using tensors. In *The 6th Scandinavian Conference on Image Analysis*, pages 244–251, Oulu, Finland, June 1989. Report LiTH–ISY–I–1019, Computer Vision Laboratory, Linköping University, Sweden, 1989.

[Press et al., 1986] W. H. Press, B. P. Flannery, S. A. Teukolsky, and W. T. Vetterling. *Numerical Recipes*. Cambridge University Press, 1986.

[Wilson and Knutsson, 1988] R. Wilson and H. Knutsson. Uncertainty and inference in the visual system. *IEEE Transactions on Systems, Man and Cybernetics*, 18(2), March/April 1988.

14. Line Extraction Using Tensors

Carl-Fredrik Westin and Hans Knutsson

CVL, LiTH

14.1 Modules

This chapter describes theories and algorithms used in the following modules: **Möbius mapping** and **Local/global line extraction**

14.2 Introduction

This chapter focuses on various aspects of representation and grouping of information. Although this theme is discussed focused on line extraction, we stress that the presented ideas have more general applicability. In the previous chapter, a coordinate free notion of tensors were used. In this chapter we need to perform tensor operations explicitly and need notations for this. Therefore, we will start with giving a short introduction to the notations in tensor theory since it may not be familiar to all readers. More complete introductions will be found in [Kay, 1988; Young, 1978; Kendall, 1977].

The work at the Computer Vision Laboratory at Linköping University has over the years produced a number of results in vision showing a clear quantum mechanical influence, e.g. the discussions of uncertainty and vision presented in Wilson and Granlund [Wilson and Granlund, 1984] and Wilson and Knutsson [Wilson and Knutsson, 1988]. It will be shown that the orientation tensor [Knutsson, 1989] introduced in the previous chapter can be used as a grouping operator for collinear/coplanar estimates. The technique used involves a type of operators commonly used in quantum mechanics; projection operators. Further, an investigation of the uniformity of some standard parameter spaces (also termed parameter mappings) is presented. The analysis shows that, to avoid discontinuities, great care should be taken when choosing the parameter space for a particular problem. A new parameter mapping well suited for line extraction, the Möbius strip parameterization [Westin and Knutsson, 1992], is presented. We describe a local-global line extraction algorithm based on this parameter mapping, where only local orientation estimates having global support is used for estimation of line segments. Most of this chapter is based on [Westin, 1991]. The procedure has similarities to the Hough Transform [Illingworth and Kittler, 1988].

14.3 Notations from tensor theory

There are two avenues to tensors, and there is a general disagreement on which is the better approach; the *component* approach or the *non-component* approach. In the component approach the underlying coordinate system is only implicit. This has the disadvantage that all the components describing the tensor change under transformation, although the tensor still is the "same". The other approach, the non-component approach, is the favourite one for people working in the mathematical community. Although this way of treating tensors is necessary for many modern applications and may give a more complete understanding for tensor theory, the component approach is easier to begin with. The component approach also has the advantage of often allowing interpretation of the calculations in terms of linear algebra. Conversion to the component approach is always necessary when algorithms are to be implemented on a computer. In order to be able to proceed, we need some notations. Let V denote a linear vector space with finite dimension and let v_i be a base for V:

$$V : \{ \mathbf{v}_1, \mathbf{v}_2, \ldots\ldots, \mathbf{v}_n \} \tag{1}$$

A vector x in V is denoted by:

$$\mathbf{x} = x^1\mathbf{v}_1 + x^2\mathbf{v}_2 + \ldots + x^n\mathbf{v}_n = \sum_{i=1}^{n} x^i\mathbf{v}_i \tag{2}$$

where x^i denotes the components of the vector.

Einstein summation convention

The summation rule is that any expression involving a twice-repeated index shall automatically stand for its sum over the values $1, 2, 3...n$ of the *repeated* index.

$$\mathbf{x} = x^1\mathbf{v}_1 + x^2\mathbf{v}_2 + \ldots + x^n\mathbf{v}_n = \sum_{i=1}^{n} x^i\mathbf{v}_i = x^i\mathbf{v}_i \tag{3}$$

The convention used here is that this rule only is applied on pair of indices of type one superscript and one subscript (one up and one down). A double sum expressed using the Einstein summation convention might look like the following:

$$a_{ij}x^iy^j \quad \longleftrightarrow \quad \sum_{i=1}^{n}\sum_{j=1}^{n} a_{ij}x^iy^j \tag{4}$$

This summation convention is very useful, but care is warranted. Compare for example the following expressions

$$a_{ij}x^iy^j = a_{ij}y^jx^i$$
$$a_{ij}(x^j + y^j) = a_{ij}x^j + a_{ij}y^j$$

with

$$a_{ij}x^iy^j \neq a_{ij}y^ix^j$$
$$a_{ij}(x^i + y^j) \neq a_{ij}x^i + a_{ij}y^j$$

It may be worth noting that the symbol $\neq$ in the last inequality actually should stand for "not even wrong", since both sides are meaningless expressions in tensor analysis.

Covariant and contravariant tensors

Tensors are based on two main ingredients. Depending on the transformation properties of a tensor, it will be categorized as being a covariant tensor, a contravariant tensor or a mixture of these two, i.e. a mixed tensor. A vector is a contravariant tensor and it has *components* with superscripts. A covector is an example of a covariant tensor. An electrical field or the gradient of a scalar field are examples of covectors. As mentioned, their transformation properties are different. A vector is expressed in units of the coordinate grid, while an electrical field is expressed in voltage *per* units of the coordinate grid. Consider the change of coordinate system resulting in doubling the value of each vector component, i.e. a rescaling of the axes. What does this bring about for the electrical field description? The electrical field expressed in its components will in the new base only have *half* of the original component values. Thus, under this particular transformation, the components of a vector is multiplied by two, while those of a covector is divided by two. This shows that the components of the two different types of elements scale in *opposite* direction under transformation.

Let V^* denote the dual space to the linear vector space V, and let ω^i denote a base for V^*:

$$V^* : \{ \omega^1, \omega^2, ..., \omega^n \} \tag{5}$$

V^*, the dual space to V, consists of all linear mappings from V to the real numbers: $V \rightarrow \mathbb{R}$. Elements in V^* are covariant tensors:

$$\mathbf{x} = x_i \omega^i \tag{6}$$

where x_i denotes the components of the element $\mathbf{x}$. Note that subscripts are used for the components in the dual space.

The metric tensor

A metric or scalar product defines distances in a vector space. The distance between two vectors $\mathbf{a}$ and $\mathbf{b}$ is:

$$d^2 = g(\mathbf{a} - \mathbf{b}, \mathbf{a} - \mathbf{b}) \tag{7}$$

where:
d denotes the distance
$\mathbf{g}$ denotes the metric tensor
$\mathbf{a} = a^i \mathbf{v}_i$ and $\mathbf{b} = b^i \mathbf{v}_i$

Writing the same expression using components gives

$$d^2 = g_{ij}(a^i - b^i)(a^j - b^j) \tag{8}$$

The standard metric in $\mathbb{R}^3$ when having a cartesian ON-base has the following components:

$$g_{ij} = \begin{pmatrix} 1 & 0 & 0 \\ 0 & 1 & 0 \\ 0 & 0 & 1 \end{pmatrix} \tag{9}$$

The scalar product between two vectors in $\mathbb{R}^3$ is denoted:

$$g_{ij}a^i b^i = \begin{pmatrix} a^1 & a^2 & a^3 \end{pmatrix} \begin{pmatrix} 1 & 0 & 0 \\ 0 & 1 & 0 \\ 0 & 0 & 1 \end{pmatrix} \begin{pmatrix} b^1 \\ b^2 \\ b^3 \end{pmatrix} = a^1 b^1 + a^2 b^2 + a^3 b^3 \tag{10}$$

The metric does not only define distances in the vector space, but also couples elements in V and V^*. For each element in V there is a corresponding element in V^* when a metric is defined. Therefore the two vector spaces V and V^* are *closely* related through the metric $\mathbf{g}$. Actually, when working with cartesian tensors, the two spaces V and V^* have a deceptive likeness. If the metric operates only on one vector

$$g(\mathbf{a}, \cdot) \quad \leftrightarrow \quad g_{ij}a^i = b_j \tag{11}$$

we see that the element produced has components with a subscript indicating that the element is a covector, i.e. a dual element:

$$g(\mathbf{a}, \cdot) \in V^* \tag{12}$$

The transformed element $g(\mathbf{a}, \cdot)$ is coupled to $\mathbf{a}$ via the metric. Of course, the basis elements, $\mathbf{v}_i \in V$ and $\omega^i \in V^*$, also have counterparts in the other domain. The dual base is often chosen so:

$$\omega^i(\mathbf{v}_j) = \begin{cases} 1 \text{ if } i = j \\ 0 \text{ if } i \neq j \end{cases} \equiv \delta^i_j \tag{13}$$

where
ω^i denotes the base for V^*, the dual base
$\mathbf{v}_j$ denotes the base for base for V
$\delta^i{}_j$ denote the Kronecker delta

If $\mathbf{v}_i$ is an ON-base the following relation holds:

$$g(\mathbf{v}_i, \cdot) \ = \ \omega^i \tag{14}$$

The tight coupling between the two domains through the metric suggests the following notation:

$$g_{ij}a^i \ = \ a_j \tag{15}$$

The effect of the transformation can be seen as "lowering" the index from a^i to a_j. The two different tensors can be viewed as the "same" tensor in a more abstract sense and as two different representations of the same entity. When working with cartesian tensors the metric is of a very simple form which makes the border between V and V^* almost invisible. Therefore the metric in many cases is referred to as a tool for raising and lowering indices (The inverse metric $g^{-1} \leftrightarrow g^{ij}$ is used for raising indices).

14.4 Geometric interpretation of the orientation tensor

The orientation tensor introduced in chapter 13.1 can be visualized by an ellipse in 2D and an ellipsoid in 3D.

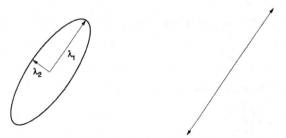

Fig. 14.1. Left: Geometric representation of a tensor as an ellipse. Right: Geometric representation of a rank 1 tensor: since the smallest eigenvalue is zero the ellipse has collapsed to a line. Two arrows have been added indicating that it is an outer product of a vector which can be seen as an "unsigned" vector.

In two-dimension it is simple to extract the direction of the eigenvectors of the orientation tensor:

Statement 1 *Let the coefficients of a symmetric positive semidefinite tensor* $\mathbf{T}$ *be:*

$$\begin{pmatrix} T_{11} & T_{12} \\ T_{12} & T_{22} \end{pmatrix} \tag{16}$$

The orientation of the eigenvectors of $\mathbf{T}$, $\hat{e}_1$ *and* $\hat{e}_2$, *can be calculated by:*

$$2\varphi_1 = 2arg(e_{1x}, e_{1y}) \ = \ arg(T_{11} - T_{22}, 2T_{12}) \tag{17}$$
$$2\varphi_2 = 2arg(e_{2x}, e_{2y}) \ = \ arg(T_{22} - T_{11}, -2T_{12}) \tag{18}$$

where φ_1 *is the angle of the eigenvector having the largest eigenvalue and* φ_2 *to the one with the least.*

A geometric interpretation of this statement can be made. The spectrum decomposition theorem allows us to write the orientation tensor as:

$$\mathbf{T} = \lambda_1 \hat{\mathbf{e}}_1 \hat{\mathbf{e}}_1^T + \lambda_2 \hat{\mathbf{e}}_2 \hat{\mathbf{e}}_2^T \tag{19}$$

When the tensor is of rank 1, only one of the eigenvalues is greater than 0 and we get:

$$\mathbf{T} = \lambda_1 \hat{\mathbf{e}}_1 \hat{\mathbf{e}}_1^T \quad \leftrightarrow \quad \lambda_1 \begin{pmatrix} e_{1x}^2 & e_{1x} e_{1y} \\ e_{1x} e_{1y} & e_{1y}^2 \end{pmatrix} \tag{20}$$

which shows the relation between the tensor of rank 1 and its only eigenvector; the tensor is the outer product of the eigenvector. In this special case, the validity of statement (1) is trivial:

$$2\varphi_1 = 2arg(e_{1x}, e_{1y}) \tag{21}$$
$$= arg(e_{1x}^2 - e_{1y}^2, 2e_{1x}e_{1y}) \tag{22}$$
$$= arg(\cos(\varphi_1)^2 - \sin(\varphi_1)^2, 2\cos(\varphi_1)\sin(\varphi_1)) \tag{23}$$
$$= arg(\cos(2\varphi_1), \sin(2\varphi_1)) \tag{24}$$

which shows that the angle of the eigenvector can be expressed directly in the tensor coefficients. This is of course true for the tensor $\lambda_1 \hat{\mathbf{e}}_1 \hat{\mathbf{e}}_1^T$ and the tensor $\lambda_2 \hat{\mathbf{e}}_2 \hat{\mathbf{e}}_2^T$ separately. The reason why this simple formula is true for a rank 2 tensor is that the two eigenvectors are *parallel* in the 2φ-domain. In figure 14.2 this relation is shown. To the left in the figure the direction of the two eigenvectors of a tensor is illustrated. To the right the corresponding double angle representations of these vectors show that they are parallel. Only their signs are different.

Fig. 14.2. Left: Geometric representation of a tensor. Right: The two eigenvectors expressed in double angle. Note that they are parallel.

The resultant vector in the 2φ-domain has apparently the same direction for all λ_2. The length of the resultant vector varies as $(\lambda_1 - \lambda_2)$. The figure also illustrates that statement 1 collapses when $\lambda_1 = \lambda_2$. However, in that case the tensor is completely round, which means that *all* vectors are eigenvectors.

14.5 The local orientation tensor as grouping operator

14.5.1 Projection operators

Before presenting the theory for orientation tensor based grouping it is helpful to touch briefly upon the concept of projection operators. Projection operators differ from ordinary rotation and reflection operators which are bijective. For a projection operator, we may have distinct vectors $\mathbf{x}_1$ and $\mathbf{x}_2$ such that $\mathbf{A}(\mathbf{x}_1) = \mathbf{A}(\mathbf{x}_2)$.

Definition 1. A linear operator $\mathbf{A}$ is said to be a projection operator if $\mathbf{A}$ is both Hermitian and idempotent, where idempotent is defined by $\mathbf{A}\,\mathbf{A} = \mathbf{A}$

If $\mathbf{A}$ is real, the Hermitian requirement reduces to requiring that $\mathbf{A}$ is symmetric. We will see that a symmetric operator on $\mathbf{R}^N$ can be decomposed into a weighted sum of projection operators (this is true for the Hermitian case as well). Consider e.g. an operator $\mathbf{A}$ with distinct eigenvalues λ_i, using the spectral decomposition theorem gives:

$$\mathbf{A} = \sum_{i=1}^{n} \lambda_i \hat{\mathbf{e}}_i \hat{\mathbf{e}}_i^T \tag{25}$$

where the vectors $\hat{\mathbf{e}}_i$ are the normalized eigenvectors of $\mathbf{A}$. It is easy to verify that all the outer products of the normalized eigenvectors are projection operators. Since the outer product produces a symmetric operator in any dimension:

$$\hat{\mathbf{e}}_i \leftrightarrow \begin{pmatrix} e_{i_1} \\ e_{i_2} \\ \vdots \\ e_{i_n} \end{pmatrix} \quad \Rightarrow \quad \hat{\mathbf{e}}_i \hat{\mathbf{e}}_i^T \leftrightarrow \begin{pmatrix} e_{i_1}^2 & e_{i_1} e_{i_2} & \cdots & e_{i_1} e_{i_n} \\ e_{i_2} e_{i_1} & e_{i_2}^2 & \ddots & \vdots \\ \vdots & \ddots & \ddots & \vdots \\ e_{i_n} e_{i_1} & \cdots & \cdots & e_{i_n}^2 \end{pmatrix} \tag{26}$$

and the operator is idempotent:

$$\mathbf{P} = \hat{\mathbf{e}}_i \hat{\mathbf{e}}_i^T \quad \text{and} \quad \mathbf{PP} = \hat{\mathbf{e}}_i \hat{\mathbf{e}}_i^T \hat{\mathbf{e}}_i \hat{\mathbf{e}}_i^T = \hat{\mathbf{e}}_i \underbrace{(\hat{\mathbf{e}}_i^T \hat{\mathbf{e}}_i)}_{=1} \hat{\mathbf{e}}_i^T = \mathbf{P} \tag{27}$$

it follows from definition 1 that $\hat{\mathbf{e}}_i \hat{\mathbf{e}}_i^T$ is a projection operator. This class of operators is commonly used in quantum mechanics [Jauch, 1968; Hughes, 1989].

14.5.2 Decomposing the orientation tensor

As mentioned, the orientation tensor which was introduced in chapter 13.1 can be used for grouping. Since the orientation tensor is symmetric, equation (25) shows that the tensor can be decomposed into a weighed sum of projection operators:

$$\mathbf{T} = \sum_{i=1}^{n} \lambda_i \hat{\mathbf{e}}_i \hat{\mathbf{e}}_i^T \tag{28}$$

where $\hat{\mathbf{e}}_i$ is a normalized eigenvector of $\mathbf{T}$.

If all the λ_i are distinct, the decomposition of $\mathbf{T}$ is unique and the projection operators are the outer product of the eigenvectors:

$$\hat{\mathbf{e}}_i \hat{\mathbf{e}}_i^T \tag{29}$$

If two or more eigenvalues are equal, there is an option in defining the eigenvectors in that subspace, and hence there is an option in the decomposition in eigenvectors. The subspace they span is, however, distinct. Say for example that $\hat{\mathbf{e}}_1$ and $\hat{\mathbf{e}}_2$ share the same eigenvalue and define an operator projecting onto this subspace:

$$\mathbf{T}_2 = \hat{\mathbf{e}}_1 \hat{\mathbf{e}}_1^T + \hat{\mathbf{e}}_2 \hat{\mathbf{e}}_2^T \tag{30}$$

Thus, decomposing the tensor in such operators gives a unique set of coefficients. Let λ_k be the distinct values, then:

$$\mathbf{T} = \sum_{i=1}^{k} \lambda_i \mathbf{T}_i \tag{31}$$

where the $\mathbf{T}_i$ are the projection operators projecting onto the different subspaces. The number of terms in the sum is equal to the rank of the tensor. We can note the one-to-one correspondence between subspaces and projection operators. Letting $\mathbf{T}_i$ act on itself easily verifies that it is a projection operator. Since all the components are orthogonal only the components acting on themselves will contribute. From equation (27), it then trivially follows that $\mathbf{T}_k$ *is* a projection operator.

Although the coefficients in the expansion in equation (31) are unique, the decomposition is not. There are other sets of projection operators spanning the tensor as well. We just chose one with orthogonal elements. An interesting choice is to decompose into projection operators having different geometrical meaning. We define the following projection operators:

$$\mathbf{T}_1 = \hat{e}_1\hat{e}_1^T \tag{32}$$
$$\mathbf{T}_2 = \hat{e}_1\hat{e}_1^T + \hat{e}_2\hat{e}_2^T \tag{33}$$
$$\mathbf{T}_i = \hat{e}_1\hat{e}_1^T + \hat{e}_2\hat{e}_2^T + \ldots\ldots + \hat{e}_i\hat{e}_i^T \tag{34}$$

The first projection operator corresponds to a one-dimensional subspace, a line. The second corresponds to a two-dimensional subspace, a plane, etc. Expressing the orientation tensor in this new basis gives:

$$\mathbf{T} = \sum_{i=1}^{n} (\lambda_i - \lambda_{i+1})\mathbf{T}_i \tag{35}$$

where the dummy variable $\lambda_{n+1} = 0$. The coefficients are trivially derived by comparing this expression with the one produced be the spectral decomposition theorem in equation (25).

Note that, although in a different context, the above tensor decomposition is the same as the one used in chapter 13.1 for classifying higher order neighbourhoods.

14.5.3 Covariant and contravariant tensors

In section 14.3 it was stated that tensors are based on two main ingredients, namely covariant and/or contravariant tensors. Since the orientation tensor has close relations to the outer product of the gradient, it can be argued that orientation is a covariant tensor of order two [Westin, 1991]. Estimating the orientation in a *simple* neighbourhood produces a tensor of rank 1, see section 13.6.3. In our context this corresponds to the projection operator $\mathbf{T}_1$ in equation (32). Letting this tensor act on its spatial position vector gives:

$$\rho = \mathbf{T}_1(\mathbf{x}) \tag{36}$$

or written in component form (the index 1 is omitted):

$$\rho_j = T_{ij}x^i \tag{37}$$

where:
T_{ij} denotes the components of the orientation tensor
x^i denotes the components of the vector defining the position in image space
ρ_j denotes the components of the produced tensor; a *covector*

Thus, the projection operator $\mathbf{T}_1$ projects the spatial position (a contravariant vector) of the estimate (a covariant tensor of order two) to a covariant vector. The fact that the output is a covector is indicated by the *subscript*, see section 14.3.

14.5.4   Lines in two dimensions

Equation (35) gives the decomposition of the orientation tensor into projection operators the following form in 2D:

$$\mathbf{T} = (\lambda_1 - \lambda_2)\hat{\mathbf{e}}_1\hat{\mathbf{e}}_1^T + \lambda_2(\hat{\mathbf{e}}_1\hat{\mathbf{e}}_1^T + \hat{\mathbf{e}}_2\hat{\mathbf{e}}_2^T) = (\lambda_1 - \lambda_2)\mathbf{T}_1 + \lambda_2\mathbf{T}_2 \tag{38}$$

As mentioned, the operation in equation (36) produces a *covector*, ρ. All collinear tensors having equal orientation are projected onto the same covector. In order to be able to visualize this covector with a vector arrow, it is transformed into its vector shape via the metric $\mathbf{g}$; $\rho^j = g_{ij}\rho_i$. It turns out that the orientation of this vector is *perpendicular* to the line to be estimated, see figure 14.3. Thus, the vector is pointing at the line coordinate being the closest to the origin.

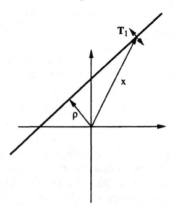

Fig. 14.3. Grouping of information

To facilitate drawing of the orientation tensor, we have in this figure used the "double arrow" notation from figure 14.1. The double arrow corresponds to the direction of the largest eigenvector. The reason for having two directions is that if $\hat{\mathbf{e}}_1$ is the eigenvector of $\mathbf{T}_1$, so is $-\hat{\mathbf{e}}_1$. The double arrow indicates these two directions. A tensor of rank 1 can therefore be seen as an "unsigned" vector.

The grouping that the tensor operation performs can be illuminated by the following derivation. The vector x in figure 14.3 can be divided into one vector perpendicular to the line of collinearity and one parallel to it:

$$x^i = \rho^i + \alpha\rho_\perp^j \tag{39}$$

where α defines the position along the line. Using this expression in the operation from equation (37);

$$T_{ij}x^j = T_{ij}(\rho^j + \alpha\rho_\perp^j) = T_{ij}\rho^j + \alpha T_{ij}\rho_\perp^j \tag{40}$$

Recalling the definition of the ideal orientation tensor as the tensor product between the covector $\perp$ to the line and itself:

$$T_{ij} = \rho_i\rho_j \tag{41}$$

gives:

$$T_{ij}\rho^j + \alpha T_{ij}\rho_\perp^j = \rho_i\rho_j\rho^j + \alpha\rho_i\rho_j\rho_\perp^j = \rho_i + 0 = \rho_i \tag{42}$$

The fact that both the orientation tensor and ρ are covariant tensors can be shown by studying their transformation properties. Suppose we change our coordinate system so the line in figure 14.3 looks like in figure 14.4.

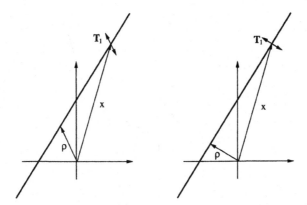

Fig. 14.4. In this figure the new transformed coordinate system is shown . If $\mathbf{T}$ and ρ were contravariant tensors, they would be transformed as in the left figure. It can be seen that they transform as the vector $\mathbf{x}$, i.e. they become more vertical. Thus, they turn in the "wrong" direction. To the right it is shown how covariant tensors would be transformed.

14.5.5 Lines and planes in three dimensions

Equation (35) shows that the decomposition of the orientation tensor into projection operators has the following form in 3D:

$$\mathbf{T} = (\lambda_1 - \lambda_2)\mathbf{T}_1 + (\lambda_2 - \lambda_3)\mathbf{T}_2 + \lambda_3\,\mathbf{T}_3 \tag{43}$$

Simple and higher rank neighbourhoods were discussed in chapter 13.1 when interpreting the 3D orientation tensor. As mentioned there, the rank of the tensor reflects the complexity of the neighbourhood. The *plane case* corresponds to what was referred to as a simple neighbourhood. The *line case* corresponds to a neighbourhood that is constant on *lines*. The orientation of the lines is given by the eigenvector corresponding to the smallest eigenvalue, $\hat{\mathbf{e}}_3$.

The projection operator $\mathbf{T}_1$ groups coplanar points and $\mathbf{T}_2$ groups collinear points. This grouping is visualized in figure 14.5 where the set of coplanar orientation tensors will be projected onto ρ_1 and the set of collinear tensors will be projected onto ρ_2. In both cases the elements are projected to the *closest* point on the plane/line to the origin.

14.5.6 Parameter spaces defined by the projection operators

Two dimensions : In 2D, the parameter mapping is two dimensional: the components of the produced covector. However, lines passing through the origin of the image coordinate system ($\mathbf{x} = 0$) are projected to the same cluster. One way of preventing the estimates from different lines from mixing is to incorporate the operator, $\mathbf{T}_1 = \hat{\mathbf{e}}_1 \hat{\mathbf{e}}_1^T$ in the mapping as well. The operator contains three different components and the produced covector contains two. This gives us a five dimensional grouping space:

$$(\rho_x, \rho_y, e_{1_x}^2, e_{1_y}^2, e_{1_x} e_{1_y}) \tag{44}$$

Three dimensions : In three dimensions, we similarly get two parameter mappings. One nine-dimensional grouping space for simple neighbourhoods:

$$(\rho_x, \rho_y, \rho_z, e_{1_x}^2, e_{1_y}^2, e_{1_z}^2, e_{1_x} e_{1_y}, e_{1_x} e_{1_z}, e_{1_y} e_{1_z}) \tag{45}$$

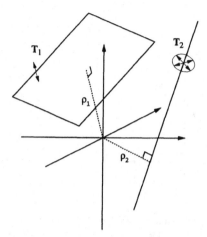

Fig. 14.5. Grouping of information in 3D.

and another nine dimensional grouping space for rank 2 neigbourhoods:

$$(\rho_x, \rho_y, \rho_z, e_{1_x}^2 + e_{2_x}^2, e_{1_y}^2 + e_{2_y}^2, e_{1_z}^2 + e_{2_z}^2, e_{1_x}e_{1_y} + e_{2_x}e_{2_y}, e_{1_x}e_{1_z} + e_{2_x}e_{2_z}, e_{1_y}e_{1_z} + e_{2_y}e_{2_z}) \qquad (46)$$

Note, however, that the parameters in these mappings are dependent. This means that only a subspace of the parameter space contains data. Unfortunately these subspaces are folded in a complex manner.

This concludes the general discussion about projection operators and grouping data. In the next section we will focus on parameter mappings for 2D lines and present a parameter mapping that unfolds the 5D-mapping presented in this section.

14.6   The Möbius strip parameterization

14.6.1   Standard parameter mappings

The reason for using a parameter mapping is often to convert a difficult global detection problem in image space into a local one. Spatially extended patterns are transformed so that they produce spatially compact features in a space of parameter values. In the case of line segmentation the idea is to transform the original image into a new domain so that collinear subsets, i.e. global lines, fall into clusters. The topology of the mapping must reflect closeness between wanted features, in this case features describing properties of a line. The metric describing closeness should also be uniform throughout the space with respect to the features. If the metric and topology do not meet these requirements, significant bias and ambiguities will be introduced into any subsequent classification process. These aspects are especially important when working with dynamic scenes. If for example, the position in the parameter space for a moving line do not change smoothly, it may be hard or impossible to keep track of them. In this section some problems with standard mappings for line segmentation will be illuminated. We start with discussing some intuitively natural mappings for line segmentation.

The Hough transform, HT, was introduced by P. V. C. Hough in 1962 as a method for detecting complex patterns [Hough, 1962]. It has found considerable application due to its robustness when using noisy or incomplete data. A comprehensive review of the Hough transform covering the years 1962-1988 can be found in [Illingworth and Kittler, 1988].

Severe problems with standard Hough parameterization are that the space is unbounded and will contain singularities for large slopes. The difficulties of unlimited ranges of the values can be solved by using two plots, the second corresponding to interchanging the axes. This is of course not a satisfactory solution. Duda and Hart [Duda and Hart, 1972] suggested that straight lines might be most usefully parameterized by the length, ρ, and orientation φ, of the normal vector to the line from the origin, the *normal parameterization*, see figure 14.6. This mapping has the advantage of having no singularities.

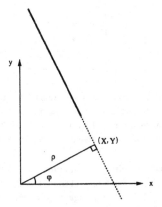

Fig. 14.6. The *normal parameterization*, (ρ, φ), of a line. ρ is the magnitude of displacement vector from the origin and φ is its argument.

Measuring local orientation provides additional information about the slope of the line or the angle φ when using the normal parameterization. This reduces the standard HT to a one-to-one mapping. With one-to-one we do not mean that the mapping is invertible, but that there is only one point in the parameter space that defines the parameters that could have produced it.

Duda and Hart discussed this briefly in [Duda and Hart, 1973]. They suggested that this mapping could be useful when fitting lines to a collection of short line segments. Dudani and Luk [Dudani and Luk, 1978] used this technique for grouping measured edge elements. Princen, Illingworth and Kittler performed line extraction using a pyramid structure [Princen *et al.*, 1990]. At the lowest level they used the ordinary $\rho\varphi$-HT on subimages for estimating small line segments. In the preceding levels they use the additional local orientation information for grouping the segments.

Unfortunately, however, the normal parameterization has problems when ρ is small. The topology here is very strange. Clusters can be divided into parts very far away from each other. Consider for example a line going through the origin in a xy-coordinate system. When mapping the coordinates according to the normal parameterization, *two* clusters will be produced separated in the φ-dimension by π, see figure 14.7. Note that this will happen even if the orientation estimates are perfectly correct. A line will always have at least an infinitesimal thickness and will therefore be projected on *both* sides of the origin. A final point to note is that a translation of the origin outside the image plane will not remove this topological problem. It will only be transferred to other lines.

Granlund introduced a double angle notation [Granlund, 1978] in order to achieve a suitable continuous representation for local orientation. However, using this double angle notation for *global* lines, removes the ability to distinguish between lines with the same orientation and distance, ρ, on opposite sides of the origin. The problem near $\rho = 0$ is removed, but unfortunately we have introduced another one. The two horizontal lines (marked a and c), located at the same distance ρ from the origin, are in the double angle normal parameterization mixed into one cluster, see figure

14.7.

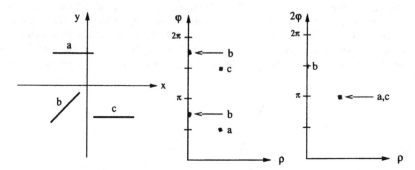

Fig. 14.7. A test image containing three lines and its transformation to the $\rho\varphi$-domain, the normal parameterization of a line. The cluster from the line at 45° is divided into two parts. This mapping has topological problems near $\rho = 0$. The $\rho 2\varphi$-domain, however, folds the space so the topology is good near $\rho = 0$, but unfortunately it is now bad elsewhere. The two horizontal lines, marked a and c, have in this parameter space been mixed in the same cluster.

It seems that we need a "double angle" representation around the origin and a "single angle" representation elsewhere. This raises a fundamental dilemma: is it possible to achieve a mapping that fulfills both the single angle and the double angle requirements simultaneously?

We have been concerned with the problem of the normal parameterization spreading the coordinates around the origin unsatisfactorily although they are located very close in the cartesian representation. Why do we not express the displacement vector, i.e. the normal vector to the line from the origin, in *cartesian* coordinates, (X, Y), since the topology is satisfactory? This parameterization is defined by

$$\begin{cases} X = x\cos^2(\varphi) + y\cos(\varphi)\sin(\varphi) \\ Y = y\sin^2(\varphi) + x\cos(\varphi)\sin(\varphi) \end{cases} \tag{47}$$

where φ is as before the argument of the normal vector (same as the displacement vector of the line).

Davis uses the $\rho\varphi$-parameterization in this way by storing the information in a cartesian array [Davis, 1986]. This gives the (X, Y) parameterization. There are two reasons for not using this parameterization. First, the spatial resolution is very poor near the origin. Secondly, and worse, all lines having ρ equal to 0 will be mapped to the same cluster.

The first problem, the poor resolution near the origin, can at least be solved by mapping the XY-plane onto a logarithmic cone. That would stretch the XY-plane so the points close to the origin get more space. However, the second problem still remains.

14.7 The Möbius strip mapping

In this section we present a new parameter space and discuss its advantages with respect to the arguments of the previous section. The topology of the mapping introduces its name, the "Möbius strip" parameterization. This mapping has topological advantages over previously proposed mappings.

The Möbius strip mapping is based on a transformation to a 4D space by taking the *normal parameterization* in figure 14.6, expressed in cartesian coordinates (X, Y) and adding a "double angle" dimension, (consider the Z-axis in a XYZ-coordinate system). The problem with the cartesian normal parameterization is, as mentioned, that all clusters from lines going through the origin mix into one cluster. The additional dimension, $\phi = 2\varphi$, separates the clusters on the origin and close to the origin if the clusters originate from lines with different orientation. Moreover, the wrap-around requirement for ϕ is ensured by introducing a fourth dimension, R.

The 4D-mapping

$$\begin{cases} X = x\cos^2(\varphi) + y\cos(\varphi)\sin(\varphi) \\ Y = y\sin^2(\varphi) + x\cos(\varphi)\sin(\varphi) \\ \phi = 2\varphi \\ R = R_0 \in \mathbf{R}^+ \end{cases} \tag{48}$$

The two first parameters, X and Y, represent the normal vector in figure 14.6 expressed in cartesian coordinates. The two following parameters, ϕ and R, define a circle with radius R_0 in the $R\phi$-subspace. Any $R_0 > 0$ is suitable. This gives a $XY\phi$-system with wrap-around in the ϕ-dimension. This 4D-mapping is a subspace of the 5D-space presented in section 14.5.6. The relation between the mappings is: $X = \rho_x$, $Y = \rho_y$ while statement (1) describes the relation between the tensor components and the double angle representation.

In the mapping above, the parameters are dependent. As the argument of the vector in the XY-plane is φ and the fourth dimension is constant, it follows that for a specific (X, Y) all the parameters are given. Hence, the degrees of freedom are limited to two, the dimension of the XY-plane. Thus, all the mapped image points lie in a 2D subspace of the 4D parameter space, see figure 14.8.

2φ

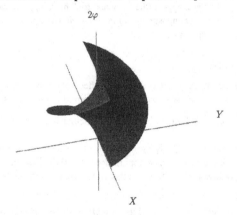

Y

X

Fig. 14.8. The 2D subspace

The 2D-surface The regular form of the 2D-surface makes it possible to find a two-parameter form for the desired mapping. Consider a $\eta\phi$-plane corresponding to the flattened surface in figure 14.8. Let

$$\rho^2 = X^2 + Y^2 = (x\cos^2(\varphi) + y\cos(\varphi)\sin(\varphi))^2 + (y\sin^2(\varphi) + x\cos(\varphi)\sin(\varphi))^2 \tag{49}$$

$$\Leftrightarrow \qquad \rho = x\cos(\varphi) + y\sin(\varphi) \tag{50}$$

Then the (η, ϕ) mapping can be expressed as

$$\eta = \begin{cases} \rho & 0 \le \varphi < \pi \\ -\rho & \pi \le \varphi < 2\pi \end{cases} \tag{51}$$

$$\phi = 2\varphi \tag{52}$$

η is the variable "across" the strip, with 0 value meaning the position in the middle of the strip, i.e. *on* the 2φ axis. The wrap-around in the ϕ dimension makes it easy to interpret the surface as a Möbius strip.

Finally, using the same test image as before, we can see that we can distinguish between the two lines on opposite sides of the origin while not dividing the cluster corresponding to the line going through the origin, see figure 14.9.

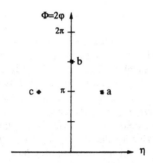

Fig. 14.9. The (η, ϕ) parameter mapping (the "Möbius strip" mapping). We can see that we can distinguish between the two lines, a and c, at opposite side of the origin at the same time as the cluster corresponding to the line going through the origin is not divided.

14.7.1 Conclusions

The name of the mapping, as mentioned above, reflects the topology of the $\eta\phi$-parameter surface, its *twisted* wrap-around in the ϕ-dimension. The proposed mapping has the following properties:

- The parameter space is bounded and has no singularities such as the standard HT parameterization for large slopes.
- The metric reflects the underlying geometry of the eatures. The mapping does not share the topological problem of the $\rho\varphi$-mapping near $\rho = 0$. In [Westin and Knutsson, 1992] it was shown that any line passing through the origin produces *two* clusters, separated by π in the φ-dimension.

Note that these properties have been achieved *without* increasing the dimension of the parameter space compared to for example the *normal parameterization* in figure 14.6.

14.8   Local estimates with global support

Before showing results from the classification of the projected data, a simple segmentation will be performed by demanding different energy support in the histogram. The test image used is found in figure 14.13. The output image from the local orientation algorithm (a 2D projection of the spatio temporal tensor) and corresponding histogram is found in the top of figure 14.10. The image is then transformed a second time. This time no new histogram is produced. The parameter coordinate obtained via the mapping is only used as a pointer, checking the energy concentration in the first histogram. If the energy concentration is greater than a specified threshold, α, the image point is left untouched (or amplified). If the energy level is less, the image point is set to zero, see figure 14.10. This is of course a very stylized example. Nevertheless, we believe that a moderate requirement of the energy concentration in the histogram can remove a lot of "noncoherent" edge points, i.e. noise. This requirement for line coherence reduces the amount of data passed to the classifier.

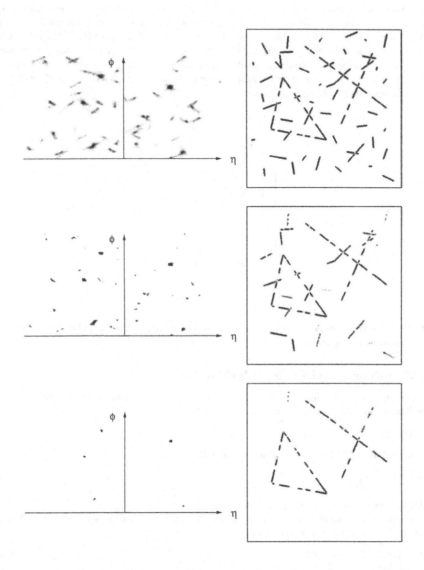

Fig. 14.10. Requiring different levels of energy support, α, in the histogram space for the orientation estimates in the image can be used for removing short segments and noise. Note that this is a pixel based method. No line segmentation has been carried out. (histogram maxvalue is scaled to 1). Top $\alpha = 0$, middle $\alpha = 0.3$ and bottom $\alpha = 0.7$.

14.9   Estimation of global line parameters

14.9.1   Segmentation of Collinear Points Into Line Groups

The resulting Möbius strip histogram is classified using a clustering procedure [Westin, 1991]. The estimates in the clusters are linked into a list structure. A sequential segmentation algorithm is used for dividing the set of collinear points in each linked list into appropriate line segments. When projecting the orientation estimates into the Möbius histogram, the elements are ordered according to a distance value, ξ, see figure 14.11, in a linked list structure. The line endings are easily found by sequentially scanning this list while searching for big jumps in this distance value. In this way, the set of collinear points is divided into subgroups corresponding to lines in the image. A control parameter defines the maximum gap that should be bridged. Line parameters are estimated from these subgroups using a minimum variance method.

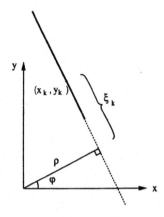

Fig. 14.11. The *normal parameterization*, (ρ, φ), of a line. ρ is the magnitude of the displacement vector from the origin and φ is its argument.

14.9.2   Fitting a Straight Line to N Given Points

In this section we derive the formulas giving the parameters of the line best fitting the measured data. Assume that we have N points

$$\mathbf{x}_i = (x_i, y_i) \qquad\qquad i = 1, ..., N \tag{53}$$

The method used is a least-squares fit to measured data.

$$\text{minimize } \sum_i d_i^2 \tag{54}$$

where d_i is the distance from the point to the estimated line.

Note that this approach is conceptually different from *linear regression*, where the set of N data points (x_i, y_i) is fit to a straight line model

$$y(x) = y(x; m, k) = m + kx \tag{55}$$

A straight line model like this assumes that the uncertainty associated with each measurement y_i is known, and that the x_i's (values of the dependent variable) are known *exactly*. To measure how well the data fits the model a chi-square merit function is used:

$$\chi^2(m, k) = \sum_i \left(y_i - m - kx_i\right)^2 \tag{56}$$

The general principle, the least-squares fit, relates to the subject of *maximum likelihood* (ML) estimators. It can be shown that the least-squares fit is a ML-estimator if the measurement errors are independent and normally distributed with a constant standard deviation.

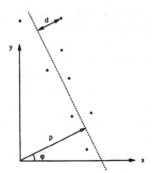

Fig. 14.12. Finding minimum variance axis by minimizing the sum of squared normal distances

However, the minimization expressed in equation (54) finds the *minimum variance axis* of the data. A minimum variance approach is advantageous when both x_i and y_i are uncertain, not only y_i, as in linear regression. The minimum variance axis can be found by investigating the covariance matrix, P, of the points.

$$\mathbf{P} = \sum_i (\mathbf{x}_i - \bar{\mathbf{x}})(\mathbf{x}_i - \bar{\mathbf{x}})^T \tag{57}$$

where
$\mathbf{x}_i \Leftrightarrow (x_i, y_i)$ is the coordinate of point number i
$\bar{\mathbf{x}} = \frac{1}{N}\sum_i \mathbf{x}_i \Leftrightarrow (x_c, y_c)$ is the average coordinate of the points

The minimum variance axis is oriented in the same direction as the eigenvector of $\mathbf{P}$ corresponding to the least eigenvalue. The direction of the eigenvector corresponding to the largest eigenvalue is of course the maximum variance axis. Simplification of equation (57) gives:

$$\mathbf{P} = \sum_i \mathbf{x}_i \mathbf{x}_i^T + \sum_i \bar{\mathbf{x}}\bar{\mathbf{x}}^T - \sum_i \bar{\mathbf{x}}\mathbf{x}_i^T - \sum_i \mathbf{x}_i \bar{\mathbf{x}}^T \tag{58}$$

$$= \sum_i \mathbf{x}_i \mathbf{x}_i^T - \sum_i \bar{\mathbf{x}}\bar{\mathbf{x}}^T \tag{59}$$

$$= \begin{pmatrix} \sum_i x_i^2 & \sum_i x_i y_i \\ \sum_i x_i y_i & \sum_i y_i^2 \end{pmatrix} - \begin{pmatrix} \sum_i x_c^2 & \sum_i x_c y_c \\ \sum_i x_c y_c & \sum_i y_c^2 \end{pmatrix} \tag{60}$$

$$= \begin{pmatrix} \sum_i x_i^2 - \frac{1}{N}(\sum_i x_i)^2 & \sum_i x_i y_i - \frac{1}{N}(\sum_i x_i)(\sum_i y_i) \\ \sum_i x_i y_i - \frac{1}{N}(\sum_i x_i)(\sum_i y_i) & \sum_i y_i^2 - \frac{1}{N}(\sum_i y_i)^2 \end{pmatrix} \tag{61}$$

Denoting the elements of the covariance matrix $\mathbf{P}$ by:

$$\begin{pmatrix} P_{11} & P_{12} \\ P_{12} & P_{22} \end{pmatrix} \tag{62}$$

allows for expressing the orientation of the eigenvectors of $\mathbf{P}$, $\hat{e}_1$ and $\hat{e}_2$, using the formulas introduced in section 14.4:

$$2\varphi_1 = 2arg(e_{1x}, e_{1y}) = arg(P_{11} - P_{22}, 2P_{12}) \tag{63}$$
$$2\varphi_2 = 2arg(e_{2x}, e_{2y}) = arg(P_{22} - P_{11}, -2P_{12}) \tag{64}$$

where φ_1 is the angle of the eigenvector having largest eigenvalue and φ_2 to the one with least. It can be worth pointing out that the key to the reduced expressions in the theorem is the representation of the *double* orientation angle. Using this double angle formula, we get a closed form of the direction of the eigenvector corresponding to the least eigenvalue:

$$2\varphi = 2arg(e_{1x}, e_{1y}) \tag{65}$$
$$= arg(P_{22} - P_{11}, -2P_{12}) \tag{66}$$
$$= arg(A, 2B) \tag{67}$$

were
$$A = \sum_i y_i^2 - \sum_i x_i^2 + \frac{1}{N}\left((\sum_i x_i)^2 - (\sum_i y_i)^2\right)$$
$$B = \frac{1}{N}\sum_i x_i \sum_i y_i - \sum_i x_i y_i$$

Thus, the following entities must be calculated:

$$\sum_i x_i, \quad \sum_i y_i, \quad \sum_i x_i y_i, \quad \sum_i x_i^2 \text{ and } \sum_i y_i^2 \tag{68}$$

It can be shown that the line minimizing the sum of *normal* distances in equation (54) is minimized by a line going through the average point (x_c, y_c). The normal vector from the origin, (x_p, y_p), to the minimum variance axis is obtained by projecting (x_c, y_c), along the axis:

$$\begin{pmatrix} x_p \\ y_p \end{pmatrix} = \begin{pmatrix} \cos(\varphi)^2 & \sin(\varphi)\cos(\varphi) \\ \sin(\varphi)\cos(\varphi) & \sin(\varphi)^2 \end{pmatrix} \begin{pmatrix} x_c \\ y_c \end{pmatrix} \tag{69}$$

If only the length of this vector, the normal distance ρ, is needed the formula is reduced to:

$$\rho = x_c \cos(\varphi) + y_c \sin(\varphi) \tag{70}$$

Dudani and Luc discussed this problem of fitting a line to a set of data points in a frequently cited paper [Dudani and Luk, 1978]. They minimized the sum of squared distances by derivation followed by finding zeros. By this technique they obtained a quadratic equation giving two solutions (ρ_1, φ_1) and (ρ_2, φ_2). They found the best of these two by calculating the square of sums and picked the one giving the least value. Note that equations (70) and (67) provide a *closed form* solution of the best fitting line. The final start and end coordinates of the line are obtained by projection the start and end coordinate in the data set on the infinite line defined by the derived parameters (ρ, φ).

 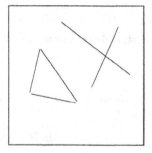

Fig. 14.13. The test image and the estimated lines.

14.9.3 Segmentation example

To illustrate how the user can influence the algorithm, here is a stylized example. By telling the algorithm to have rather large acceptance on interruptions of the line and not accept too short lines we get the result in figure 14.13. More results from this algorithm can be found in [Westin, 1991].

References

[Davis, 1986] E. R. Davis. Image space transforms for detecting straight edges in industrial parts. *Pattern Recognition Letters*, Vol 4:447–456, 1986.

[Duda and Hart, 1972] R. O. Duda and P. E. Hart. Use of the Hough transform to detect lines and cures in pictures. *Communications of the Association Computing Machinery*, 15, 1972.

[Duda and Hart, 1973] R. O. Duda and P. E. Hart. *Pattern classification and scene analysis*. Wiley-Interscience, New York, 1973.

[Dudani and Luk, 1978] S. A. Dudani and A. L. Luk. Locating straight/lines edge segmentgs on outdoor scenes. *Pattern Recognition*, 10:145–157, 1978.

[Granlund, 1978] G. H. Granlund. In search of a general picture processing operator. *Computer Graphics and Image Processing*, 8(2):155–178, 1978.

[Hough, 1962] P. V. C. Hough. A method and means for recognizing complex patterns. U.S. Patent 3,069,654, 1962.

[Hughes, 1989] R. I. G. Hughes. *The structure and interpretation of quantum mechanics*. Harvard University Press, 1989. ISBN: 0-674-84391-6.

[Illingworth and Kittler, 1988] J. Illingworth and J. Kittler. A survey of the Hough transform. *Computer Vision, Graphics and Image Processing*, 44, 1988.

[Jauch, 1968] J. M. Jauch. *The Foundations of Quantum Mechanichs*. Reading, MA:Addison-Wesley, New York, 1968.

[Kay, 1988] David C. Kay. *Schaum's outline of theory and problems of tensor calculus*. McGraw-Hill, 1988.

[Kendall, 1977] D. E. Bourne P.C. Kendall. *Vector Analysis and Cartesian Tensors*. Van Nostrand Reinhold (UK), 1977. ISBN: 0 442 30743 8.

[Knutsson, 1989] H. Knutsson. Representing local structure using tensors. In *The 6th Scandinavian Conference on Image Analysis*, pages 244–251, Oulu, Finland, June 1989. Report LiTH–ISY–I–1019, Computer Vision Laboratory, Linköping University, Sweden, 1989.

[Princen *et al.*, 1990] J. Princen, J. Illingworth, and J. Kittler. A hierarchical approach to line extraction based on the Hough transform. *Computer Vision, Graphics, and Image Processing*, 52, 1990.

[Westin and Knutsson, 1992] C-F Westin and H. Knutsson. The Möbius strip parameterization for line extraction. In *Proceedings of ECCV–92, LNCS–Series Vol. 588*. Springer–Verlaag, 1992.

[Westin, 1991] C-F Westin. Feature extraction based on a tensor image description, September 1991. Thesis No. 288, ISBN 91–7870–815–X.

[Wilson and Granlund, 1984] R. Wilson and G. H. Granlund. The uncertainty principle in image processing. *IEEE Transactions on Pattern Analysis and Machine Intelligence*, PAMI–6(6), November 1984. Report LiTH–ISY–I–0576, Computer Vision Laboratory, Linköping University, Sweden, 1983.

[Wilson and Knutsson, 1988] R. Wilson and H. Knutsson. Uncertainty and inference in the visual system. *IEEE Transactions on Systems, Man and Cybernetics*, 18(2), March/April 1988.

[Young, 1978] Eutiquio C. Young. *Vector and Tensor Analysis*. Dekker, 1978.

Part IV

Active Gaze Control and 3D Scene Description

Exploiting control of processing to adapt the system to the current goals and context has always been a major theme in the VAP project. Planning discussions during the first year soon led the consortium to the conclusion that it is equally important to control the image formation process. It is fairly obvious that a system can gain a considerable power from being able to direct the direction of gaze. However, changing gaze direction soon leads to a requirement to adapt the camera optics to focus on objects which are near or far, and to modify the aperture according to the level of ambient light. The ability to change the focal length of lenses allows a system to see context or detail, according to its need. The ability to change the camera vergence angle permits the system to optimise its cameras for looking near or far. Thus it was decided to develop camera head systems with motorised focus, aperture, zoom and vergence .

Motorised control of focus, aperture, zoom and vergence was discovered to provide capabilities which are much more important to perception than simply optimising the camera configuration. Control of focus provided a simple and robust means to measure the distance to objects using techniques for depth from focus. These techniques provide a depth measurement which is often more precise than stereo. More importantly, they also provide a simple and effective separation of figure and ground. Control of aperture permits the depth of field to be opened or closed according to the needs of the system to perceive a narrow or wide range of depths. Stereo vergence was also discovered to play a fundamental role in separation of figure and ground, as well as to provide a fast reliable means to discover the ground plane in order to organise perceptions in 3D space. Depth from focus and stereo are complementary techniques whose combination provides a robust and rapid form of direct 3D perception.

Chapter 15 describes the KTH head-eye system. This system provides computer control of 13 degrees of freedom for experiments in primate and machine vision. Extremely fast motors and clever control engineering have provided a binocular head which has become the reference for the state of the art. The motors for pan and tilt of the individual cameras provide eye movements approaching 180 degrees per second with a precision of 50,000 steps per resolution (or 0.0072 degrees per step). This permits the head to mimic human saccadic movements. The cameras are equipped with motorised zoom lenses with a range of focal lengths from 12.5 mm to 75 mm. The head is connected to a transputer based control system which has permits the construction of layers of basic visual behaviours such as automatic vergence, focusing, stabilisation and tracking. An experiment is described in which co-operating processes for focus and vergence are coupled to a stabilisation/tracking process based on the use of correlation computed using a cepstral filter.

Chapter 16 describes a layered control architecture for binocular camera heads. This control architecture was developed at LIFIA for the SAVA system, employing a lightweight binocular head (6 degrees of freedom) mounted on a robot arm serving as a neck (6 degrees of freedom) mounted on a mobile robot (2 degrees of freedom), for a total of 14 degrees of freedom. The camera control unit has been built using a hierarchical control system composed of three levels. At the lowest level are standard PID motor controllers tuned to the dynamics of each actuator. At the second level is a device controller which estimates and controls a 3-D fixation point. Fixation point estimation and control have been achieved by extending a standard 6 axis Cartesian arm controller by the addition of 2 additional axes: the offset from the coordinates to the head coordinates, and the 2D angle and distance to the intersection of the optical axes in the gaze plane. The device controller has been designed to control an idealised "virtual head". All information specific to the configuration and dynamics of the head are contained in a "translator" procedure which transforms the virtual head to a specific set of actuators. In this way, the head controller can be mapped onto other binocular heads such as the KTH head. The third layer is composed of perceptual actions which reactivity drive the state of the virtual head. Perceptual actions have been defined for reactive control of vergence, focus, fixation and tracking along linear features in 3D.

Chapter 17 describes the processes used to build and maintain a description of the 3D structures in a scene using active control of a binocular camera head. This process was implemented in the standard SAVA module described in parts I and II above. The system is novel in that it separates the stereo reconstruction process into several parts:

1)   Reactive control of vergence using a coarse to fine algorithm based on phase
2)   Estimation of the 3D fixation point (described in chapter 16)
3)   Stereo matching within the horopter based on simple measures of similarity of the position and orientation of edge segments
4)   Reconstruction of 3D edges in an coordinate system which is intrinsic the object
5)   Dynamic calibration of the stereo camera models based on objects in the scene

Edge segments are obtained from the 2D tracking and description modules. Similarity of edge segments is determined by computing a Mahalanobis distance for segments position, orientation and length. Matched segments are used to build 3D segments based stereo camera models which are dynamically estimated using reference frames defined by objects in the scene.

Stereo reconstruction using depth is extremely sensitive to the precision of the estimated values of the intrinsic camera parameters. Constantly changing the focus, aperture and vergence using reactive techniques has made it impossible to use classical calibration for the camera parameters. Cumbersome and time consuming set-up means that calibration can not be performed "on the fly" as the system operates. We have developed a system for dynamically calibrating stereo cameras to reference frames defined by objects in the scene. This technique allows standard stereo techniques to reconstruct a 3D description in a coordinate system which is intrinsic to the objects in the scene.

Such an "object-based" reference frame is invariant to small movements of the head and thus makes registration for updating the model a trivial process. This calibration process is described in chapter 17.

A 3D description in terms of edge segments is quite complex. The group at KTH have developed a more synthetic representation for modelling the 3D form of objects based on GEONS. This system, described in chapter 18, processes edge chains and interest points provided by the image processing module in order to develop a GEON based description in both a top-down and a bottom up manner. This technique exploits knowledge of context to predict faces which are used to organise contours. When the system is unable to predict shape based on context, the system attempts to interpret the GEONS as faces which are then grouped into objects. This object are used to build and maintain a description of the contents of the scene in terms of 3D forms.

15. The KTH Head-Eye System *

Jan-Olof Eklundh, Kourosh Pahlavan and Tomas Uhlin

CVAP, KTH

Abstract. The KTH-head is the result of adapting a cost effective mechanical design to the fundamental mechanical qualities present in primate head. In the article it is argued that if the goal of research in computational vision is considered to be that of understanding seeing systems, then we need to develop systems with anthropomorphic features. The empirical nature of such research also requires systems by which one can perform experiments in real time and in interaction with the environment.

By an analysis based on observations of the human visual system we describe a design of such a system. Details of its implementation and performance are also provided. To demonstrate that the system meets the requirements derived from our analysis, we finally describe a set of experiments with our head involving gaze control and the integration of primary ocular processes.

15.1 Introduction

The goals of research in computational vision can and have been stated in many different ways. It is generally being agreed upon that the field is concerned with systems having the capability of processing and analyzing visual information. However, there is less agreement on what this actually means.

Much effort has over the past two decades been devoted to the problem of reconstructing a scene from images of it. Although this work has provided us with valuable knowledge, it only indirectly addresses the problem of developing a computational theory of a seeing system. An important reason for this, is that in comparison to the only existing seeing systems we know of, i.e. the biological systems, one overlooks the fact that the latter systems look at the *world* rather than at *prerecorded images* and that their visual processing is related to a *task*. These observations have been extensively discussed in recent work on active or animate vision, see e.g. [Baj85; Bal89; Alo90], and we shall develop them further below. Let us first state that *we* consider the problem of the design of seeing systems as being the essence of research in computational vision. Some explaining remarks on this statement are needed.

Obviously one cannot talk about a seeing system without defining what tasks it is going to solve. If we are to construct a system for achieving certain goals in, say, a manipulation or a navigation task, we can utilize the specific restrictions in the situation to obtain a high-performing solution. Such a specialization can also be observed in biological vision. There are species displaying very high performance in particular visual tasks, in fact often much higher performance than that of humans in the corresponding task. On the other hand, the visual system of the primates have a remarkable flexibility. In the way that these systems combine range of applicability with performance, they seem to outperform most other seeing systems.

* Part of this work has been perfomed within the ESPRIT-BRA project "Vision as Process, VAP" (BR-3038). The support from the Swedish National Board for Technical and Industrial Development, NUTEK is also gratefully acknowledged.

Although such an argument is rather vague, we see it as our primary motive for defining our goal as that of studying computational systems behaving like the human (or primate) visual system. Of course, this does not imply any claims of predictions about biological systems, since those are not being observed. However, it does imply that much of the inspiration must come from biological vision.

Given that our goal is to develop a computational theory for seeing systems in the sense outlined above, we have to ask ourselves the question how such studies can be undertaken. Although the field hopefully can be given a theoretical foundation, it is obviously empirical: validating or falsifying theories can only be done experimentally. This is the basic reason why there exists a need to devise sensor systems with anthropomorphic features, like the one we are going to describe in this article. It provides us with the means of performing the necessary empirical research.

By taking a closer look at how biological vision works, one can derive more specific motivations for, and requirements on, such a sensor. First, it allows us to study vision as part of behaviors and in interaction with the environment, which as argued by e.g. [Alo90; Bal91] is crucial and also different from the reconstruction based vision paradigm. Secondly, one can investigate problems in real-time. Even though this poses difficulties for the computations, it allows us to exploit time to obtain feedback and more information from the environment and to focus the attention on important spatial and temporal events. Notably, real-time here refers more to the fact that we consider the visual analysis as a process running *in time* than as something running at a particular clock frequency. However, the processing speed, as well as the mechanical speed of the system, by necessity correspond to the speed of the events in the world. An underlying hypothesis, for which especially Ballard and his co -workers have made a good case, is that the computations in such a scenario can be implemented by both simple and robust algorithms.

These arguments about what the research in computational vision is all about, and how we can approach these problems, led us to devise a head-eye system with several anthropomorphic features. In the remainder of this article, we shall describe its design, thereby also elaborating on the motivation of specific features. A picture of the head is shown in Figure 15.1.

We shall finally exemplify some of its performance to demonstrate that the system actually constitutes not only a necessary but also a useful research tool in computer vision.

15.2   What active vision imposes on the design

The design of a head-eye system for active vision is in a natural way dependent on what we put into this notion, usually only vaguely defined. In our view, there are two basic aspects to take into account.

The first one is that active vision means active integration of primary processes like vergence, focus, stabilization and tracking. Existing head-eye systems use such processes in their control and therefore some kind of integration exists in all such systems. An active vision system is not just an optomechanical device feeding a computer and carrying out the commands from the computer. The degree of integration, i.e. the extent to which the loops are closed, is a crucial feature of such a system as well. As a consequence, the issue of real-time processing and control becomes central.

Of course, one could argue here that the processes given above are based on specific assumptions about the visual system. If we had a spherical visual sensor shaped like an onion with layers of transparent sensor elements which could cover the whole world, we would possibly not build a head-eye system. However, this seems to be at least a highly impractical solution, and presently we believe that we are forced to look at the world the way nature has evolved to do it.

Returning to real-time processing and integration we feel that these factors determine the behaviors one can obtain. The degree of complexity by which an animal, observing the world, reacts towards its environment, is based on its capabilities to capture the events in the region of interest in its surrounding. Real-time vision is actually addressing this issue. The faster and more elaborately the visual (generally the sensory) system reacts to the surrounding environment, the more evolved will the primary reactive tasks be. The factor that determines the required speed, i.e. what constitutes real-time, is of course, the speed of events to be observed in the environment.

The nature of the processes we want to integrate will be related to the architecture chosen. Biological study of vertebrates shows that those not engaging vergence, also lack foveal vision with a non-uniform retina. The same is true about the elaborate vision of many birds engaging multiple foveas for different kinds of vergence [AlK85].

Fig. 15.1. The KTH-head.

We have like Krotkov [Kkv87], Ballard [Bal91] and others chosen to design a binocular head-eye system. There is no clear argument for this being the only possible solution, besides the fact that the knowledge about mammalian vision provides us with valuable insights into how such a system could work. With this choice, vergence as well as foveation become natural features.

The second aspect of active vision that influences our design is that the system should be purposive. This means that it needs a behavioral engine, a set of different and sometimes contradictory interests, that could result in a specific interaction with the surrounding environment. It might seem as if the problem has come to a bifurcation point. The question is now if our system is a reactive system, interacting with the environment only if something happens, or an active system, taking initiatives to explore the scene even in the absence of events.

We argue that a reactive system represents nothing but the primary and involuntary behaviors of an active system. The existence of an active system is, in other words, preconditioned by its reactive primary processes, but its activeness is preconditioned by its set of interests. The important conclusion of this is that real-time gaze control is also a necessary feature, as has been stressed by e.g. Ballard.

In summary, we have argued that the fact that active vision implies both purposiveness and an integration of primary processes suggests that an active vision system should have the features of real-time processing, high-speed gaze control, binocularity and foveation. These features are indeed proposed by other researchers as well and probably not at all controversial. However, we shall in the next sections consider other design features on which there is less agreement.

15.3   Heads on shoulders vs. heads on arms!

In nature, heads are expected to be on shoulders and not on arms. This playing with words is actually referring to two different approaches in building head-eye systems. Some researchers are, often for good reasons, interested in putting head-eye systems on robot arms. The head we have designed is, on the contrary, put on shoulders, i.e. a platform. This is mainly because we are presently not interested in the study of structure from ego-motion, but there are also other reasons behind our construction.

When we talk about a "head", we include the neck movements as part of its function. Hence, 3 of the degrees of freedom in a robot arm actually do the job the neck would do. However, as we shall see in our discussion about eye movements, there are limitations in many of the previously constructed arm-mounted head-eye systems that actually make them function more like pairs of eyes than as heads with eyes. More practical limitations also arise when one is forced to adapt to the kinematics and the command language of a robot devised for some other or at least more general purpose.

The outline drawings of the front and side views of our head-eye system is illustrated in Figure 15.2. The neck part, the base-line rig and the eye modules are marked in the figure. These modules will be discussed separately later. We then also analyze what degrees of freedom are actually needed in order for a head-eye system to be a "decent" head, in the sense of our introductory remarks.

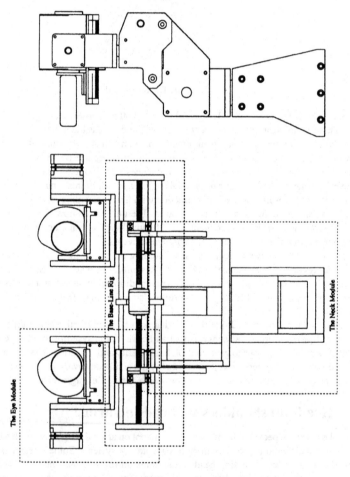

Fig. 15.2. An outline drawing of the KTH-head.

15.4   On eye movements and the KTH-head

There is an extensive literature on eye movements in human beings, see e.g. [Car88; Ybs67]. Somewhat less attention seems to have been paid to the study of neck movements and body movements in connection with vision. This might be the major source of confusion in the design of head-eye systems, because the eye movements in existing head-eye systems are actually hybrids of eye and neck movements, consequently none of them really.

Since our purpose in this section is to elucidate the difference between eye movements and neck movements, we start by characterizing them.

Eye movements are different from other movements affecting visual cues in their quality that they actually do not affect the cues! By this we mean that apart from the fact that fovea centralis is occupied by images from different parts of the objects/scene, no geometrical distortion is generated in the image by normal eye movements. So is not the case with neck movements and other body movements. On the contrary, neck movements are supposed to change the perspective of the images and generate parallax and they do it well thanks to the eccentric position of the spline. In a sense, body movements and neck movements are serving the same tasks, although there still are functional differences between them.

It might be pointed out that the idea behind previous head-eye designs has been to study the geometrical inter-ocular relations rather than imitating the complicated mechanical structure of mammalian heads. We are convinced that the layout of the biological solutions to mechanical problems concerning vision serves a very fundamental purpose: simplifying the task of seeing. This issue is beyond the scope of this article, but in [PUE92a] we discuss it in more detail. The comment is here simply clarifying the reasons behind the approach chosen in our design of the head.

In the KTH-head, the eye modules[2] are separate units, capable of doing their own tilt and pan rotations. The eye modules are also capable of carrying motorized zoom lenses with a drift of optical center up to 90 mm. The lenses are always rotated about the lens center, whatever the focal length or accommodation distance is. This property is implemented through two dedicated motors that position the lenses on the basis of a pre-calibrated ("learned") look-up table. The rotation about the lens center guarantees the least possible distortion of the image and the camera geometry is coarsely converted to the geometry of a pin-hole camera. This is an essential feature of our design.

The displacement of the lens center is a result of changes in optical parameters like focal length, focusing distance and even the iris opening. For example, the focal length of a thin lens is longer for smaller iris openings. An interesting example where this occurs in nature is the biological evidence of animals whose iris openings deform their crystalline lenses.

Figure 15.3 (a) demonstrates a wire model of an eye with 3 mechanical degrees of freedom: superior-inferior rotation (tilt), lateral-medial rotation (pan) and superior-inferior oblique rotation (cyclotorsion). This model is not a rigid model, hence difficult to implement precisely. The model, especially with the spherical retina, is an example of a mechanical construction which simplifies the computational tasks. Let us elaborate this a bit. When tilting a certain amount of degrees by pulling and pushing superior and inferior wires, the lateral and medial wires are not affected at all; meaning that no transformational computations are needed for bringing a point with certain coordinates on the spherical retina to the fovea centralis. All rigid designs are subject to some kind of more or less complex kinematics, caused by the fact that rotation about one axis, also causes rotation of the other axis.

Figure 15.3 (b) illustrates an eye module in the KTH-head. Two of these modules form the pair of binocular eyes on the head. Although the human oculomotor system constrains the motion of the two eyes, the eye modules in the KTH-head can move totally independently. Hence, the control

---

[2] We use the word eye rather than camera, unless this could create confusion, because the word camera has an unnecessary technical connotation in our context

system implemented in software handles the coordination of the two eyes. Earlier designers of head-eye systems have often chosen to link the pan movement of the eyes mechanically, and used a common rig for tilt. There are several drawbacks with such approaches. The systems can only verge symmetrically[3] and consequently the design converts the research issue of eye movements in general, into the issue of symmetric vergence in particular. These systems can never calibrate themselves on-line, they can never "fixate" and their mechanical linkage system is often a source of error. Besides, the solution is neither cheaper nor simpler than the solution engaging separate motors.

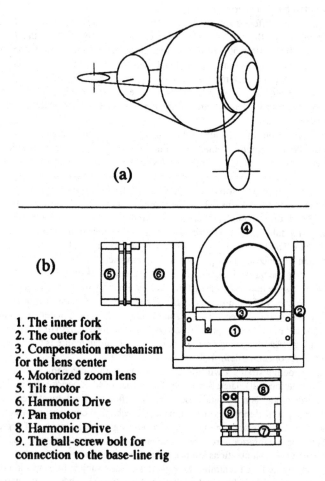

1. The inner fork
2. The outer fork
3. Compensation mechanism for the lens center
4. Motorized zoom lens
5. Tilt motor
6. Harmonic Drive
7. Pan motor
8. Harmonic Drive
9. The ball-screw bolt for connection to the base-line rig

Fig. 15.3. (a) The wire model of a mammalian eye. (b) An eye module in the KTH-head

Talking about fixation, it is worth adding some remarks. Fixation is not only a mechanism to bring the binocular pair of regions of interest together. It is also keeping them in the same position

---

[3] It is of course possible to make complex mechanical cam sets that can even manage non-symmetric vergence, as it is possible to make mechanical computers!

under both ego-motion and the motion of the object. This is exactly what happens when human beings are fixating at a point. While looking at the point, whatever movement the subject undergoes, due to neck movement, body movement or even movement of a vehicle carrying him, the fixation at the point remains in effect. Symmetric vergence cannot do this, unless an extra degree of freedom is added to the system.

This kind of fixation is what we call stabilization. It is psycho-physically established that the vergence and pursuit motions in amplitude and speed yield the same characteristics.

We shall below return to the details of our eye module design.

15.5   Neck movements

As mentioned earlier, the major difference between an eye movement and a neck movement is that with eyes rotating about the optical centers, the eye movement will not change the geometry in the image, while the neck movement will. Hence parallax information can be extracted by neck movements only.

We have assumed that the number of degrees of freedom in the neck should be three. In mammals, with flexible spline mechanism, the situation is actually a bit more complicated. However, we have judged this construction as too elaborate to model and concentrated on implementing the qualities of the human neck that are most important for vision, namely the movements at the top of the spline. Hence we have contended ourselves with two degrees of freedom here, pan and tilt, much in line with what other systems have. However, it is important to note that our particular design allows eccentric motions with a constant geometry and hence a proper study of depth from motion; at least to some extent.

Neck movements should be studied from two different viewpoints. Fixation can be done on moving objects, or it can be done by a moving subject. In both cases, neck movements are very essential in the process of fixation. In the former case the movements help the observer to keep the angle of vergence reasonably close to symmetry, and in the latter case, the neck movements make it possible to continue fixation on an object although the observer is moving. In other words, fixation *in general* is meaningless without involving neck movements.

The participation of neck movements in the process of tracking is another story. Here the problem is how and when the neck movements cooperate with the eye tracking. Some simple ad hoc model can be used, e.g. by inspiration from some psychological experiments. The problem is, however, that the coordination of neck and eye movements seems to be much more complicated than the solution that intuition and experimental psychology can offer today.

In a tracking experiment we have done–and we will explain it later in this article–the neck movements are weighted by the amount of eye rotations. The more asymmetric vergence, the higher the neck involvement. The approach helps to keep a reasonably symmetric vergence on the dynamic fixation point. However, as the case is with eyes, there are very complicated behavioral patterns in the saccadic movements of the neck which seem to be influenced directly by attentive mechanisms, rather than eye movements. It is beyond the scope of this article to discuss this issue further and in fact, the literature indicates that there is still much to be discovered about the complicated cooperation of eye and neck movements.

15.6   Detailed technical description

We shall now describe the eye module, the base-line rig and the neck module of the KTH-head in more detail.

15.6.1   Eye module–detailed description

Each eye module is built of two ⊔-formed forks–an inner tilt fork and an outer pan fork. The inner fork rotates vertically in the outer one. The motor which has the task of rotating the inner fork is fixed to the outer fork via a harmonic drive. A sliding tray, the compensation mechanism for the lens center, is attached to the inner fork's lower plate. the camera is attached to this very sliding mechanism and the lens is screwed to the C-mount standard ring of the camera while its body is resting on the slide plate.

The outer ⊔-formed fork is coupled to the local pan motor via another harmonic drive gear-box. The practical meaning of this construction is that doing an asymmetric vergence towards a point in the scene, results in different tilt angles in the two eyes because of the dependency of the tilt angle on the pan angle. This is not the case in the human eyes. In human eyes the rectus muscles are in charge of tilt and pan movements of the eye. There are four of these: the superior and inferior rectus muscles take care of the tilt movement and the lateral and medial rectus muscles handle the pan movement. The difference appears in the non-rigid construction of the muscles.

The general impression one gets from the mechanics of human eyes is that everything is put together so that the least distortion and irregularity is achieved. The mechanism is simply helping the image acquisition system in such a way that the relations between the images are as simple as possible.

The muscle construction allows the tilt and pan movements to be independent of each other and consequently whatever the pan angle is, the two eyes will have the same angle when fixating at a point in the scene, even if the vergence is done asymmetrically. It can be shown that implementation of such a mechanism is impossible with a rigid construction.

Having bulky zoom lenses, as the case is here, a rigid construction is however to prefer; instead the tilt angle of the following eye should be regulated to adjust itself with the dominant eye. In human eyes, the epipolar line going through the center of the image, is rotated to the horizontal position by the cyclotorsion movement of the eye.

The last part in the eye module to describe here is the compensation mechanism. As mentioned earlier this mechanism is a sliding square table which is attached to the inner ⊔-formed fork and slides forward and backward by means of a 4-phase stepper motor rotating a bolt along a screw. The motor is equipped with a linear gear-box which reduces the input axis with a factor of 1:20. The output axis of the gear-box is not a shaft but a bolt. The two ends of the screw is fixed at the two sides of the sliding plate. The motor is attached to the gear-box and the gear-box is attached to the inner fork, i.e. when the motor rotates, the output of the gear-box, the bolt, is rotated and the fixed screw moves linearly forward and backward, depending on the direction of the rotation. A linear slide serves as the device filling the gap between the gear-box and the table; it simply guarantees an accurate linear translation.

The motors for pan and tilt are 5-phase stepper motors stepping 1000 half-steps/revolution. A Harmonic Drive on the motor shaft reduces this resolution with a ratio of 1:50. The final resolution will consequently be 50,000 steps/revolution. The 5-phase stepper motors are practically resonance-free and have a good holding torque. The ones used here can theoretically be run at a step rate of 100 KHz (6000 rpm in half-step mode).

An important issue in eye movements is the stability and speed of the module. When standing still the module should not vibrate until the next saccade is ordered and a saccadic movement should be quite fast in a real-time system.

The cooperative nature of the rotations of the two eyes requires partly dependent movements and at the same time independent ones. In order to make this point more clear, eye movements in a human being can be used as an example. The eyes always move together, even if they are closed. From our experimental studies for implementing vergence, it becomes clear that this cooperative action saves a lot of time for verging the eyes.

The eye movements should yet be independent since either eye could have the task of focusing at the point of interest. This is simply because the feature which is interesting in the image from one eye might not exist in the image of the other eye.

The result of this discussion is that eye modules should be constructed so that they are mechanically independent in their motion strategy, while the control system keeps them in a cooperative relation to each other.

The discussion above led to the design of the eye units as independent modules which are mechanically independent of one another, but at the same time coordinated and synchronized by the control system. As pointed out earlier, there has been other solutions implemented on earlier heads made by different research groups, where a coordination of the eyes is done by means of rigid mechanical links or cams. This was possible because a symmetric vergence was accepted. The approach may be regarded as a simplified model of head/eye movements, but in fact the resulting systems behave neither like a head nor like a pair of eyes. At the same time, certain head constructions have involved independent eye mechanisms, but still use them with symmetrical vergence. These constructions often do not employ an integrated neck module.

The major argument here is that eye and neck movements realize two basically different tasks. The eye movements, which in turn are divided into different classes of movements depending on the amplitude and the speed of rotation, could be thought of as changes based on the scope of the present image; i.e. the observed part of the scene at the time. Small saccades, tremors, larger correction saccades following vergence movements, etc. are all following this scheme. Neck movements however, are the ones which serve as an action of widening the bandwidth of these motions. Following fast moving objects, the study of depth from motion, etc. can be classified as tasks of these movements.

The idea here is not to divide each class of motion in a dependent category and beforehand state their tasks. It is simply to establish the differences in head and eye movements and avoid the confusion.

The details of the mechanical structure of an eye module is shown in Figure 15.3 (b).

Motors for pan and tilt have a resolution of 50,000 steps/revolution. That is, one pulse on each motor drive results in a rotation of 0.0072 degree or 0.000126 radian ($<$ 26 arc seconds); this is a resolution well suited to the resolution of a good CCD chip and a large focal length, the worst case scenario with present facilities.

When trying to find the zero disparity points in an image, it is very important to be able to do very small eye movements which are even smaller than a pixel. The repeatability and high resolution of the eye movements in the KTH-head allows such delicate rotations. These fine displacements are still over 10 times larger than the ones in a human eye, but so is the case even in a comparison of pixels in a CCD camera and cones in human eye's fovea centralis.

The maximal speed of each rotation in the KTH-head's eyes is dimensioned to be half the speed of human saccades (assuming medium heavy lenses), which is good enough if the computational bottlenecks are taken into account.

Since the head as such is deliberately designed so that different lenses can simply be mounted on it, lenses are not discussed here.

The connection of eye modules to the base-line cage is implemented by passing the base-line rail through the pan gear-box housing. The module is also coupled to the base-line ballscrew by means of the bolt fastened to the same part of the body of the eye module.

15.6.2 The base-line cage

Changing the base-line is a special feature of the head. Unlike animals, our head can change the length of the base-line. This is an experimental feature of the head and is designed to allow a flexible approach to the study of stereopsis.

It is important to point out that the rotation of the eyes about the lens centers guarantees that changing the length of the base-line does not generate any side effects. The base-line is defined as the line between the two lens centers. If the eyes do not rotate about their lens centers, their rotations result in a foreshortening or prolonging of the base-line.

The rail bars together with the supporting bars and side plates build a cage which joints the neck to the eye modules. The eye modules can slide along this cage symmetrically. The base-line motor, which has two shafts each connected to a high precision ballscrew, one left-threaded and the other right-threaded, is located in the middle and in front of the cage. Since the screws coupled to the motor are threaded in different directions, any rotation of the motor shaft results in a displacement of one eye module to the right while the other eye module is displaced to the left and vice versa (symmetric displacement of the eye modules.)

There are two more plates–small arms–attached to the two supporting bars. These plates are located in the middle part of the supporting bars so that they surround the tilt motor of the neck part in between. One arm is freely rotating with a ball-screw mounted in the neck part; the other one is coupled to the output shaft of the Harmonic Drive of the neck tilt.

In order to compensate for some of the amount of the torque occurring, when tilting the bulky cage, balance weights can be mounted on the rear part of the cage arms. The speed of 10 KHz, presently applied to the neck, does not require such balance weights, but higher speeds or faster acceleration ramps require them. All other degrees of freedom are tested successfully at much higher speeds. The system is however, as mentioned before, designed for half the speed of human saccades (about 25 KHz pulse frequency) and in the long run, the system is subject to mechanical damages at higher speeds.

The base-line cage can contain several modules like a third eye module for peripheral vision or ear modules and similar experimental accessories. The stroke length of the base-line is min 150 mm and max 400 mm.

The wider the base-line, the larger the dynamic forces. Although the torque variations appearing in the tilt movement could be compensated for by adding weights, these additional weights cannot be a solution for dynamic forces generated by the pan movement of the neck. The use of Harmonic Drives does, however, decrease the effect of dynamic forces drastically. Nevertheless small oscillations are inevitable.

15.6.3   The neck mechanism

The neck module in the KTH-head gives two degrees of freedom. These two rotations are the tilt and the pan movements of the neck. The tilt rotation mechanism lies on the pan plate which in turn is coupled to the pan motor shaft. The pan motor, in its turn, is hidden in the base of the head. Compared to the motors in the eye modules and the base-line motor, the neck motors have a larger torque to deal with. Therefore they are larger and cannot accelerate as fast as the other motors. The reduction on the motor shaft is however still the same (1:50), although the Harmonic Drive can manage a larger load.

In this context it should be mentioned that the same phenomenon that appeared in the structure of the eye modules comes up even here. Depending on whether the pan motor is attached to the output shaft from the tilt gear-box or vice versa, the coordination of the system will be different; again because the absolute position of the motor on the top is dependent on the degree of rotation of the motor in the bottom. No matter which of pan or tilt movements is at the top, the other one becomes dependent of it. This drawback of the rigid mechanism of the neck, should be compensated for by the control system, as the case was with the eye modules.

The rotation axis of the pan mechanism goes through the rotation axis of the tilt motor. The cross point is then the origin of the coordinate system of the head.

The tilt motor shaft is directly jointed to an emergency break which is quickly activated at power down, that is, a sudden power failure does not result in any damages to the head.

The neck module per se is very stable, but the burden of the head makes its oscillation amplitudes larger than the ones in the eyes. As mentioned above, the wider the base-line is, the larger the oscillations become. The main source of oscillations can however be a poor and unstable/undamped base ("shoulders"). The base plate of the head is therefore equipped with proper fixture holes for stable mounting of the head on a rigid base. We believe that the oscillations still are much less than the ones in the previous head constructions, and in practice negligible.

15.7   Performance

In order to give a better idea about what performance the KTH-head has, some data about it are briefly presented here:

– General data:
  Total number of motors: 15
  Total number of degrees of freedom: 13
  Number of mechanical degrees of freedoms in each eye: 2
  Number of optical degrees of freedom in each eye: 3
  Number of degrees of freedom in neck: 2
  Number of degrees of freedom in the base-line: 1
  Top speed on rotational axes (when all motors run in parallel): $180°/s$
  Resolution on eye and neck axes: 0.0072°
  Resolution on the base-line: 20 μm
  Repeatability on mechanical axes: virtually perfect
  Min/max length of the base-line: 150/400 mm
  Min/max focal length: 12.5/75 mm
  Weight including the neck module: about 15 Kg
  Weight excluding the neck module: about 7Kg

– Motors:
  7 5-phase stepper motors on the mechanical axes
  2 4-phase stepper motors for keeping the optical center in place
  6 small 4-phase stepper motors on optical axes, 3 on each lens

The soft and hard configuration of the control scheme is discussed later in a section of its own.

At least in computer vision, the performance of an algorithm or a head-eye system, do not mean much, if they are not experimentally tested. Therefore, we will also briefly demonstrate some experimental results achieved by the system.

15.8   The control system

The control system of the KTH-head is based on the ability to change all mechanical and optical parameters in parallel. A flexible implementation of such a specification requires soft and hard configurations which can achieve such process parallelism.

The KTH-head intensively uses stepper motors. Although the older generations of stepper motors were bulky and offered resolutions easily matched by DC motors, the newer generation of stepper motors, especially the 5-phase stepper motors used in this construction are very accurate machines, yielding high resolution, speed and torque. Stepper motors require a device called "indexer". This device takes as input some kind of information about number of steps, speed and alike; as output it generates a pulse train which is the input to the motor drives.

For sake of high flexibility in the control of motors, the indexer is implemented in software. That is, the pulse train is generated by the software through writing ones and zeros to a port. This way the process of ramping up and down in motors has become very flexible and the application programmer is quite free to implement a variety of control schemes for the motors.

Running motors in parallel is not the only satisfactory base for a flexible control scheme. The interprocess communication of the motor processes is even more important. Quite often, the motors need to synchronize themselves with each other. They also need to keep a direct and continuous communication with the image processing units when run in short closed loops, e.g. when image stabilization/gaze holding is active.

Since the primary ocular processes implemented on the head, i.e. the fundamental processes defining the reactive behavior of the system, are also following this parallel scheme, a common solution in form of a network of transputers, with each node dedicated to a specific task, seems to be a proper approach.

The transputer node dedicated to the control of optical and mechanical parameters contains one process for each motor. All these processes on the transputer are run in pseudo-parallel. Although this is not a pure parallelism, the processes are in effect run in real parallelism. The reason is simply the fact that all processes write to common pools whose values are read by another sampling process. The sampling process samples the new value of the pool at a rate faster than the step length of the fastest motor, so that the system doesn't experience the pseudo-parallelism as an obstacle. Besides, the transputer's computational power is used optimally.

The indexer implemented in software can take as command a target position or speed variations on each axis. The former is appropriate when a saccade is needed, the latter when smooth pursuit or tracking is under execution. In the experiments presented later in this article, the vergence process takes advantage of all these kinds of movements.

Not all movements of the head are independent. On the contrary, they often have very complicated connections and dependencies. For example the eye movements, although asymmetric, are either converging or diverging; neck movements keep the fixation valid, unless a sudden saccade in the two eyes is also accompanying it. The complexity of these connections, which in our system are implemented as separate processes, requires a total mechanical independence in each module[4].

The observant reader has certainly noticed that we have tried to isolate the motor control system from the closed loop control of the motoric system. The former, i.e. how the motor services are carried out, is discussed here; the section describing the implemented primary ocular processes demonstrates examples on the latter.

Figure 15.4 illustrates how the motoric processes are configured. As depicted in the figure, each process is linked to a command distribution process from one side, and a sampling process which gathers their pulse output from the other side. The processes share in a set of common data like ramp tables, global positions and interrupt flags.

Two movements in the system are so dependent of others that they simply cannot be considered as degrees of freedom. These movements are the ones compensating for the loci of focal points. The position of the focal points in a lens is a function of both focal length and accommodation distance[5]. This means that the compensation mechanism is activated as soon as the focal length or the focusing distance are ordered to be changed in the lenses.

---

[4] The approach chosen by evolution. From a physiological point of view, eyes are totally independent of each other, although the oculomotor system binds them together so that a specific behavior in the movements is generated.

[5] It is actually a function of aperture size too. The contribution from this variable is however small and we have not taken it into account when calibrating the system.

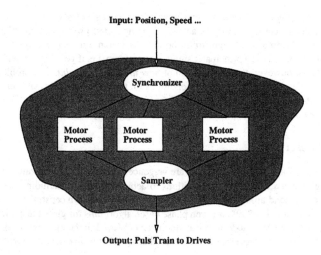

Output: Puls Train to Drives

Fig. 15.4. The scheme of motor processes in the KTH-head.

15.9 Basic experiments

In the previous sections we have from an argument about the issues of computational vision and about some properties of mammalian vision, proposed a design of a head-eye system. We have also described a system developed and implemented on the basis of these principles. The goal of this work is, of course, to provide a research tool useful in our attempt to study seeing systems in the sense outlined in our introduction. The work reported here is only a first step in that direction, but to substantiate the claim that we have such a system, we need to show by experiments that our system has the properties we have listed. Single elementary functionalities, the primary processes, are indirectly implied by the performance data. Still we shall describe how they can be obtained in real-time under computer control. As we pointed out in Section 15.2, a more important aspect is that these primary processes indeed can be integrated into more complex behaviors. We shall also describe some experiments demonstrating this.

Obviously, the very fact that these experiments occur in real-time implies that a written description never will offer the full flavor of them; they have to be seen. However, by presenting some recorded images, we shall illustrate what goes on. In particular, these experiments indicate that with our deliberate design with a high number of degrees of freedom, we can indeed obtain a systematic build-up of behaviors, by integrating primary processes. Moreover, they demonstrate that we can concentrate on the visual behaviors and their related eye and head movements, without various considerations about what kind of movements the mechanical system can achieve, since our system is directly designed to be a head-eye system with a large number of independent degrees of freedom. We feel that the control problems therefore are posed in their right context and that already our first experiments show that they can be handled.

Before starting a description of the experiments, it is in its place to say some words about kinematics and inverse kinematics in head-eye systems. Many researchers have had the problem of complex kinematics in their head-eye systems which actually is caused by the fact that they do not rotate the cameras about the right point. Moreover, they have not used the neck movement (if present) in the way described above, i.e. as a cyclopean motion. We have no complex kinematics in our system and whenever we choose a point in the world to look at, the positioning error is of subpixel magnitude, something that has been shown during extensive tests.

The problem of inverse kinematics is actually a complicated problem in robotics. However, manipulative robotics is totally different from robotics for vision. Finding the proper trajectory of the movement of a *robot arm* is a principally significant problem, i.e. if you want to avoid collisions. In head-eye systems, however, the question is not *how to reach the point*. The problem is, as we have underlined earlier, that there are proper conditions for the vergence angles and certain physical constraints on the body/neck, and we want to find the best *view* out of the situation. This is what our system's eye-neck coordination (a part of the stabilization process) does.

15.9.1 Brief description

The basic, or primary functions of our system are vergence, focusing and stabilization/tracking

We have implemented a number of algorithms to achieve vergence, including methods based on correlation along a band around the epipolar line and the well-known cepstral method proposed by Yeshurun and Schwartz [YeS89] and even phase based algorithms for global disparity detection. The spatial and phase based methods all run at video rate (25 Hz). For the moment we shall however not elaborate on the vergence process, since we shall discuss it below in connection with integration of processes.

A stabilization process has also been implemented. It keeps track of the foveal image of each eye (camera) and generates motor commands so that the central part of the image, i.e. the region of interest, is stable. The method is again a minimization using steepest descent, but the process also involves an $\alpha - \beta$ tracker (see [DeF90]) for motion prediction. The result is a very smooth real-time stabilization, which in the present implementation on a single transputer manages speeds up to $60°/s$ for arbitrary patterns. The sequence of images in Figure 15.5 depicts the results from an earlier version of the stabilization process, when the hand in the middle of the frame moves around and in depth. Figure 15.6 is the same algorithm optimized for tracking. Here, the vision part of the algorithm is less interesting (in this example it is tracking the brightest part of the image), instead, the control algorithm and image acquisition system have been dedicated more time for faster pursuit. The algorithm is executed at double video rate using both interlace fields (50 Hz).

The final basic mechanism is that of focusing. This process keeps the foveal image of the eye in question in focus, that is, it focuses in the vicinity of the center of the image. This process has also been implemented in several ways. The best one seems to be one that maximizes gradient magnitude, a finding stemming from Tenenbaum [Ten70], often called the tenengrad method. The focusing process is described in detail in [Hor91]. Figure 15.7 illustrates the effect of stabilization on focusing on moving objects. The smooth evaluation function is a necessity for making focusing on moving objects possible. Later in this section, we will also briefly discuss the benefits of using variance methods for blur detection and the application of the method in our later experiments. For details about different criteria for sharpness see [Kkv87].

Another experiment showing the functionality of our system concerns dynamic event detection. The system fixates some point in a static environment when a moving object appears peripherally. The system then instantly fixates the point where the change is detected, presently defined as the centroid of the moving area. Figure 15.8 shows an example.

The final experiments we shall describe demonstrate in a more conclusive way that our head-eye system actually meets the design criteria we stated in the beginning of the article. These experiments, accounted for in detail in [PUE92a], aim at showing that cooperating independent primary processes can achieve complex behaviors in a robust way. The task is to do dynamic vergence with support from the focusing and stabilization processes. The process configuration is given in Figure 15.9.

Each eye here has a loop of primary processes of its own. The two eyes are then coordinated by the vergence process. The vergence process sends the proper motoric commands.

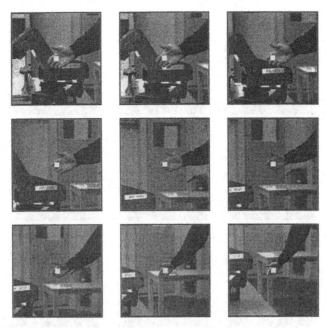

Fig. 15.5. This sequence depicts the stabilization on the hand in the center of the image. The order is from top left to bottom right. The hand is moving in the scene and the camera is adjusted to keep the image of the hand in the center (marked with a □). The images are real images redirected from the output of the frame-grabber. Every 25th image is shown here. The speed of the hand here is about 20°/s. The images here are from the right camera; the same process is executed on the other camera's images and both are coordinated by the vergence process. In newer experiments evolving a common representation of both left and right images the speed has been incresed to up to 60°/s.

In Figures 15.10, 15.11, 15.12 and 15.13, it is shown how the process without and with accommodation obtains one or a multitude of matches in trying to verge on a repetitive wallpaper pattern. Figures 15.15 and 15.16 illustrate the same thing on an indoor scene with more structural complexity, which makes the problem even easier. Of course, a repetitive pattern of high enough frequency will cause false minima, but even then the accommodation process decreases ambiguity considerably, see Figure 15.14. The reader is referred to [PUE92a] for more details.

In some recent experiments, we have changed the process configuration so that a simple common representation of the images from left and right cameras is the input to the stabilization process. That is the stabilization is performed on the superimposed images and the symmetric disparity with respect to the common line of sight of the two cameras in the superimposed image and the variance of the gray-levels are used to feed the accommodation and vergence processes. The superiority of this approach is in its virtue that the variance is computed along with disparity detection and the integration of blur and disparity cues is performed implicitly.

This integration can be executed very quickly and therefore the fixation procedure can be stable even under dynamic conditions. Figure 15.17 depicts the sequence of the dynamic stabilization on the superimposed images recorded under operation and in real time. The details of this work can be found in [PUE92b; PUE92c].

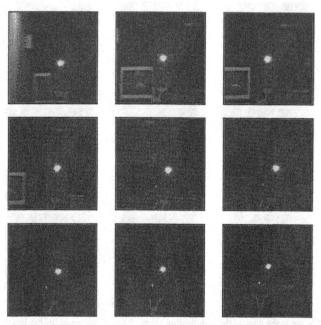

Fig. 15.6. Every fifth image from the tracking sequence is depicted here. The sequence is from top left to bottom right. The images here contain both interlaces, but in order to increase the flow rate of the images, the image processing is done on each interlace field separately. The deviation from the center of images is because of sudden acceleration/decelerations. This deviation is compensated for by the α-β tracker. The image processing is simple and just looks for the flash lamp.

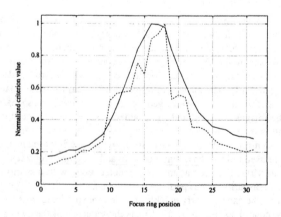

Fig. 15.7. The evaluation function for focusing without stabilization (dashed curve), and with stabilization (solid curve). Stabilization results in a smooth curve free from local minima.

Fig. 15.8. The motion of the person walking in the scene results in a temporal difference. The white band is the positive difference and the black one is the negative difference.

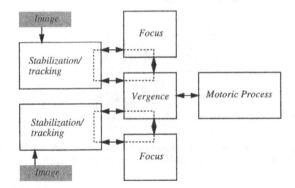

Fig. 15.9. Process configuration in the implementation without common representation of leeft and right images. The meeting point for the processes dedicated to the left and the right eye (the two rings), is the vergence process.

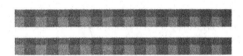

Fig. 15.10. The repetitive pattern without the band limits of accommodation. The band (top). The pattern square superimposed on the best match (bottom). The match here, represented by the smallest value, is false.

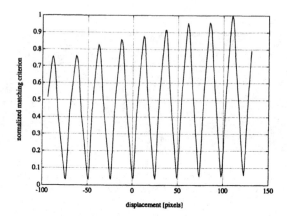

Fig. 15.11. The evaluation function for matching. A good match here is represented by minimum points. As shown in the figure, without the dynamic constraints from the focusing process, there are multiple solutions to the problem. The obtained match here is false.

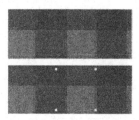

Fig. 15.12. The repetitive pattern with the band limits provided by accommodation. The band (top). The pattern square superimposed on the best match (bottom); here the correct match.

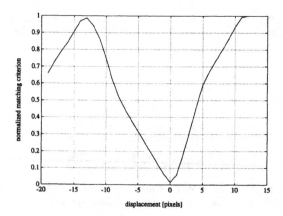

Fig. 15.13. The evaluation function for matching. A good match here is represented by the minimum point. To be compared with the curve not using the accommodation limits.

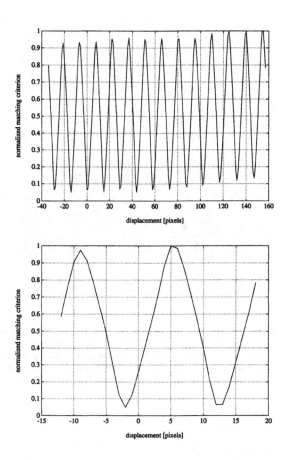

Fig. 15.14. The same experiment with a higher frequency. The graph at the top illustrates the result of matching along the epipolar band. The one at the bottom illustrates the area confirmed by accommodation. The one at the bottom, in spite of the two minima, detects a correct fixation. However, a potential source of false matching (the second minimum) exists. The focal parameters are by no means ideal here; the angle of view is larger than 20°. The displacement is defined to be 0 at the center of the image. Note, however, that the cameras are at different positions in the two cases. Hence the displacements do not directly relate.

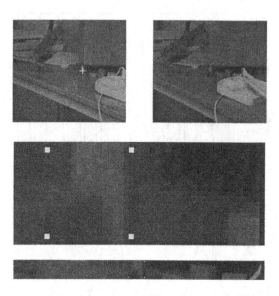

Fig. 15.15. The left and right images of a stereo pair (top). The point marked with a cross (+) in the left image, is the next fixation point. The epipolar band and the superimposed correlation square on it are also shown. In this case, the matching process alone could manage the fixation problem (the stripe in the bottom). The cooperation has however, shrunk the search area drastically (the stripe in the middle).

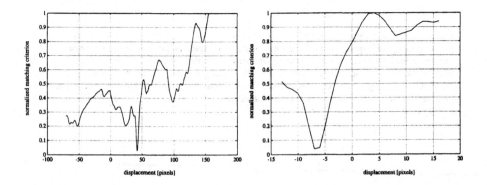

Fig. 15.16. The graph to the left illustrates the result of matching along the epipolar band. The one to the right illustrates the interval of search suggested by the accommodation process. See also the note in the caption of Figure 15.14.

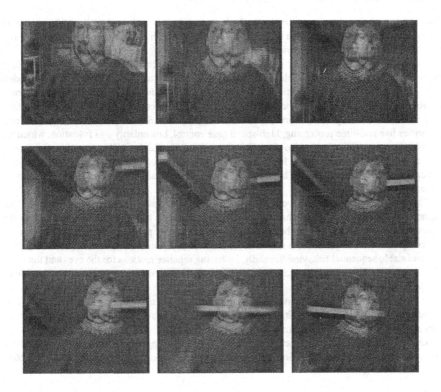

Fig. 15.17. A sequence of dynamic tracking of a moving person in real-time. The movement is at a speed of about $60°/s$. Every fifth frame is depicted from top-left to bottom-right. The left and right images here are superimposed and the tracking is performed on this common representation. The vergence process running in parallel, brings the images of the object of interest in the middle to gether. In order to see the symmetries, the fusing vergence movements are disabled. The background can be used as the speed reference. When tracking is precisely performed, the interlace effect is totally removed from the center of the images and is maximal in the rest of the images.

15.10   Conclusion and discussion

We have argued that if we consider the goal of research in computational vision as being that of understanding seeing systems, then we need to have a system with anthropomorphic features, or one that at least can imitate aspects of mammalian vision.

To obtain computational models of systems that even in a weak sense can achieve the combination of performance and flexibility displayed by the human visual system, we suggest that these models should operate under similar conditions. Among other things this implies that the system should be able to look at the world rather than at prerecorded images, that they should involve both purposiveness and integration of primary involuntary processes. Such behaviors are obtained from properties like real-time processing, high-speed gaze control, binocularity and foveation, which we therefore propose to include. Also other researcher have argued for such properties, foremostly Ballard [Bal91]. However we have furthered the analysis to let it determine our design of a head-eye system in a more complete way. This analysis as well as the final design of our system have been presented in the article.

Our design has two features worth underlining. First, the system has a large number of independent degrees of freedom, mechanical and optical. This is in line with the assumption that the behaviors we want to achieve can be modeled as sets of primary processes working in cooperation and competition and in parallel. As shown in some of our experiments, this does not exclude purposive and stable sequential behavior. Secondly, by having separate motions for the eyes and the neck, and in particular, by rotating the eyes about the lens centers, we avoid a number of control problems that potentially could arise.

It can be mentioned that the system is modular and its design allows removal as well as addition of features. For instance, since suitable foveated sensors are unavailable, one can add one or two cameras to acquire peripheral information.

Our current work aims at demonstrating how our design supports the study and development of a number of visual behaviors. Some early and later results can be found in [PUE92a; PUE92b; PUE92c].

References

[AlK85] M. A. Ali, M. A. Klyne. *Vision in Vertebrates*, Plenum Press, New York, 1985

[Alo90] Y. Aloimonos. *Purposive and Qualitative Active Vision*, Proc. DARPA Image Understanding Workshop, 1990, 816-828

[Baj85] R. Bajcsy. *Active Perception vs Passive Perception*, Proc. 3rd IEEE Workshop on Computer Vision, Bellair, 1985

[Bal89] D. H. Ballard. *Animate Vision*, Proc. 11th IJCAI, Detroit, 1989

[Bal91] D. H. Ballard. *Animate Vision*, Artificial Intelligence, 48, 1991, 57-86

[Car88] R. H. S. Carpenter. *Movements of the Eyes*, Pion Press, London, 1988

[DeF90] R. Deriche, O. Faugeras. *Tracking Line Segments*, Proc. 1st ECCV, Antibes, 1988 259-268

[Hor91] A. Horii. *Cooperative Focusing and Stabilization in the KTH-Head*, KTH, Stockholm, 1991 (to appear)

[Kkv87] E. P. Krotkov. *Exploratory Visual Sensing for Determining Spatial Layout with an Agile Stereo System*, University of Pennsylvania, PhD thesis, 1987

[PUE92a] K. Pahlavan, T. Uhlin, J. O. Eklundh. *Integrating Primary Ocular Processes*, Proc. 2nd ECCV, 1992

[PUE92b] K. Pahlavan, T. Uhlin, J. O. Eklundh. *Dynamic Fixation*, 4th ICCV, submitted, 1992

[PUE92c] K. Pahlavan, T. Uhlin, J. O. Eklundh. *Fixation by Active Accommodation*, Proc. SPIE 92, Boston, 1992

[Ten70] J. M. Tenenbaum. *Accommodation in Computer Vision*, Stanford University, PhD thesis, 1970

[Ybs67] A. Yarbus. *Eye Movements and Vision*, Plenum Press, New York, 1967

[YeS89] Y. Yeshurun, E. L. Schwartz. *Columnar Image Architecture: a Fast Algorithm for Binocular Stereo Segmentation*, IEEE Trans. PAMI, Vol. 11, Jul 1989, 759-767

16. Hierarchical Control of a Binocular Camera Head

James L. Crowley, Philippe Bobet and Mouafak Mesrabi
LIFIA, INPG

16.1 Introduction

This chapter describes a hierarchical control system for an active binocular camera head. The first section reviews the background of work on active camera control. The second section reviews principles of layered control. In the third chapter, a head level controller is presented. This head level controller permits the other modules to obtain an estimate of the Cartesian position of the head and of its gaze point, and to control the head position and gaze position. The final section discusses the reflexive control of aperture, focus and vergence.

16.1.1 Objectives of the Camera Control Unit

The development of binocular camera heads and an integrated vision system has opened a line of cooperation between the scientific communities of biological vision, machine vision and robotics. This chapter is concerned with a robotics problem posed by such devices: How to organize the control architecture. We will argue for a layered control architecture in which a "gaze point" may be commanded by an external process or driven by simple measurements of information from the scene.

The camera control unit serves as the system interface and the device controller for the active binocular head. The camera control unit is designed to be a portable interface. It provides an ability to estimate and command the state of a binocular sensor in a manner which is relatively independent to the actual geometry of the sensor.

The camera control unit involves control loops at three levels. At the lowest level are classical motor controllers that are on board the robot. At the second level is a device level controller that estimates the position of the head and the position of a 3D gaze point defined as the intersection of the optical axes. At the third level are perceptual actions such as tracking along a contour or keeping an object at the gaze point.

This reports begins by describing some of the background behind the SAVA-II camera control unit. It then describes the head-level device controller. The final section describes reflexive control of aperture, focus and convergence.

16.1.2 Binocular Heads within the VAP Consortium

In the initial year of the VAP project, it became apparent that actively controlled binocular stereo provided low-complexity and inexpensive algorithms for separating figure from ground and for determining the 3D position and orientation of objects in a scene. As a consequence, the consortium has developed four different approaches actively acquiring binocular stereo data.

The group at KTH has desitgned and built a table top binocular head. The KTH head optimizes speed and precision, at the cost of weight. It also contains a number of redundant degrees of freedom in order to permit a variety of experiments in active stereo.

The group at AUC has constructed a head which is similar to that of KTH. The AUC head is mounted on a ceiling mounted X-Y table able to move around an area of about 3 meters by 3 meters. At University of Linkoping, a Stardent graphics computer was used to build a simulator for a pair of cameras mounted on a robot arm in a simulated room. The Linkoping group can control their simulated binocular camera system for experiments in active camera control. At LIFIA we have constructed a binocular head to be mounted on a six-axis robot manipulator which itself mounted on a mobile robot. Because of weight restrictions, the LIFIA head design minimizes weight and maximizes precision and simplicity at the expense of speed.

The device level of the software system described in this report is designed to sufficiently general to apply to any of the VAP heads. In order to explain this system we need to contrast the two most extreme designs from LIFIA and KTH.

The KTH head has a rotating base, giving a pan angle which we will call γ_h. This is followed by a tilt platform, whose angle we will call β_h. The two cameras are mounted on a screw mechanism which permits their base line to be modified. Each camera is then mounted on a small platform that may tilt (β_l and β_r) and that may pan (α_l and α_r). Each camera has controllable zoom, focus and aperture.

The LIFIA head is mounted on a "neck" which is in fact a six axis AID arm mounted on the Robotsoft mobile platform. The neck permits us to command the position and orientation of the camera in coordinates which are relative to the position of the mobile robot. The mobile robot provides us with an estimate of its position and orientation in a world coordinate system. The arm is mounted at a point which is midway between the power wheels of the vehicle. This point serves as the origin for both the vehicle and arm coordinate systems.

Fig. 16.1 Configuration of the Six Axes Manipulator which serves as a neck for the LIFIA Head.

The configuration of joints in the AID arm is described is shown in figure 16.1. The joint ϕ_4 corresponds to the pan axis γ_h on the KTH head. The axis ϕ_5 corresponds to the tilt axis, β_h, on the KTH head. The axis ϕ_6 is redundant with vergence angles of the two cameras. We also call this axis α_h. An important problem is controlling these redundant axes.

The mobile robot vehicle controller, the arm Cartesian controller and the six axes of the LIFIA head are all accessible to the SAVA system from a SAVA Camera Control module.

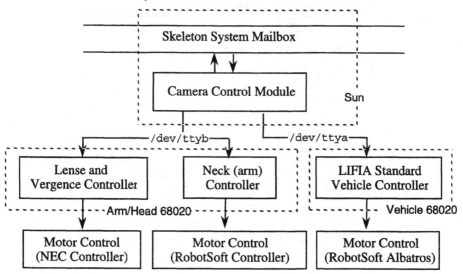

Fig. 16.2 The SAVA camera control process communicates with an on-board vehicle controller and an on-board controller for the arm and binocular head.

16.2 Principles of Layered Control

The control system for a robotic device may be organised as layers of control loops, where layers are defined by the abstraction of the control data and the cycles time of the control loop [12]. Robot vehicles and robot arms often have layers of control loops at four level: <u>Motor</u>, <u>Device</u>, <u>Actions</u>, and <u>Tasks</u>. We have found that such a layered control architecture maps quite naturally onto the control of a binocular head, as shown in figure 16.3.

<u>Motor Level</u> The motor level is concerned with control parameters defined by the motor shaft. Sensor (typically optical encoders) provide information in terms of position and angular speed of the motor. Commands are generated in terms of motor position, speed and perhaps acceleration. Typical control cycle times for robotics are on the order of 1 to 10 milliseconds. The motor level is typically controlled by a form of PID Controller.

<u>Device Level</u> The device level is concerned with the geometric and dynamic state of the entire device. Control cycles for robotics applications are typically on the order of 10 to 100 milliseconds.

For device independence, it is useful to design a controller for an idealized "virtual device". Our virtual head is based on controlling an "gaze-point" defined as the intersection of the optical axes. The mapping from the virtual head to a particular mechanical head is performed by a translation layer between the device controller and the motor controllers. The device level also permits any of the axes of the virtual head to be directly controlled.

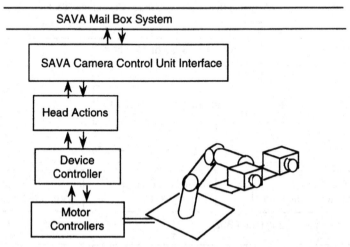

Fig. 16.3 A Layered Architecture for Control of a Binocular Head in the SAVA System.

Action Level: The action level concerns procedural control of the device state based on measurements taken from sensor signals. An action will drive the device through a sequence of states. The action level for a binocular head involves control of head motion and optical parameters. The action level often involves control cycles of 0.1 to 1.0 seconds.

Task Level: A task level controller has a goal expressed in terms of a symbolic state. The task level controller chooses actions to bring the device and the environment to the desired state. The selection of actions is based on a symbolic description of the preconditions and results of actions, as well as a description of the current state of the device and environment. This leads us to propose a control cycle composed of three phases: Evaluation (of state), Selection (of an action), and Execution (of an action).

16.3 "Head Level" Device Controller

This section describes the control system for a binocular head. It begins with the device level controller and then describes the virtual head. This is followed by the techniques for estimating the position of the gaze point with respect to the head.

16.3.1 Device Level Protocol

One of our design goals is that the head controller provide a general control protocol which can be easily transported to different mechanical configurations. For this reason, we have defined our device controller to command a "virtual head".

The head controller has four components: Protocol Interpretation (1), State Estimation (2), Command Generation (3) and Translation (4). These four components are illustrated in figure 16.4 and described in the following sections.

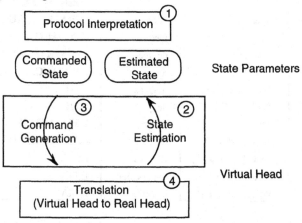

Fig 16.4 Components of the Binocular Head Device Controller

16.3.2 A Virtual Head

The virtual head is an idealized mechanical structure which should be general enough to map onto the kinematic structure of any of the physical heads. Any information which is specific to a particular head should be included in the "translator". The motor/encoder axes which may be driven, and their physical configuration is defined as a "composition" of the binocular heads used by the VAP consortium. The units of all measurements are translated from encoder counts to degrees (or meters) in the translator.

The commands for each of the axis of the virtual head are made available at the protocol level. In addition, this level offers the control of a head-centered "interest-point". This interest point is defined by the 3D position at which the optical axes of the two cameras intersect. This can pose a problem in the case of heads with independent tilt for each camera. In such a case, there is no guarantee that the optical axes will intersect. The interest point is then taken as the midpoint of the shortest line between the two optical axes.

Table 16.1 lists the axes for the virtual head, as well as their units. The set of heads on which an axis is present is listed on the left. The LIFIA head is mounted on a 6 axis arm. The last three axes of this arm form a wrist which may modeled as three rotations (γ_h, β_h, α_h) about a common point. These three rotations correspond to the last three axes of the arm (ϕ_4, ϕ_5, ϕ_6). For the moment let us consider that the first three axes serve simply to place this rotational center at a location (X_h, Y_h, Z_h). In the case of the KTH and AUC heads, the head is composed of a base, described by (X_h, Y_h, Z_h) at which their is a pan axis followed by a tilt axis β_h. The same model may be used for these two configurations if we see the KYH and AUC heads as having $\alpha_h = 0$.

Table 16.1 List of Axes in the KTH, LIFIA and AUC heads.

	Axis	Units Heads Used
	Focus	meters LIFIA, KTH, AUC
Aperture	degrees	LIFIA, KTH, AUC
Zoom	meters	KTH, AUC
Vergence	degrees	LIFIA, KTH, AUC
Baseline	meters	KTH, AUC
Tilt Left Camera	degrees	KTH
Tilt Right Camera	degrees	KTH
Pan (ϕ_4 or γ)	degrees	LIFIA, KTH, AUC
Tilt (ϕ_5 or β)	degrees	LIFIA, KTH, AUC
Axes $\phi_1, \phi_2, \phi_3, \phi_6$	degrees	LIFIA
Cartesian Position	meters, deg	LIFIA
Vehicle Position	meters, deg.	LIFIA

In order to accommodate any of these axes, a device table is defined. This device table is built up from a structure called an "Axis".

```
typedef struct  _axis
{    int enabled= FALSE;      /* Boolean: Is this axis physically present?   */
     int pos = 0;             /* Current position */
     int command = 0;         /* Current Desired position */
     int min;                 /* Minimum Value */
     int max;                 /* Maximum Value */
     int conversion_base;     /* for conversion to to/from encoder counts */
     int conversion_factor;   /* for conversion to to/from encoder counts */
     char units[16];          /* Units (Meters, degrees, etc) */
     char name [32];          /* Name of this axis */
} AXIS;
```

The device table is a dynamically allocated array of AXIS structures.

```
AXIS  *  head [21];
```

Each axis structure is dynamically allocated from an initialization file. For the version of the program which is down-loaded to the on-board M 68020 processor, this initialization is performed by a procedure. In either case, initialization defines which axes are present, their units, the initial value for that axis, the conversion factor from encoder counts, and their maximum and minimum values. Subsequent access to that axis may either be based on the index of the entry in the table, or by association with the axis name. Association is provided by the function get_axis_index:

```
int  index = get_axes_index(char * name)
```

This function matches the string "name" to the name file of each of the non-NULL entries in the head table. An exact match returns the index number. A failure to match returns -1.

A command to change the value of an axis will first determine the index of the axis to be changed, and then set the commanded value for that axis to the new value. Individual axes may be commanded by an absolute or incremental move using the commands which resemble those of the motor controller described above. An important difference is that all commands are in terms of the device's units (meters or degrees).

M[ove] <Axis Name> <Value>   /* Set the desired absolute axis value */
B <Axis Name> <Value>   /* Set the desired incremental value */
P <Axis Name>   /* Get the position of the axis */

Absolute and incremental moves change the variable "command" for the axis. At the end of each cycle, the translator scans the list of axes and updates the current position. Whenever the commanded position is different from the current position, the commanded value is transformed to encoder counts and a move is issued to the motor controller. A command to an axis which is not currently in the head table will trigger a negative acknowledgement.

16.3.3 Estimating the Position of the Gaze Point

The important state component for a binocular head is the gaze point, defined by the intersection of the optical axes. Knowledge of the state of the neck permits us to transform the gaze point to 3D coordinates whose origin is at the base of the head. A major role of the head controller is estimating the current gaze point (process 2 in figure 16.4) and controlling the current gaze point (process 3 in figure 16.4). We will start by defining the gaze point in polar coordinates whose origin is midpoint on the baseline between the two cameras. We will then develop the transformation to a fixed reference frame centered at the base of the head.

Let us derive formulas for determining the gaze point within an "eye centered" 2D polar coordinate system. This eye centered coordinate system has its origin mid-way between the optical centers of a pair of stereo cameras. Let us define the X axes as coincident with the baseline, and the Y axis as perpendicular to the baseline and in the plane defined by the optical axes. Let the separation of the cameras be a distance 2B so that the location of the optical centers are the defined as the points (B, 0) and (-B, 0). Furthermore, let the optical axes be located in the (X, Y) plane with angles of α_l and α_r.

The equation of the left optical axis in the plane defined by the base line and the optical axis is:

$$X \, Sin \, (\alpha_l) - Y \, Cos \, (\alpha_l) + B \, Sin(\alpha_l) = 0.$$

The right optical axis is described by: $X \, Sin \, (\pi - \alpha_r) - Y \, Cos \, (\pi - \alpha_r) - B \, Sin(\pi - \alpha_r) = 0.$

Since $\operatorname{Sin}(\pi - \alpha_r) = \operatorname{Sin}(\alpha_r)$ while $\operatorname{Cos}(\pi - \alpha_r) = -\operatorname{Cos}(\alpha_r)$, the left equation reduces to

$X \operatorname{Sin}(\alpha_r) + Y \operatorname{Cos}(\alpha_r) - B \operatorname{Sin}(\alpha_r) = 0$.

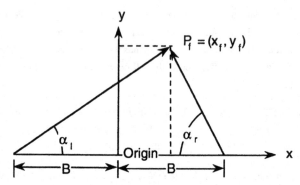

Fig. 16.5 The gaze point P is the intersection of the optical axes.

The position of the fixation point, defined by the intersection of the optical axes, can be calculated as the sum and difference of these two equations. This gives

$$X = B \, \frac{\operatorname{Cos}(\alpha_l)\operatorname{Sin}(\alpha_r) - \operatorname{Sin}(\alpha_l)\operatorname{Cos}(\alpha_r)}{\operatorname{Cos}(\alpha_l)\operatorname{Sin}(\alpha_r) + \operatorname{Sin}(\alpha_l)\operatorname{Cos}(\alpha_r)} = B \, \frac{\operatorname{Sin}(\alpha_l - \alpha_r)}{\operatorname{Sin}(\alpha_l + \alpha_r)} \qquad (1)$$

$$Y = \frac{2B\operatorname{Sin}(\alpha_l)\operatorname{Sin}(\alpha_r)}{\operatorname{Cos}(\alpha_l)\operatorname{Sin}(\alpha_r) + \operatorname{Sin}(\alpha_l)\operatorname{Cos}(\alpha_r)} = 2B \, \frac{\operatorname{Sin}(\alpha_l)\operatorname{Sin}(\alpha_r)}{\operatorname{Sin}(\alpha_l + \alpha_r)} \qquad (2)$$

In the case of symmetric vergence angles, $\alpha = \alpha_l = \alpha_r$ the two equations become:

$X = 0$

$$Y = \frac{2B\operatorname{Sin}(\alpha)}{2\operatorname{Cos}(\alpha)} = B \, \operatorname{Tan}(\alpha)$$

In polar coordinates, (D_C, α_C), the vergence angle to the gaze point can be expressed as

$D_C = \sqrt{X^2 + Y^2}$

$$\alpha_C = \operatorname{Tan}^{-1}\!\left(\frac{Y}{X}\right) = \operatorname{Tan}^{-1}\!\left(2 \, \frac{\operatorname{Sin}(\alpha_l)\operatorname{Sin}(\alpha_r)}{\operatorname{Sin}(\alpha_l - \alpha_r)}\right) \qquad (3)$$

Equation 3 describes a gaze point as if at the end of a telescopic stick which we can extend and pointed within a plane. We can solve for the position of the end of this stick in the scene using the position and orientation of the head and the state of the binocular head. The head "state" parameters on which the gaze point depend are the Distance (D_C), the azimuth gaze angle or pan (α_g) and the elevation gaze angle, or tilt (β_g). The gaze azimuth angle, α_g, is defined as the sum of the head pan α_h and a common vergence angle α_c.

$$\alpha_g = \alpha_h + \alpha_c$$

Both the polar and Cartesian forms of the gaze point are stored in a data structure that defines the head "state". The Cartesian values are computed from the polar values. Commanded values may be set by messages from other processes. The difference between a commanded value and a current position triggers the translator to call a device specific procedure to move the necessary real axes.

In order to estimate and control the fixation point in 3D, the six axis arm controller was extended by two additional axes. Axis number seven is the fixed offest (distance and direction) between the tool coordinates defined by the end of link number six and the head coordinates. Link number eight is defined by the common gaze angle, α_c, and the distance to the fixation point, D_c. Thus at each instant the head controller returns an estimate of the fixation point relative to the base of the siz axis manipulator.

Command of the fixation point involves controlling an under-constrained system of motors. Our solution has been to define a hierarchy among the axes. If the commanded fixation point can be reached by panning the cameras or tilting the head, then only these joints are moved. If camera pan or head tilt reach a specified limit then the next axes in the hierarchy (head pan or axe 3 of the arm are commanded and the lower level axes are moved to compensate. When these axes reach soft limits, the lower arm axes and finally the mobile base are brought into play.

16.3.4 Commanding Fixation

The combination of pan, tilt and vergence can be thought of as the ability to point a virtual telescopic "stick" into the world. The parameters for this stick are its length (D_c), azimuth (pan or α_g), and elevation (tilt or β_g). These three virtual axes define an important component of the head "state". Their values are specified with respect to a head centered coordinate frame. Their actual decomposition into real axes depends on strategies which are device specific.

In addition to a polar expression for the gaze point, it is useful to have a Cartesian expression. Transformation from Cartesian to polar form is quite simple. We consider that the pan and tilt axes of the head are located at Cartesian coordinates (X_h, Y_h, Z_h). The the position of the gaze point is determined by.

$$X_g = X_h + D \, Cos(\alpha_g + \alpha_h)$$

$$Y_g = Y_h + D \, Sin(\alpha_g + \alpha_h)$$

$$Z_g = Z_g + D \, Sin(\beta_g + \beta_h)$$

Both the polar and Cartesian forms of the gaze point are stored in a C structures that defines the head "state". This structure is defined as an array of V_AXES structures:

```
typedef struct  _v_axis
{    int enabled= FALSE;       /* Boolean: Is this axis physically present?  */
     int pos = 0;              /* Current value */
     int command = 0;          /* Current Desired position */
     int min;                  /* Minimum Value */
     int max;                  /* Maximum Value */
     char units[16];           /* Units (Meters, degrees, etc) */
     char name [32];           /* Name of this axis */
} V_AXIS;
```

The polar representation is stored in structure called "polar state".

```
struct polar_state {      /* The virtual head state */
    V_AXIS depth;         /* distance to vergence point */
    V_AXIS pan;           /* Composite pan angle */
    V_AXIS tilt;          /* Composite tilt angle */
}

struct cart_state {       /* Cartesian representation of the head state */
    V_AXIS    X;          /* X component of gaze point */
    V_AXIS    Y;          /* Y component of gaze point */
    V_AXIS    Z;          /* Z component of gaze point */
}
```

The current value for each of these V_AXES is determined by the translation procedures (4) in Figure 4. The Cartesian values are computed from the polar values. Commanded values may be set by messages from other processes. The difference between a commanded value and a current position triggers the translator to call a device specific procedure to move the necessary real axes.

16.3.5 The Command Interpreter

The command interpreter responds to the commands get and set. These commands may refer to any of the real or virtual axes of the head.

The formats for a "get" commands are:

(get axis <Axis-name>)
(get polar_state)
(get cartesian_state)

The interpreter decides it action based on the second keyword. For a request for an axis, the axis name is matched to the list of axes names. If the value does not matched, a message "axis undefined" is returned. Otherwise, the value is determined from the position parameter of the axis and returned. For a request for the polar_state or the cartesian_state, the values of the state variables are returned.

The set command has the formats:

(set axis <inc/abs> <axis-name> <value>)
(set polar_state <inc/abs> <parameter-name> <value>)
(set cartesian_state <inc/abs> <parameter-name> <value>)

The field <inc/abs> indicates whether the new value is incremental (inc) or absolute (abs). The value is written into the variable "command". In the next cycle, the translator notes the difference between the commanded and current values and issues the necessary move commands.

16.4 Reflex Level Control of Ocular Parameters

This section concerns action level control of the ocular parameters of aperture, focus, and convergence. This control constitute device level reflexes which have several roles:

1) To accommodate changes in the image quality due to command of the gaze point,

2) To provide an error signal for directing the gaze point, and

3) To provide an estimate of the position of objects in the scene.

In the first case, we assume that the horopter (field of view at or near zero disparity) is commanded by some external processes. Ocular reflexes for focus and aperture permit the system to maintain a suitable image quality despite these changes. Reflex control of vergence places the interest point on the scene point which is generating an image signal, providing a measure of scene position relative to the head.

In the second case, the gaze point may be directed by some external process to a scene location which is in front of or behind a surface. The discordance between the commanded value and the accommodated value constitutes an event which can signal to the system that the expectation was incorrect.

The measures which are described below are all based on smoothed versions of the image produced by a binomial pyramid. We have found that with a multiple resolution image description provides a computational support which renders our image measures both more stable and more efficient. Our multiple resolution representation is computed using a fast "optimal S/N" binomial pyramid [Chehikian-Crowley 1991]. That is, the image is convolved with a cascade of binomial filters based on the kernel [1 2 1], to produce a set of 12 re-sampled images, numbered 1 to 12. The standard deviation for the level k of this pyramid is given by an exact formula:

$$\sigma_\kappa = \sqrt{2^k}$$

This section describes preliminary experiments with image measurement for controlling the focus, aperture, and vergence. We restrict ourselves to a very brief description of the ocular reflexes for focus, aperture and vergence.

16.4.1 Focus

It is well known that focus can be controlled by the "sharpness" of contrast. The problem is how to measure such "sharpness". In [Krotkow 1987] we can find a description of several methods for measuring image sharpness. Horn [Horn 1968] proposes to maximize the high-frequency energy in the power spectrum. Jarvis proposes to sum the magnitude of the first derivative of neighboring pixels along a scan line [Jarvis 1983]. Schlag [Schlag 1983] and Krotkov [Krotkow 1987] propose to sum the squared gradient magnitude. Tenenbaum and Schlag compare gradient magnitude to a threshold and sum uniquely those pixels which are above a threshold. The problem is then the choice of such a threshold. We have found that such a measure performs poorly. After experiments with several measures, we have found our best results with the sum of gradient magnitude, without the use of the threshold.

We measure image gradient at the level five or our low-pass pyramid, providing a binomial smoothing window with a standard deviation of $4\sqrt{2}$. Gradient is calculated using compositions of the filter [1 0 −1] in the row and column directions. By default the "region of interest" is at the center of the image, but this region may be placed anywhere in the image by a message from another software module or the user. Local extrema in the gradient magnitude are summed within the region of interest. An initialize command causes focus to look for a global maximum in this sum. Subsequently, the reflex action seeks to keep the focus at a local maximum.

We note that this measure exhibits a plateau around the proper focal value. This region corresponds to the "depth of field". Reducing the aperture will enlarge the depth of field and thus enlarge this plateau. Figure 16.6 shows an example of the average gradient magnitude within a a region of interest computed for an image of a cube, as a function of focus.

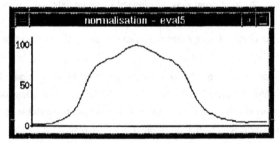

Fig. 16. 6 Example of the average gradient magnitude within a a region of interest computed for an image of a cube, as a function of focus.

16.4.2 Control of Aperture

It is quite simple to servo the aperture using the intensity level of a single pixel at a low resolution level. A pixel in a low resolution pyramid represents a form of average light level within a region of the image. For a smaller region of interest, we can select the pixel at a higher resolution level. For larger region, we select a pixel at a lower resolution level. This is the simplest example of an ocular reflex, made possible by control of the imaging system. Unfortunately, when the pixel falls on a dark area of the scene, it can lead the system to open the aperture completely, saturating the rest of the image.

The optimum contrast occurs when the dynamic range of the signal is largest. If we consider each pixel as a random variable, we can form two estimators: The variance of the signal, and the range of minimum to maximum gray level. We have found that the maximizing the variance of pixels in the region of interest provides the most robust estimator. This estimator has the following advantages:

1) It characterizes the ambient light given for different zones of the image while one does not fall on a dark patch of the image,
2) It does not exhibit a plateau of values. It results in a single peak, leading to a precise value.
3) It provides an automatic compensation between intensity and focus.
4) It is stable for different images of the same scene with the same lighting without modifying the parameters.

We can also observe that because we apply the operator to a reduced window, the measure is computed in a very short time. Figure 16.7 shows the values for this measure over a range of aperture settings.

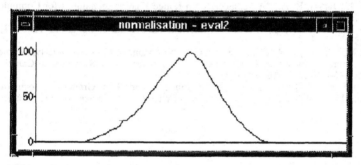

Fig. 16.7. maximum variance within a region of interest, as a function of the aperture.

16.4.3 Control of Convergence

As described above, when we specify the 3D interest point, the system computes the necessary camera angles so as to bring the optical axes to intersect at the specified gaze point. This corresponds to a simple perceptual action which might be called

Look-at <3D point>

Our collaborators at University of Linköping have found that a very robust measure for convergence is provided by the difference in Phase of the correlation of an even and odd filters with the image. Westelius and his colleagues [Westelius et al. 1991] demonstrated vergence control of a simulated head using even and odd gabor filters. The in collaboration with them, we have determined that a reasonable approximation of the phase may be obtained using filters [1 0 -1] and [1 -2 1]. These filters are used in our system to compute the first and second derivatives. While the phase measured in this way is not linear with position, it does seem to be monotonic.

We exploit the multiple resolution pyramid to converge on an object in a coarse to fine manner. An image row and an initial column positions in the two cameras are selected for convergence. We measure the phase at this row in the two cameras at level 9 of our pyramid ($\sigma = 2^{4.5} = 16\sqrt{2}$). The phase provides a shift in each image. This shift is then used to compute the column for the next higher resolution level. The process is repeated at each level. The final shift in each image is converted from pixels to encoder counts for the vergence motors and pan motors. The sum of the shift is used to compute a pan motion for axe 6. The difference is used to compute a symmetric vergence angle for the two vergence motors. The result is that the head stays symmetrically verged on an edge of the largest object which covers the selected scan line. The process repeats for each pair of images which are taken by the image acquisition module.

Bibliography

[Chehikian-Crowley 1991] A. Chehikian et J. L. Crowley, "Fast Computation of Optimal Semi-Octave Pyramids", Scandinavian Conference on Image Analysis, August 1991.

[Horn 1968] Horn, B. P. K., "Focussing", MIT Artificial Intelligence Lab Memo No. 160, May 1968.

[Jarvis 1983] Jarvis, R. A., "A Perspective on Range Finding techniques for Computer Vision", IEEE Trans. on PAMI 3(2), pp 122-139, March 1983.

[Krotkov 1987] Krotkov, E., "Focusing", International Journal of Computer Vision, 1, p223-237 (1987).

[Schlag 1983] Schlag, J., A. C. Sanderson, C. P. Neumann, and F. C. Wimberly, "Implementation of Automatic Focussing Algorithms for a Computer Vision System with Camera Control", CMU-RI-TR-83-14, August, 1983.

[Westelius et al. 1991] Westelius, C. J., H. Knutsson, and G. H. Granlund, "Focus of Attention Control", SCIA-91, Seventh Scandinavian Conference on Image Analysis, Aalborg, August 91.

17. Active 3D Scene Description in Object Reference Frames

James L. Crowley and Philippe Bobet
LIFIA, INPG

This chapter describes a real time 3D vision system which uses stereo matching of edge segments to reconstruct object centered 3D descriptions of objects. We describe a new technique which permits stereo cameras to be calibrated to a reference frame defined by any object on which we can reliably find six points. This approach provides the scene to image transformation matrices by direct observation, that is, without matrix inversion. The approach can be extended to update the transformations to account for motion and change in the optical parameters of the cameras.

A reflex measure is used to maintain active stereo convergence on an object. As a result it is possible to limit stereo matching to a horopter region in which disparity is nearly null. This permits stereo matching to be performed by a simple similarity measure based on a Mahalanobis distance. Stereo matching is also made simple by tracking stereo matches. Tracking of 2D segments is provides temporal continuity of segments. This continuity is exploited to track 3D correspondences, and thus use existing correspondences as a basis for finding additional correspondences. Thus system uses real time edge tracking to lock onto stereo matches for reconstruction.

Reconstruction is performed in an object centered coordinate frame. This assures the 3D data can be fused from a range of viewing positions, even when the viewing position is poorly known. The mathematics for reconstructing 3D segments from stereo matches is presented and a parametric representation for 3D vertical segments is then described. This representation describes 3D segments as a vector of parameters with their uncertainty.

17.1. System Organization

The stereo system is organized as a pipeline of relatively simple processes, as illustrated in figure 17.1. Stereo images are acquired and filtered by the image acquisition and processing module to produce edge segments. These edge segments are transmitted to two tracking and description modules, one for each of the two cameras. The 3D module requests edge segments from the tracking modules. It then uses a similarity measure to maintain a list of possible correspondences with their confidence factors. The most confident correspondence are selected for 3D reconstruction. Reconstructed segments are then sent to a 3D scene maintenance process.

Classic research in stereo often involves using images from cameras which have been carefully set up and calibrated, in many cases with a process which requires hours. Unfortunately, the scene to image transformation is changed whenever the camera is moved or the focus is changed. In an active continuously operating vision system these parameters are continually changing, this classic approach can not be used. Thus we have been forced to develop a new approach which permits the camera calibration to be continually maintained by observing the scene. In the following section we present a technique for calibrating to an object based reference by direct observation.

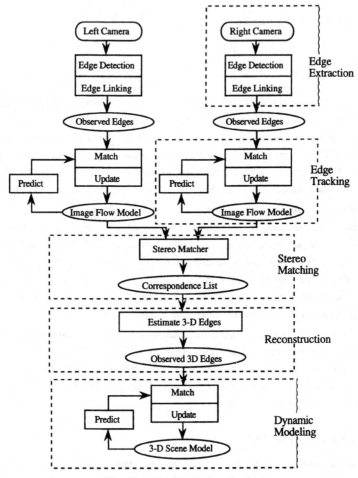

Fig. 17.1 The System Organization

17.2. Correspondence Matching within the Horopter

Stereo correspondence is a classic vision problem. In its most general form, such stereo correspondence matching has an computational complexity which is exponential. To minimize computational costs, constraints are used to limit the possible matches that are considered. An active binocular camera head provides an extremely strong constraint. The intersection of the optical axes defines a region of 3D space for which the stereo disparity is (approximately) zero. This region of space is called the "horopter". The horopter defines the 3D region of interest may be thought of as a filter for separating figure and ground.

In this section we describe a system for stereo matches within a horopter. This system assumes that the cameras have been verged on a region of interest either by a-priori information or by a reflexive response. We first provide an overview of the system. We then describe a similarity measure for edge segments from the left and right image flow models, based on a simple Mahalanobis distance.

17.2.1. System Overview

The stereo matching process is illustrated in figure 17.2. Line segments are acquired from the flow models from the left and right images. Each segment from the left flow model is compared to the segments from the right flow model. For each pair of segments a similarity measure is computed based on the position, length and orientation of the segments. The most similar segment from the right image is selected as matching. No effort is made to avoid double use of a segment from the right image, or to assure a globally consistent match. Instead, consistency is maintained based on the results of the 3D reconstruction.

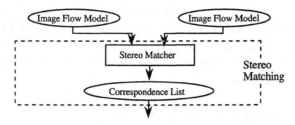

Fig. 17.2 The stereo matching process maintains a correspondence list.

The 3D reconstruction algorithm exploits the segment identities which are maintained by tracking. In this way, if a pair of segments matched the previous time, then this pair of segments is immediately associated with a 3D segment.

17.2.2. Similarity Function for Edge Segments

Each segment in the left image is compared to each segment in the right image using a similarity function based on a form of Mahalanobis distance, as illustrated in figure 17.3.

An edge segment is described by parameters $(x_p, y_p, h_p, \theta_p)$. Each parameter has an associated uncertainty, $(\sigma_x, \sigma_y, \sigma_\theta, \sigma_h)$. The Mahalanobis distance for each parameter is given by square of the difference divided by the square of the uncertainty. Unlikely matches are eliminated comparing the absolute value of the difference to a threshold. The similarity measure is given by:

$$\text{Dist} = \left(\frac{x_p - x}{\sigma_x}\right)^2 + \left(\frac{y_p - y}{\sigma_y}\right)^2 + \left(\frac{h_p - h}{\sigma_h}\right)^2 + \left(\frac{\theta_p - \theta}{\sigma_\theta}\right)^2$$

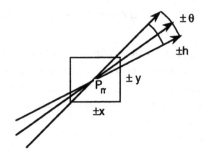

Fig. 17.3 Similarity with respect to a prototype segment.

For each segment in the left image, the similarity function is computed for all segments in the right image. If the distance is less than a threshold, then the ID of the right segment is added to a list of possible matches, in the order of the distance measure. For each segment in the left image, the closest segment in the right image is selected for 3D reconstruction.

17.3. Computing 3D Structure in an Object Based Reference Frames

For each stereo correspondence, a 3D edge segment is obtained. These 3D segments are then integrated into a 3D composite scene model as an independent observation. This section describes the process for constructing a model of the the 3D structure of an object within its own "intrinsic" reference frame. The intrinsic reference frame forms a view invariant description for fusing multiple observations and for recognition.

17.3.1. Process Overview and Notation

The 3D inference process is described in figure 17.4. It begins by correcting the end points so that only the overlapping parts of segments are reconstructed. The position of end-points within the object based reference frame are then computed, along with the uncertainty for the end points. Segment parameters are then computed from the end-points.

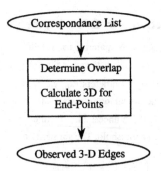

Fig. 17.4 The Inference of 3D Form from the the correspondence of Edge Segments.

In homogeneous coordinates, a point in the scene is expressed as a vector:

$${}^{s}P = [x_s, y_s, z_s, 1]^{T}$$

The index "s" raised in front of the letter indicates a scene based coordinate system for this point. The origin and scale for such coordinates are arbitrary. A point in an image is expressed as a vector:

$${}^{i}P = [i, j, 1]^{T}$$

The projection of a point in the scene to a point in the image can be approximated by a three by four homogeneous transformation matrix, ${}^{i}_{s}M$. This transformation models the perspective projection with the equation:

$${}^{i}P\, w = {}^{i}_{s}M\ {}^{s}P$$

or

$$\begin{bmatrix} w\ i \\ w\ j \\ w \end{bmatrix} = {}^{i}_{s}M \begin{bmatrix} x_s \\ y_s \\ z_s \\ 1 \end{bmatrix} \tag{1}$$

The variable w captures the amount of "fore-shortening" which occurs for the projection of point ${}^{s}P$. This notation permits the pixel coordinates of ${}^{i}P$ to be recovered as a ratio of polynomials of ${}^{s}P$. That is

$$i = \frac{w\ i}{w} = \frac{{}^{i}_{s}M_1 \bullet {}^{s}P}{{}^{i}_{s}M_3 \bullet {}^{s}P} \qquad j = \frac{w\ j}{w} = \frac{{}^{i}_{s}M_2 \bullet {}^{s}P}{{}^{i}_{s}M_3 \bullet {}^{s}P} \tag{2}$$

where ${}^{i}_{s}M_1$, ${}^{i}_{s}M_2$, and ${}^{i}_{s}M_3$ are the first, second and third rows of the matrix ${}^{i}_{s}M$, and " $\bullet$ " is a scalar product.

It is common to model cameras with a pin-hole model expressed mathematically as a projective transformation. It must be stressed that this is only an approximation. Real lenses and cameras do not have a unique projection point, nor a unique optical axis. One way to model such errors is by adding an unknown random vector, U_M, which accounts for the difference in pixel position.

$$ {}^{i}P\, w = {}^{i}_{s}M\ {}^{s}P + U_M$$

17.3.2. Computing 3D Structure From Stereo Correspondences

Let ${}^{L}_{s}M$ and ${}^{R}_{s}M$ represent the transformations for the left and right cameras in a stereo pair. Let ${}^{L}_{s}M_1$, ${}^{L}_{s}M_2$, and ${}^{L}_{s}M_3$ represent the first, second third rows of the ${}^{L}_{s}M$, and ${}^{R}_{s}M_1$, ${}^{R}_{s}M_2$ and ${}^{R}_{s}M_3$ represent the first, second third rows of the ${}^{R}_{s}M$. Observation of a scene point, ${}^{s}P$, gives the image points ${}^{L}P = (i_L, j_L)$ and ${}^{R}P = (i_R, j_R)$. From equation 2 we can write.

$$i_L = \frac{{}^{L}_{s}M_1 \bullet {}^{s}P}{{}^{L}_{s}M_3 \bullet {}^{s}P} \qquad\qquad i_R = \frac{{}^{R}_{s}M_1 \bullet {}^{s}P}{{}^{R}_{s}M_3 \bullet {}^{s}P}$$

$$j_L = \frac{{}_S^L M_2 \bullet {}^S P}{{}_S^L M_3 \bullet {}^S P} \qquad\qquad j_R = \frac{{}_S^R M_2 \bullet {}^S P}{{}_S^R M_3 \bullet {}^S P}$$

With a minimum of algebra, these can be rewritten as

$$({}_S^L M_1 \bullet {}^S P) - i_L \, ({}_S^L M_3 \bullet {}^S P) = 0 \qquad ({}_S^R M_1 \bullet {}^S P) - i_R \, ({}_S^R M_3 \bullet {}^S P) = 0 \qquad (3)$$

$$({}_S^L M_2 \bullet {}^S P) - j_L \, ({}_S^L M_3 \bullet {}^S P) = 0 \qquad ({}_S^R M_2 \bullet {}^S P) - j_R \, ({}_S^R M_3 \bullet {}^S P) = 0 \qquad (4)$$

This provides us with a set of four equations for recovering the three unknowns of $^S P$. Each equation describes a plane in scene coordinates that passes through a column or row of the image, as illustrated in figure 17.5. The two equations from the left image describe a two planes which form a line projecting from the pixel (i_L, j_L) to the scene points. The equation containing i_R from the right image describes a vertical plane passing through i_R. The intersection of this plane with the line from the left image is the scene point which we wish to recover.

We can solve for a 3D point with two equations from the left camera and one from the right, or equally, using one equation from the left and two from the right. When the projection matrices are exact, these points are identical. Unfortunately, because of errors in pixel position due to sampling and image noise, the projection of the rays from the left and right camera do not necessarily meet at a point.

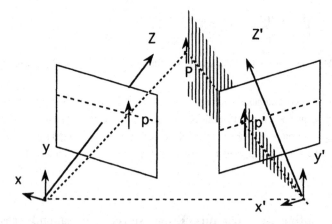

Fig. 17.5 Computation of a Scene Point by Stereo Projection

Let us define point $^S P_L$ as the point obtained using two equations from the left camera and one equation from the right camera. Let us define, $^S P_R$, as the scene point obtained from two equations in the right and one in the left. A more precise scene point can be obtained from the mid-point of the 3D segment which joins these two points:

$$^S P = {}^S P_L + \frac{{}^S P_L - {}^S P_R}{2}$$

Stereo reconstruction produces errors which are proportional to the distance from the origin. By placing the origin on the object to be observed, such error may be minimized. Computing the matrix $_s^i M$ for a pair of cameras permits a very simply method to compute the position of points in the scene in a reference frame defined by the scene. Dynamically developing the transformations for the left and right images permit objects in the scene to be reconstructed independent of errors in the relative or absolute positions of the cameras.

17.3.3. Computing the Overlapping Part for Matching Segments

Stereo projection is performed on the segment end-points. Because segment length is not always reliable, we must first determine the overlapping part of the corresponding segments. This requires computing the epipolar lines for the end-points of each of the segments.

The equation of the epipolar lines for each of the stereo images may be solved from the correspondence of three points [Faugeras et. al. 92]. Junctions defined by the intersection of a pairs of corresponding line segments are used to define these points. For image coordinates (i, j), the equation of an epipolar line in an image has the form:

$$A i + B j + C = 0$$

This can be written as an inner product of a line coefficient vector $^i L$ and the transpose of an image point, $^i P$.

$$^i P^T \bullet {}^i L \; = [i, j, 1] \begin{bmatrix} A \\ B \\ C \end{bmatrix} = 0 \tag{5}$$

Let us use the indices L and R for the left and right images. What we need is the 3 by 3 matrix which will map a point in one image, $^L P$ or $^R P$, into the line equation for the other image, $^R L$ or $^L L$. Let us call these matrices $_R^L B$ and $_L^R B$.

$$^L L_i = {}_R^L B \, ^R P_i \qquad\qquad\qquad {}^R L_i = {}_L^R B \, ^L P_i \tag{6}$$

In order to determine the coefficients of $_R^L B$ we compose matrices for three points in the left image and their corresponding points in the right images.

$$^L P = \; [^L P_1 \cup {}^L P_2 \cup {}^L P_3] \qquad {}^R P \; = \; [^R P_1 \cup {}^R P_2 \cup {}^R P_3]$$

By plugging the equations of 6 into 5 we can then write the equations 7.

$$^L P \, _R^L B \, ^R P = 0 \qquad\qquad\qquad {}^R P \, _L^R B \, ^L P = 0 \tag{7}$$

We then solve equation 7 to obtain the coefficients for $_R^L B$ and $_L^R B$.

For each corresponding pair of line segments, S_L and S_R, the epipolar line is computed for each of the end-points. If an epipolar line intersects the line segment, then the segment end-point is replaced by this intersection point, as shown in figure 17.6.This technique works well for vertical segments, but can become unstable for segments which are nearly aligned with the epipolar lines.

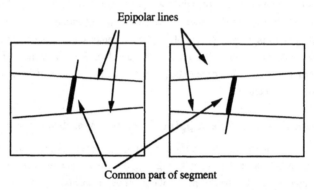
Fig. 17.6 Determining the Common Part of a Segment

This operation produces point correspondences which may be used to compute 3D structure. In order to calculate the 3D we need to have a coherent pair of matrices $^L_S M$ and $^R_S M$ which describe the projection from the same scene reference frame into the left and right cameras. We obtain these matrices by observing a set of six (or four) corresponding scene points.

17.4. Calibration to an Object-based Reference Frame

The intrinsic parameters are the parameters which are independent of camera position and orientation. They are typically listed as the "center" of the image, defined by the intersection of the optical axis with the retina, and the ratio of pixel size to focal length in the horizontal and vertical directions [Tsai 87]. Most stereo algorithms estimate depth from the head to scene points and then use the intrinsic parameters to reconstruct the scene. It is often overlooked that the pin-hole model is only a rough approximation for the optics of a camera. For a real camera, the intrinsic parameters are extremely sensitive to the setting for focus and aperture and even to small perturbations in retina mounting due to vibration! It is not surprising that most 3D vision systems work require a long careful camera calibration and then only work for a short period.

In an active head, in which focus and vergence are continually changing, such an approach is completely impractical. Such a system requires a method by which the calibration is obtained and maintained by direct observation of objects in the scene. Our search for such auto-calibration led us to consider the work of Mohr [Mohr et. al. 91] on the use of projective invariants. In particular Mohr and his collaborators have shown that the cross ratio can be used to construct a scene based reference frame, in which the objects the scene provide the reference coordinate system. Such an approach abandons the use of the camera intrinsic parameters, and measures the form of objects directly in a scene based reference frame. Mohr proposes a projective reference frame for 2D data based on four points. The extension to three dimensions leads to a projective reference based on five points, viewed over multiple images.

Koenderink and Van Doorn [Koenderink-Van Doorn 89] have observed that four fiducial marks ought to be sufficient to define a scene based reference frame for structure from motion or stereo. They have attempted to "stratify" the problem into a three stage process. They define an affine "projection" from an image reference to the image based on two views of four points. They then use this affine transformation to recover the position of points in the scene. In the second phase, they apply a Euclidean metric to the resulting structure by imposing a rigidity constraint. In the final phase, a third view removes possible ambiguous interpretations. Koenderink and Van Doorn argue that the most important and the most difficult part of the problem is the first phase, and that affine transformations provide a simple solution. They develop a mathematical model for recovering 3D form in an affine basis.

Sparr has shown how arbitrary reference points can be used to define an affine basis for reconstruction [Sparr 92]. In such a coordinate frame, the shape of a collection of points is represented in terms of "affine coordinates" provided by a sub-configuration of points. In the case of each of these techniques, the objects observed in the scene provide the reference frame in which objects are reconstructed. With such a technique, an object provides its own reference frame. In the techniques described below, we will first develop an affine basis and then show that an orthogonal basis is a special case which results when the reference points form a set of orthogonal vectors.

We have found that a robust 3D vision system may be constructed using the objects in a scene to calibrate the cameras. With this technique, the cameras are calibrated by fixating on any known set of 6 points. Calibration is then updated continually by tracking the image position of points as optical parameters are adjusted or as the camera is moved.

We begin by showing how an orthographic transformation matrix from affine scene coordinates to image coordinates can be obtained from the observation of four non-coplanar points. These four points define a set of three basis vectors whose lengths become the units of measure in the system. If the three vectors are orthogonal, then the reference frame provides an orthogonal basis.

We show that the orthographic projection matrix can be completed to obtain the full perspective transformation by the observation of an additional two points whose position is known relative to the first four points. This permits us to use any known object containing six identifiable points to define a reference frame for 3D reconstruction. If the size of the object is known, then the units of reference frame can be deduced.

17.4.1. Calibrating an Orthographic Projection from Scene Object to Image

Any four points in the scene which are not in the same plane can be used to define an affine basis. Such a basis can be used as a scene based coordinate system (or reference frame). One of the four points in this reference frame will be taken as the origin. Each of the other three points defines an axis, as shown in figure 17.7. On an arbitrary object, these axes are not necessarily orthogonal.

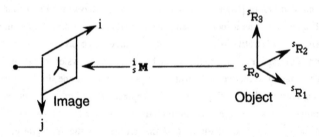

Fig. 17.7 Any four points in the scene define a scene based coordinate system

A simple way to exploit this idea is to use any four non-coplanar points to define an orthographic projection from an affine reference frame in the scene to the image. Let us designate a point in the scene as the origin for a reference frame. By definition,

$${}^{s}R_o = [0, 0, 0, 1]^T$$

Three axes for an affine object-based reference frame may be defined by designating three additional scene points as:

$${}^{s}R_1 = [1, 0, 0, 1]^T$$

$${}^{s}R_2 = [0, 1, 0, 1]^T$$

$${}^{s}R_3 = [0, 0, 1, 1]^T$$

The vector from the origin to each of these points defines an axis for measuring distance. The length of each vector defines the unit distance along that vector. These three vectors are not required to be orthogonal. The four points may be used to define an affine basis by the addition of a constraint that the sum of the coefficients be constant [Sparr 92]. We note that when the points are the corners on a right parallel-piped (a box), then they can be used to define an orthogonal basis and the additional constraint is unnecessary.

Let the symbol $\cup$ represent the composition of vectors as columns in a matrix. We can then represent our affine coordinate system by the matrix ${}^{s}R$.

$${}^{s}R = [{}^{s}R_1 \cup {}^{s}R_2 \cup {}^{s}R_3 \cup {}^{s}R_o] = \begin{pmatrix} 1 & 0 & 0 & 0 \\ 0 & 1 & 0 & 0 \\ 0 & 0 & 1 & 0 \\ 1 & 1 & 1 & 1 \end{pmatrix}$$

The projection on these four points to the image can be written as four image points ${}^{i}P_o$, ${}^{i}P_1$, ${}^{i}P_2$, and ${}^{i}P_3$. These image points form an observation of the reference system, represented by the matrix ${}^{i}P$, where the term w_o has been set to 1.0.

$${}^{i}P_w = [{}^{i}P_1 w_1 \cup {}^{i}P_2 w_2 \cup {}^{i}P_3 w_3 \cup {}^{i}P_o] = \begin{pmatrix} w_1 i1 & w_2 i2 & w_3 i3 & io \\ w_1 j1 & w_2 j2 & w_3 j3 & jo \\ w_1 & w_2 & w_3 & 1 \end{pmatrix}$$

This allows us to write a matrix expression.

$${}^i_w P = {}^i_s M \, {}^s R$$

The reference matrix ${}^s R$ has a simple inverse, which can be solved by hand.

$${}^s R^{-1} = \begin{pmatrix} 1 & 0 & 0 & 0 \\ 0 & 1 & 0 & 0 \\ 0 & 0 & 1 & 0 \\ -1 & -1 & -1 & 1 \end{pmatrix}$$

Inverting this matrix allows us to write the expression:

$${}^i_s M = ({}^i P_w) \, {}^s R^{-1} = \begin{pmatrix} w_1 i_1 & w_2 i_2 & w_3 i_3 & i_0 \\ w_1 j_1 & w_2 j_2 & w_3 j_3 & j_0 \\ w_1 & w_2 & w_3 & 1 \end{pmatrix} \begin{pmatrix} 1 & 0 & 0 & 0 \\ 0 & 1 & 0 & 0 \\ 0 & 0 & 1 & 0 \\ -1 & -1 & -1 & 1 \end{pmatrix}$$

or

$${}^i_s M = \begin{pmatrix} w_1 i_1 - i_0 & w_2 i_2 - i_0 & w_3 i_3 - i_0 & i_0 \\ w_1 j_1 - j_0 & w_2 j_2 - j_0 & w_3 j_3 - j_0 & j_0 \\ w_1 - 1 & w_2 - 1 & w_3 - 1 & 1 \end{pmatrix} \qquad (8)$$

Having performed the inversion of ${}^s R$ by hand, there is no need to compute an inverse when the system is calibrated.

The problem with equation 8 are the values w_1, w_2 and w_3. Two approaches are possible. One may employ the approximation $w_1 = w_2 = w_3 = 1$, yielding an orthographic approximation to the projective transformation. The magnitude of the error for such an approximation is proportional to the change in distance from the camera and inversely proportional to the focal length of the camera lens.

The orthographic approximation can provide a usable approximation for points near the reference object. For example, such an approximation was employed by artists before the effects of projection were discovered. In a case where the reference object is unknown, an approximate 3D reconstruction using orthographic projection can be constructed in terms of four non-coplanar reference points.

Alternatively, we may seek to determine the full perspective transformation by solving a set of linear equations to determine w_1, w_2 and w_3. Solving for this vector requires three additional constraints, or the observation of one and a half additional points whose position is known with respect to the first four points. To obtain the perspective transformation from equation 8 we must solve for w_1, w_2 and w_3. Solving for these three variables requires 3 independent equations, or the observation of the image coordinates for one and a half scene points. Let us define two known scene points as ${}^s R_4$ and ${}^s R_5$.

$${}^s R_4 = [x_4, y_4, z_4, 1]^T$$

$${}^s R_5 = [x_5, y_5, z_5, 1]^T$$

Note that we require that the position of the points ${}^s R_4$ and ${}^s R_5$ be known with respect to the four reference points. Using equation 2, we can write four equations with three unknowns.

$$i_4 = \frac{{}^i_s M_1 \bullet {}^s R_4}{{}^i_s M_3 \bullet {}^s R_4} = \frac{(w_1 i_1 - i_0)x_4 + (w_2 i_2 - i_0)y_4 + (w_3 i_3 - i_0)z_4 + i_0}{(w_1 - 1)x_4 + (w_2 - 1)y_4 + (w_3 - 1)z_4 + 1} \tag{9}$$

$$j_4 = \frac{{}^i_s M_2 \bullet {}^s R_4}{{}^i_s M_3 \bullet {}^s R_4} = \frac{(w_1 j_1 - j_0)x_4 + (w_2 j_2 - j_0)y_4 + (w_3 j_3 - j_0)z_4 + j_0}{(w_1 - 1)x_4 + (w_2 - 1)y_4 + (w_3 - 1)z_4 + 1} \tag{10}$$

$$i_5 = \frac{{}^i_s M_1 \bullet {}^s R_5}{{}^i_s M_3 \bullet {}^s R_5} = \frac{(w_1 i_1 - i_0)x_5 + (w_2 i_2 - i_0)y_5 + (w_3 i_3 - i_0)z_5 + i_0}{(w_1 - 1)x_5 + (w_2 - 1)y_5 + (w_3 - 1)z_5 + 1} \tag{11}$$

$$j_5 = \frac{{}^i_s M_2 \bullet {}^s R_5}{{}^i_s M_3 \bullet {}^s R_5} = \frac{(w_1 j_1 - j_0)x_5 + (w_2 j_2 - j_0)y_5 + (w_3 j_3 - j_0)z_5 + j_0}{(w_1 - 1)x_5 + (w_2 - 1)y_5 + (w_3 - 1)z_5 + 1} \tag{12}$$

Provided that no five of our six points are coplanar, these four equations can be solved to obtain the values of w_1, w_2 and w_3. The full projection matrix, ${}^i_s M$, can then be obtained from equation 8.

When the positions of the points ${}^s R_4$ and ${}^s R_5$ are known in advance, the solution can be structured to yield the full perspective transformation by direct observation, without matrix inversion. To illustrate this, let us consider the problem of calibrating ${}^i_s M$ by observation of 6 vertices of right parallel-piped.

17.4.2. Calibrating to a Right Parallelpiped

Consider a reference frame defined by six points on a right parallelpiped, as shown in figure 17.8. Point ${}^s R_0$ defines the origin. Points ${}^s R_1$, ${}^s R_2$ and ${}^s R_3$ define the unit vectors for the X, Y and Z axes. Points ${}^s R_4$ and ${}^s R_5$ permit the full projective transformation to be recovered. Points ${}^s R_0$, ${}^s R_1$, ${}^s R_2$ and ${}^s R_3$ are defined above as:

$${}^s R_0 = [0, 0, 0, 1]^T$$

$${}^s R_1 = [1, 0, 0, 1]^T$$

$${}^s R_2 = [0, 1, 0, 1]^T$$

$${}^s R_3 = [0, 0, 1, 1]^T$$

Points ${}^s R_4$ and ${}^s R_5$ are given by:

$${}^s R_4 = [1, 0, 1, 1]^T$$

$${}^s R_5 = [0, 1, 1, 1]^T$$

Substituting ${}^s R_4$ and ${}^s R_5$ into equations 10 through 13 gives:

$$(i_4 - i_1) w_1 + (i_4 - i_3) w_3 - (i_4 - i_0) = 0 \tag{13}$$

$$(j_4 - j_1) w_1 + (j_4 - j_3) w_3 - (j_4 - j_0) = 0 \tag{14}$$

$$(i_5-i_2)\, w_2 + (i_5-i_3)\, w_3 - (i_5-i_0) = 0 \qquad (15)$$

$$(j_5-j_2)\, w_2 + (j_5-j_3)\, w_3 - (j_5-j_0) = 0 \qquad (16)$$

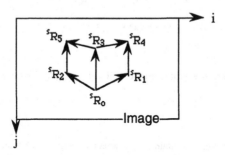

Fig. 17. 8 The full projective transform can be computed directly from the observation of 6 known scene points. Point ${}^{s}R_0$ defines the origin. Points ${}^{s}R_1$, ${}^{s}R_2$ and ${}^{s}R_3$ define the unit vectors of the three scene dimensions. Points R_4 and ${}^{s}R_5$ permit the full projective transformation to be recovered.

The coefficients w_1 and w_3 can be had from equations 17 and 18, that is from observation of ${}^{s}R_4$. The coefficients w_2 and w_3 can be had from equations 19 and 20, that is from observation of ${}^{s}R_5$. From the point ${}^{s}R_4$ we obtain:

$$w_1 = \frac{(i_4-i_0)\,(j_4-j_3) - (i_4-i_3)\,(j_4-j_0)}{(i_4-i_1)\,(j_4-j_3) - (i_4-i_3)\,(j_4-j_1)} \qquad (17)$$

$$w_3 = \frac{(i_4-i_0)\,(j_4-j_1) - (i_4-i_1)\,(j_4-j_0)}{(i_4-i_3)\,(j_4-j_1) - (i_4-i_1)\,(j_4-j_3)} \qquad (18)$$

While, from the point ${}^{s}R_5$ we obtain:

$$w_2 = \frac{(i_5 \; i_0)\,(j_5-j_3) - (i_5-i_3)\,(j_5-j_0)}{(i_5-i_2)\,(j_5-j_3) - (i_5-i_3)\,(j_5-j_2)} \qquad (19)$$

$$w_3 = \frac{(i_5-i_0)\,(j_5-j_2) - (i_5-i_2)\,(j_5-j_0)}{(i_5-i_3)\,(j_5-j_2) - (i_5-i_2)\,(j_5-j_3)} \qquad (20)$$

The fact that the equations are over-constrained poses a small problem. If the image position of points ${}^{s}R_4$ and ${}^{s}R_5$ are not perfectly measured, the resulting solution for w_1, w_2 and w_3 will not be consistent. However, it is inevitable that the position of the reference points will be corrupted by small random variations in position, if for no other reason, because of image sampling. If we simply compute w_1 and w_3 from ${}^{s}R_4$ and then compute w_2 from ${}^{s}R_5$, this inconsistency can yield an imprecise solution for the position of 3D points. We can turn this problem to our advantage by exploiting the redundancy of the last half of a point to correct for random errors in the image position of the reference points.

The classic method for minimizing the inconsistency in reference point position is to compute a mean-squared solution. Faugeras and Toscani [Toscani-Faugeras 87] present a direct method to minimize the sum of the error between the projection of calibration points and their observation. From equation 5.2, for each calibration point ^sR_k and its image projection ^iP_k, we can write:

$$(^i_sM_1 \bullet {}^sR_k) - i_k\,(^i_sM_3 \bullet {}^sR_k) = 0 \qquad (^i_sM_2 \bullet {}^sR_k) - j_k\,(^i_sM_3 \bullet {}^sR_k) = 0$$

For N non-coplanar calibration points we can write a linear system of 2N equations of the form:

$$A\;{}^i_sM = 0.$$

where the rank of A is 11. The problem is to find a matrix ^i_sM which best minimizes a criterion equation

$$C = \|\,A\,{}^i_sM\,\|$$

We use Lagrange multipliers to obtain a least squares value for ^i_sM which minimizes C. We will refer to this below as the "mean square technique", denoted "msq" in the tables of experimental results below. To compare the mean square solution with the direct technique, we have experimented with reconstruction of a an aluminium cube with a side of 20cm.

In our experiments, the image size is 512 by 512 pixels. A standard left handed image coordinate system is used in which the origin is the upper left hand corner, positive i (columns) is to the left, and positive j (rows) is down. The images used for this example are shown in figure 17.9.

Fig. 17.9 Stereo Images of a 20 cm Calibration Cube at a distance of 1.2 meters.

For the left image, the vertices of the cube were detected at:

$$^LP_0 = (228, 481) \qquad\qquad ^LP_1 = (347, 351) \qquad\qquad ^LP_2 = (77, 374)$$
$$^LP_3 = (229, 223) \qquad\qquad ^LP_4 = (354, 107) \qquad\qquad ^LP_5 = (69, 125)$$

Equations 17 through 20 give a solution for $\vec{w}$ of:

$$w_1 = 0.917610, \quad w_2 = 0.858158, \quad w_3 = 1.052614$$

By the direct method we then obtain

$$\begin{array}{c}L\\{}_s M\end{array} = \begin{pmatrix} 147.589396 & -146.422112 & -11.048572 & 228.000000 \\ -101.081043 & -84.764543 & -269.732889 & 481.000000 \\ 0.082390 & 0.059453 & -0.052614 & 1.000000 \end{pmatrix}$$

Using the least squares technique, the matrix for the left image ${}_s^L M$ is computed as:

$$\begin{array}{c}L\\{}_s M\end{array} = \begin{pmatrix} 148.016122 & -146.716244 & -12.239302 & 228.149911 \\ -100.417731 & -85.159763 & -270.607106 & 481.003325 \\ 0.084301 & 0.058403 & -0.056504 & 1.000000 \end{pmatrix}$$

For the right image, the vertices of the cube were detected at:

$^R P_0 = (212, 464)$ $^R P_1 = (343, 332)$ $^R P_2 = (74, 360)$

$^R P_3 = (197, 208)$ $^R P_4 = (337, 88)$ $^R P_5 = (52, 116)$

With the direct method, from equation 17 through 20 we obtain

$$\begin{array}{c}R\\{}_s M\end{array} = \begin{pmatrix} 158.141055 & -132.711116 & -26.996216 & 212.000000 \\ -105.729358 & -78.270296 & -268.666055 & 464.000000 \\ 0.079128 & 0.071471 & -0.060894 & 1.000000 \end{pmatrix}$$

Using the least squares technique, the matrix for the right image ${}_s^R M$ is computed as:

$$\begin{array}{c}R\\{}_s M\end{array} = \begin{pmatrix} 158.066763 & -132.620333 & -26.745194 & 211.958839 \\ -105.863649 & -78.136621 & -268.493161 & 464.002612 \\ 0.078734 & 0.071856 & -0.060038 & 1.000000 \end{pmatrix}$$

As a check, we detected the image positions of the point $^S R_6 = [1, 1, 1, 1]^T$ and construct the 3D position of this point by a stereo solution. The image of this point is detected at:

$^L P_6 = (200, 23)$ $^R P_6 = (193, 11)$

Solving for the 3D position with the stereo technique using all four equations as described above gives

Dist	method	X	Y	Z
0.007559	direct	1.004628	1.005564	0.997816
0.010183	msq	0.992905	0.993915	1.004042

These reconstructed points are expressed in units defined by the side of the cube. One multiplies by 20 to obtain centimeters. We can observe that for this example, the direct calculation gives an error of about 0.7%, while the mean square technique gives about 1% error. The error is due to both the sampling interval of the pixels and imprecision of detection of the corners. Although the direct solution happened to perform best in this example, the mean square technique tends to give an error with the lowest average value. Such a tendency is made evident by a systematic exploration of the precision of the two techniques.

17.4.3. Correcting Calibration by Tracking Image Points

In this section we develop techniques for correcting the projective transformation matrix by tracking image points. We compare an affine correction obtained by tracking three points to a full projective correction obtained by tracking four points. The results show that an affine correction is sufficient for focus and aperture, while the full projective correction gives slightly better results for camera vergence.

A modification to the lens of a camera will induce a transformation on the image position of points in the scene. In theory, any movement in the 3D position of the principal point (projection or stenope point) will result in an image transformation which depends on distance to the point. Correcting such a transformation requires a 3 x 3 projective correction matrix, which may be obtained by tracking the image positions of four points. When the movement of the projection point is very small with respect to the focal length, the effects of the distance to the scene point may be much smaller than random pixel noise. In such a case, the transformation may be approximated by a 3 by 3 affine correction matrix obtained by tracking image coordinates of any three points.

In this section we develop an exact technique to compute the 3 by 3 projective correction. We then experimentally measure the precision of this correction for the cases of change in focus, aperture and vergence angle. Our results show that that stereo convergence is best corrected by a 3 by 3 projective correction matrix obtained by tracking four points. Changes in focus and aperture may be corrected by a 3 x 3 affine transformation, obtained by tracking three points. Movements of the head require tracking the scene position of points.

Let us define the positions of points in image 1 (before the change) as ^1P_i and in the image 2 (after the change) as ^2P_i. We seek to model the transformation of these points by a 3 by 3 homogeneous transformation ^2_1D.

$$^2P_i\, w = {}^2_1D\, {}^1P_i$$

or

$$\begin{bmatrix} w\, i_2 \\ w\, j_2 \\ w \end{bmatrix} = {}^2_1D \begin{bmatrix} i_1 \\ j_1 \\ 1 \end{bmatrix}$$

Where w is the homogeneous variable. Let us choose three non-colinear points in the first image $^1P_0$, $^1P_1$ and $^1P_2$. These points can be used to compose a matrix, ^1P.

$$^1P = [\,^1P_1 \cup {}^1P_2 \cup {}^1P_0\,] = \begin{pmatrix} i_{11} & i_{21} & i_{01} \\ j_{11} & j_{21} & j_{01} \\ 1 & 1 & 1 \end{pmatrix}$$

Let us designate the corresponding points in the second image as $^2P_0$, $^2P_1$ and $^2P_2$ and use these to compose the matrix ^2P.

$$^2P = [\,^2P_1 \cup {}^2P_2 \cup {}^2P_0\,] = \begin{pmatrix} i_{12} & i_{22} & i_{02} \\ j_{12} & j_{22} & j_{02} \\ 1 & 1 & 1 \end{pmatrix}$$

The transformation from image 1 to image 2 is defined by the homogeneous matrix $^2_1\mathbf{D}$ and written as

$$^2\mathbf{P}_w = {}^2_1\mathbf{D}\ {}^1\mathbf{P} \tag{21}$$

As above, the matrix $^2\mathbf{P}_w$ contains the scaling coeficients w_1 and w_2. To solve for $^2_1\mathbf{D}$ we invert $^1\mathbf{P}$ to obtain:

$$^2_1\mathbf{D} = {}^2\mathbf{P}_w\ {}^1\mathbf{P}^{-1} \tag{22}$$

In cases where the transformation from image 1 to image 2 is a composition of a translation, rotation and scale change, the scaling coefficients are given by $w_1 = w_2 = 1$ and $^2_1\mathbf{D}$ has the form of an affine transformation. We will call this an affine correction matrix $^2_1\mathbf{D}_a$. For such a transformation, $^2_1\mathbf{D}$, may be computed from any three points which are not in a straight line in image 1 for which a correspondence is known in image 2 by inverting the matrix $^1\mathbf{P}$.

$$^2_1\mathbf{D}_a = {}^2\mathbf{P}\ {}^1\mathbf{P}^{-1} \tag{23}$$

When $^2_1\mathbf{D}$ is not affine, the vector w_1 and w_2 must be computed. We can use equation 23 to obtain two equations for the two unknowns, by observing the change in position of a fourth point. This can be formulated using a technique similar to that used in section 17.2. Unlike section 17.2, the solution is exact. The only requirement is that the image positions of the four corresponding points are available before and after the transformation. Tracking may be used to determine the correspondence [Crowley et al. 89].

Can focus be corrected with an affine correction matrix $^2_1\mathbf{D}_a$ or is the full projective form required? To investigate, we placed our cube at a distance of about 105 cm and set the focus to the corresponding distance and calibrated. The images used for calibration are shown in Figure 17.10 and are the same used for the first example presented above.

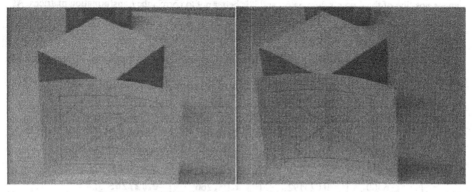

Fig. 17.10 Image of cube with X in front, used for focus experiments.

After our initial calibration, we placed a paper with an x in front of the cube as shown in figure 17.10. We stepped the focus through 7 positions, while tracking 4 points on the "X". We corrected the calibration matrices using both the affine and projective correction methods, and then used the stereo images to reconstruct the upper three corners of the cube. As a control, we also reconstructed without correction. We computed the average distance between the observed position of the three points and the true position. The results are shown in Table 17.1.

Table 17.1 Results of reconstruction of the top three points of cube, using an affine correction matrix (D_a) and a projective correction matrix (D_p) to compensate for focus. The cube is placed a distance of approx 105 cm from the camera head. The most precise value is indicated in bold. The fourth line is the initial calibration position.

Encoders	distance	no-correction	D_a	D_p
1	58 cm	0.092609	0.044556	**0.031695**
3000	72 cm	0.105037	**0.010446**	0.014922
6000	95 cm	0.032241	**0.007682**	0.008848
7000	105 cm	0.010183	-	-
8000	115 cm	0.030358	**0.051334**	0.066590
10000	162 cm	0.066598	**0.039397**	0.061256
14000	539 cm	0.155521	0.093470	**0.077903**

Table 17.1 indicates that the affine and projective correction matrices give very similar precision in reconstruction. We were surprised to observe that the affine correction matrix was more precise for small changes in focus.

For changes in aperture, the situation is more delicate, as illustrated by the following experiment. Using the same scene of the cube, we stepped the aperture of the left camera through 5 position near the middle of its setting. We then performed a 3D reconstruction for the seven visible corners of the cube and measured the average distance between the reconstructed and true points. The results are shown in Table 17.2. For small changes in aperture, both the affine and projective correction matrices are vanishingly close to identity. In fact, in such a case, the error due to quantization of the image position of the 4 points dominates, and it is better not to correct the transformation matrices. As the aperture opens, pixels begin to be shifted, and the projective correction began to give better results.

Table 17.2 Average Reconstruction error for change in aperture, with no correction, affine correction and projective correction.

Encoders	no-correction	D_a	D_p
5400	**0.016365**	-	-
5800	**0.016365**	0.016873	0.016877
6000	**0.016365**	0.016873	0.016877
6200	**0.022030**	0.022405	0.022409
6400	0.023192	0.032266	**0.021265**

In our stereo head the vergence rotational axis is nearly aligned with the principal point. In this case, we have speculated that the transformation due to vergence is closely approximated by a simple translation in the image, modelled by an affine correction matrix. Experiments show that this expectation is incorrect.

Table 17.3 shows results from an experiment in which the head was calibrated to a cube at a distance of 120 cm, and the left camera was then slowly turned. The encoder values and angle of the left camera are shown in the first two columns. The other three columns show the average error for reconstructing the 7 visible corners of the cube without correction ("no-correction"), with an affine correction D_a and with a projective correction, D_p. The affine correction was computed by tracking corners 1, 2 and 4. The projective correction was computed by tracking points 1, 2, 4, and solving for w_1 and w_2 with corner 5.

Table 17.3 Average precision obtained for reconstruction of the 7 points of the cube when correcting for vergence with an affine transformation D_a and a projective transformation D_p.

Encoders	angle	no-correction	D_a	D_p
2800	84.45°	0.010183	-	-
2950	85.20°	0.513628	0.019190	**0.018208**
3100	85.96°	1.405333	0.022504	**0.016943**
3250	86.71°	1.774919	0.031842	**0.019300**
3400	87.47°	2.117045	0.034800	**0.012026**

Without a correction, the reconstruction error grows rapidly (21% for a change of 3°). Both the affine and projective correction matrices limit this error to a few percent. However, the average error for the affine transformation continues to grow as a function of the vergence angle, while the error is roughly constant for the projective correction. Our conclusion is that an affine transformation provides an acceptable correction for small vergence angles, but that the most precise correction requires a projective correction matrix.

One might ask how the choice of points influences the precision of the correction. To measure this, we have performed an experiment in which the left camera was converged while tracking all six points of the cube. We then computed the correction matrix with all possible sets of four of the six points. Average 3D reconstruction precision ranged from 1.3% error to 6.2% error. More importantly, the reconstruction precision was proportional to surface of the area enclosed by the four points.

In this section, we have shown that a 3 by 3 correction matrix obtained by tracking image points can be used to correct for changes in camera vergence, focus and aperture. In the next section we show how a correction matrix obtained by tracking scene points can corrected for movements in the cameras.

17.4.4. Experiments in Precision of Reconstruction

Let us illustrate the precision of our technique with an experiment. Using the image in figure 17.9, we computed calibration matrices for the left and right cameras using the two techniques described above (direct and msq). We then placed a box of sugar next to the calibration cube, as shown in figure 17.11 and reconstructed the corners of the box using the matrices determined by the three techniques. The six visible corners of the sugar box are listed as points S_0 through S_5. The 3D error, measured as a percentage of the side of the cube, are shown for each of the 6 points.

Fig 17.11 Left and right images of cube and sugar box used for Table 17.4.

The first thing that we can observe is that each of the two techniques produces the most precise result for half of the points. The largest error was on the order of 10% for point S_1 using the direct method, while the smallest was on the order of 1% for point S_2 also using the direct method. None-the-less, our conclusion from these and many other experiments is that computing the calibration matrix using the mean-square technique gives a slight improvement in precision at a slight increase in computational cost. The direct method provides a 3D solution which is less precise but easier to program.

Table 17.4 Errors for reconstructed corners of sugar box using both techniques. All distances are in units defined by the side of the calibration cube (20cm). The most precise values are indicated in bold.

Point	Real Position	direct	msq
S_0	(0, −0.325, 0)	**0.0175**	0.0185
S_1	(4.75, −0.325, 0)	0.1065	**0.0948**
S_2	(0, 0, 0)	**0.0131**	0.0523
S_3	(0, −0.325, 0.95)	**0.0240**	0.0549
S_4	(0.475, −0.325, 0.95)	0.1082	**0.0372**
S_5	(0, 0, 0.95)	0.0750	**0.0549**

17.4.5. Keeping Scene Coordinates Locked on a Reference Object

The projective transformations ${}_s^L M$ and ${}_s^R M$ are rigidly attached to the cameras. If the head moves, the reference system is translated and rotated. We require a method to keep these matrices locked onto the object centered reference frame as the head moves. This is possible by constructing a correction matrix by observing the 3D position of the four reference points.

Let us designate the current coordinate system as 1. Moving the cameras from position 1 to 2 can have the effect of translating and rotating the reference frame. We can express the resulting matrix as a product of the previous matrix and a transformation.

$${}_2^L M = {}_1^L M \, {}_2^1 T \qquad\qquad {}_2^R M = {}_1^R M \, {}_2^1 T$$

To shift the reference back to the object, we require the inverse transformation, ${}_1^2 T$. We reconstruct the reference points in the new reference frame, 2, using the current calibration matrices ${}_2^L M$ and ${}_2^R M$. We then compose the four points into a matrix

$${}^2 R = \left[{}^2 R_1 \cup {}^2 R_2 \cup {}^2 R_3 \cup {}^2 R_0 \right]$$

We note that in the original reference frame, these same points represent the origin and the three unit vectors, expressed as:

$${}^1 R = \begin{pmatrix} 1 & 0 & 0 & 0 \\ 0 & 1 & 0 & 0 \\ 0 & 0 & 1 & 0 \\ 1 & 1 & 1 & 1 \end{pmatrix}$$

To calculate ${}_1^2 T$ we write

$${}^1 R = {}_1^2 T \, {}^2 R$$

We obtain the correction matrix by inverting ${}^2 R$.

$${}_1^2 T = {}^1 R \, {}^2 R^{-1} \qquad\qquad\qquad (24)$$

Table 17.5 show the results of an experiment with locking the reference frame to an object. We calibrated to a cube, and then translate and/or rotated the cube while tracking four of the corners. We reconstruct the new positions of the tracked corners and compute the correction matrix. We then update the calibration matrices and compute the corrected position of the 3D points. Error is measured by average distance between the reconstructed points and the true positions. The cube was translated in the x,y plane and rotated about the z axis by carefully moving it around on a table. The correction matrix was computed with points R_0, R_2, R_4, and R_5. The average error in reconstructing all seven visible cube corners is shown in table 17.5.

Table 17.5 The cube is displaced by 10 cm in x and then in y. It is then rotated by 16.7 degrees. The average error in reconstruction is shown before and after correction. An error of 0.5 represents half a length of the cube, or 10 cm.

scene	x, y (cm)	θ_z	No Correction	Corrected
2	10.0, 0.0	0.0°	0.500999	0.010139
3	0.0, 10.0	0.0°	0.496915	0.013008
4	0.0, 10.0	16.7°	0.420312	0.065857

To show that the technique works as well when the head is translated and rotated, we performed the same experiment while moving the head translated roughly 10cm and rotated about 10 degrees.

Table 17.6 Correction for three head movements.

scene	No Correction	Corrected
2	0.487617	0.014577
3	0.267168	0.005398
4	0.430925	0.020571

17.5. Refining a 3D Model with Multiple Observations

A description of the structure of the scene may be build up from multiple observations using techniques from estimation theory. Our system is based on the techniques presented in [Crowley-Ramparany 87] and developed in [Crowley et. al. 92a]. We limit our presentation here to a brief overview.

17.5.1. Representation for 3D Lines Segments

The scene model and observations are expressed as 3D segments represented by a pair of end-points, P_1 and P_2, along with their 3x3 covariances, C_1 and C_2, as shown in figure 17.12.a. In addition, the midpoint, direction and length are computed from the end-points. As with 2-D segments presented in chapter 2, these redundant parameters are used for matching and for perceptual grouping.

The primary 3D segment parameters are as follows.

P_1, P_2 The end points the segment (x, y, z).

C_1, C_2 The 3x3 covariances of the end points.

CF   The confidence of segment.

ID: The Identity of the segment (from the 2-D Models).

Each 3D segment also contains a number of redundant parameters shown in figure 17.12.b

P   The center point of the segment (x, y, z).

C_p The 3x3 covariance of the center point .

ΔP_s An un-normalized direction vector.

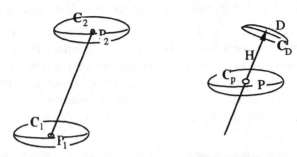

a)End-Point Representation   b) Midpoint-Direction-Length

Fig. 17.12 Representation for a 3D segment and its Uncertainties.

The parameters and uncertainty of a 3D segment are refined by multiple observations. Correspondence is furnished directly by the segment ID's, and is verified by comparing segments for similar orientation, co-linearity and overlap. The 3D segment parameters are then updated using the Kalman filter equations [Crowley et al 92a].

17.5.2. Estimating 3D Form by Multiple Observations

Calibrating to object centered coordinates permits 3D observations to be fused with a simple form of Kalman filter. Because segments are reconstructed in the same reference frame, the observation equation, which expresses the transformation from model to observation, reduces to an identity matrix.

$$
{}^{o}_{m}H = \begin{pmatrix} 1 & 0 & 0 \\ 0 & 1 & 0 \\ 0 & 0 & 1 \end{pmatrix}
$$

Updating the parameters in the 3D composite model follows directly from the Kalman filter equations [Crowley et. al. 92a]. Let us refer to P_o, C_o as an observed end point and P_m, C_m as a model end-point. For each corresponding pair of end-points, a Kalman gain vector is computed.

$$
K := C_m [C_o + C_m]^{-1} \tag{25}
$$

The end-point and its covariance are then updated by

$$
\hat{P}_m := P_m + K [P_m - P_o] \tag{26}
$$

$$
\hat{C}_m := C_m + K C_m. \tag{27}
$$

Each time a segment is observed, the confidence factor is incremented by one up to a maximum value. When the segment is not observed during an update, the confidence factor is decremented. If the CF of a segment drops below 1, the segment is removed from the composite model. Thus a segment must be present in at least 5 observations in order to be considered reliable enough to be preserved in the model.

17.5.3. Transferring the Scene Coordinates to a New Reference Frame

Our system is continually changing its fixation point. With each change it is convenient to change the reference frame for reconstruction as well. The simplest way to do this is to empty the 3D model, chose a new reference object and then reconstruct the model with respect to this new object. In some cases it is be desirable to transform the 3D model to a new reference frame. If the object is known, it is possible to completely recalibrate to it. However, it is also possible to "hop" the reference frame to any set of four non-coplanar points, by constructing a four by four homogeneous coordinate transform.

Let us designate the the original reference frame as "O" and the new reference frame as "S". To transform the reference frame, we need the homogeneous transformation from O to S: $_O^S T$. Such a transformation is easily constructed by observing the 3D positions of four new reference points in the current reference frame. Let us call these four reference points expressed in the original coordinate system: $^O R_0$, $^O R_1$, $^O R_2$ and $^O R_3$. To obtain $_S^O T$ we compose a matrix with the 3D position of these reference points.

$$^O R = [\,^O R_1 \cup \,^O R_2 \cup \,^O R_3 \cup \,^O R_0\,]$$

We note that in the new reference frame, these same points will represent the origin and the three unit vectors, represented by

$$^S R = \begin{pmatrix} 1 & 0 & 0 & 0 \\ 0 & 1 & 0 & 0 \\ 0 & 0 & 1 & 0 \\ 1 & 1 & 1 & 1 \end{pmatrix}$$

To calculate $_S^O T$ we write

$$^O R = \,_S^O T \,^S R$$

and then note that $^S R$ has a trivial inverse (as in section 2.) Thus $_S^O T$ is given by

$$_S^O T = \,^O R \,^S R^{-1} = \,^O R \begin{pmatrix} 1 & 0 & 0 & 0 \\ 0 & 1 & 0 & 0 \\ 0 & 0 & 1 & 0 \\ -1 & -1 & -1 & 1 \end{pmatrix} = \begin{pmatrix} x_1-x_0 & x_2-x_0 & x_3-x_0 & x_0 \\ y_1-y_0 & y_2-y_0 & y_3-y_0 & y_0 \\ z_1-z_0 & z_2-z_0 & z_3-z_0 & z_0 \\ 0 & 0 & 0 & 1 \end{pmatrix} \tag{28}$$

Thus the reference frame can be moved to any set of four non-coplanar points whose position is known (or observed) in the original reference frame by a direct formula.

Once the transformation matrix is constructed it is used to transform each of the end-points in the 3D model, as well as their uncertainty:

$$\hat{P}_O := \,_S^O T \, P_s$$

$$\hat{C}_O := \,_S^O T \, C_s \,_S^O T^T$$

The transformation is also used to update the calibration matrices to the new reference frame. Let ^L_OM and ^R_OM be the results of calibrating to the reference frame "O" as described above. We note that post-multiplying our calibration matrices by a homogeneous matrix ^O_ST transforms the calibration matrices to the new reference frame.

$$^L_SM = {}^L_OM\,{}^O_ST \quad \text{and} \quad {}^R_SM = {}^R_OM\,{}^O_ST$$

17.6. Conclusions

In this chapter we have presented an active 3D vision system which reconstructs the 3D structure in an object-centered "intrinsic" reference frame. This system is capable of real time response and continuous operation because we have separated the stereo problem into several simple components. These include:

1) The use of a coarse to fine convergence reflex to place the fixation point on an object.

2) The use of a proprioceptive measurements from encoders on the head motors to estimate the 3D position of the fixation point with respect to the head.

3) The use of a novel direct calibration technique to reconstruct the structure of an object in its own "intrinsic" coordinates.

4) The use of 2-D tracking to correct the camera calibration for changes in focus, convergence and shift in the retina or lense due to vibration.

5) The use of 3D tracking to keep the camera calibration locked onto the object.

6) The use of a simplified form of Kalman filter to integrate observations of object structure.

Real time response is possible because stereo matching is limited to a region of the scene around the fixation point defined by the horopter. A person can not read all of a book at the same time, nor can he see all of his visual world at the same time. At a macroscopic level perception is a serial process. The serial nature is a consequence of the computational complexity of stereo correspondence. A horopter serves to limit stereo data to a quantity which can be treated in real time. Such an approach also produces a simple form of separation of figure and ground.

An attempt to perform numerical 3D measurements with an active stereo head using classical techniques encounters a fundamental problem. The classical approach involves a time consuming camera calibration phase before 3D vision is possible. Such calibration is destroyed by changing the focus, aperture or stereo convergence angle. The calibration matrices can be destroyed by vibration as the system is moved. We have solved this problem by developing a technique by which cameras are calibrated by direct observations of points in the scene. We have also developed techniques which permit the calibration matrices to be updated to correct for changes in focus, aperture, convergence or movement of the head. We have also presented a technique which permits the reference frame to be hopped to arbitrary set of four non-coplanar points in the scene. Calibration to intrinsic object coordinates has a side effect of making the 3D fusion process very simple by producing 3D descriptions which are in a view invariant reference frame.

A number of improvements to the system are currently underway. The system currently operates using only edge segments. In the near term, we intend to extend 2D tracking and 3D reconstruction to operate on elliptical arcs. The code for extracting arcs is already in place in the system. Tracking requires a method to estimate (and calibrate) the precision of the parameters of the arcs. Extension to 3D will require a method to estimate the 3D position and orientation of an ellipse (or circle) from the stereo correspondence of 2D ellipses. We are currently working on a solution to this problem.

In the long term, we feel that edge points are an inadequate representation for the visual appearance of 3D structure. We have experimented with multi-resolution visual primitive for end-stops, spots, line elements, and corners. We are interested in methods for directly measuring 3D surface curvature from its 2D appearance. We believe that such approaches will lead to a much more robust description of the visual world.

Bibliography

[Crowley-Ramparany 87] Crowley, J. L. and F. Ramparany, "Mathematical Tools for Manipulating Uncertainty in Perception", AAAI Workshop on Spatial Reasoning and Multi-Sensor Fusion", Kaufmann Press, October, 1987.

[Crowley et. al. 92a] J. L. Crowley, P. Stelmaszyk, T. Skordas and P. Puget, "Measurement and Integration of 3D Structures By Tracking Edge Lines", International Journal of Computer Vision, July, 1991.

[Crowley et. al. 92b] J. L. Crowley, P. Bobet and C. Schmid, "Auto-Calibration by Direct Observation of Objects", Submitted to the Journal of Image and Vision Computing, August, 1992.

[Faugeras-Toscani 86b] O. D. Faugeras and G. Toscani, "The Calibration Problem for Stereo. Computer Vision and Pattern Recognition, pp 15-20, Miami Beach, Florida, USA, June 1986.

[Faugeras 92] O. D. Faugeras, "What can be seen in three dimensions with an un-calibrated stereo rig", The Second European Conference on Computer Vision (ECCV-2), St. Margherita, Italy, May 1992.

[Koenderink-Van Doorn 89] J. Koenderink and A. J. Van Doorn, "Affine Structure from Motion", Technical Report, Universtiy of Utrecht, Oct. 1989.

[Mohr et al. 91] R. Mohr, L. Morin and E. Grosso, "Relative Positioning with Poorly Calibrated Cameras", in Applicatiosn of Invariance in Computer Vision, DARPA-ESPRIT Workshop, Rejavik Iceland, March 1991. (Also available as LIFIA-IMAG Technical Report RT 64, April 1991.)

[Ramaparany 89] Ramparany, F., "Perception Multi-sensorielle de la Structure Geometrique d'une Scene", Thèse de Doctrat, INPG, Feb 1989.

18. Geometric Description and Maintenance of 3-D Objects in an Active Vision System *

Jan-Olof Eklundh, Göran Olofsson and Mengxiang Li

CVAP, KTH

18.1 Introduction

A hypothesis underlying the VAP project is that efficient high level processes for scene interpretation and object recognition are feasible in an active and continuously operating vision system. In particular it is suggested that the use of spatial *and* temporal contexts together with a purposive approach alleviates several of the problems that typically are associated with high level vision, like combinatorially growing complexity of the scene description and excessive computing times. In the purposive approach expectations play a central role and everything appearing in the scene need not be completely interpreted.

In the work so far we have considered recognition of objects of known categories and maintenance of these interpretations over time. To allow for partial interpretations both fully recognized objects and other information not necessarily tied to any specific object are maintained. On the other hand objects that have been verified and are being maintained can be represented parsimonuously.

We have investigated several different geometric representations. Reasoning methods have been developed in view of the specific conditions in an *active vision system*. They have been evaluated through experiments in an *integrated vision system* on simple indoor scenes with both known and unknown objects.

To make the problems and their attempted solutions more concrete for this system we have specified a test scenario used throughout VAP - a breakfast table. In this scenario there is a *ground plane*, i.e. the table top, on which objects are sitting. The objects can belong to a priori known categories, like cups, plates and packages, or be of unknown types. The objects within a category are chracterized by their shapes, but these shapes can vary. Objects can appear and disappear and also change places and the view of the scene may not always contain the whole table top. The scenario hence represents a semi-structured environment containing, but not entirely consisting of, manufactured objects appearing on a ground plane. Geometric descriptions have been considered as particularly appropriate for representing such structures.

The first aim of our work was to be able to perform recognition of known objects upon request. We wanted to explore the potential gain in using a top-down interpretation of image data alone and particularly in a system of active vision where the flow of information can be controlled from frame to frame. For the breakfast table scenario this means that we control the attention of the system to different parts of the scene for interpretation of specific objects. A typical request could be "Find a cup in that region".

---

* The work has been perfomed within the ESPRIT-BRA project "Vision as Process, VAP" (BR-3038). The support from the Swedish National Board for Technical and Industrial Development, NUTEK is also gratefully acknowledged.

Secondly we wanted to perform experiments on the maintenance of the previous interpretations. We then assumed that not only would the observer move but that there would also be object motion in the scene. The objective has not been to perform any advanced tracking or motion analysis. We have rather concentrated on making the interpretations stable with respect to small changes in position, orientation and view. In our experimental scenario we have had objects moving and also made objects appear, disappear and reappear into the view, while the our system maintain the interpretations. The project presently involves no study of manipulation, hence these events have been simulated.

After some early investigations we decided to represent the real objects in terms of primitives. We hypothesize the existence of these primitive objects rather than the real object instances themselves. The particular primitives we have used are *Geometric Ions*, GEONS [Biederman85].

Since maintenance of the interpretation plays an important role we have included time aspects in our representations. Time stamps are used to record the formation and confirmation of any hypothesis allowing the computation of a recency factor in the updating of any particular piece of evidence/hypothesis.

Our work builds on several contributions to the fields of scene interpretaion, especially those by Biederman [Biederman85] and Dickinson et al [Dickinson90; Dickinson91; Dickinson92a; Dickinson92b]. However, the emphasis on vision as an active procss implies several additions to these approaches.

18.2   Geometric Representations

As mentioned above our system should be able to represent generic objects in a semi-structured environment. Generally such objects will be man-made and in principle possible to describe geometrically, qualitatively as well as quantitatively. Quantitative representations use parameteric 3-D models of the objects intended for costly fitting of image feature data. The qualitative approach involve no such reconstruction of the 3-D model. The latter is to be preferred in an active recognition situation as the processing of each frame should be minimized. However, the parametric models may be better suited for tracking and manipulation. For a deeper investigation of the two representation principles and their qualities we refer to the report [VAP-ir.c.1.1]. In this there is also further motivation for our choice of a qualitative approach presented below.

The representation scheme we have chosen is based on the work by Biederman [Biederman85]. In his theory for *recognition by components* he suggest a representation based on certain volumetric primitives called GEONS. The hypothesis is that recognition and model indexing can be performed from silhouettes, if this representation is used. The subset of GEONS we have used is shown in Figure 18.1.

The GEONS are primitive objects, qualitatively distinct in geometric shape. Their visual appearance are determined by their faces and how they are connected and the silhouettes. Basically they are characterized by their contours being straight or curved. We have developed thechniques to perform this characterization which will be described below. Dickinson has demonstrated an effective way to segment these primitives into their qualitatively distinct aspects, faces and face features, organised and related in graphs [Dickinson90; Dickinson91; Dickinson92a; Dickinson92b]. Also Pentland has presented work along the same lines[Pentland86].

Quantitative descriptions are needed to refine the representations of the geometric primitives, that is the GEONS. They include parameterizations of the primitives themselves and also metric constraints on size and curvature, number of faces etc.

Such geometric descriptions are feasible for object representation as well as for representing intermediate results and unrecognised scene structures. Given that we use GEONS as a basis for our object modelling, we need to allow for partial interpretations of them. Dickinson uses qualitative

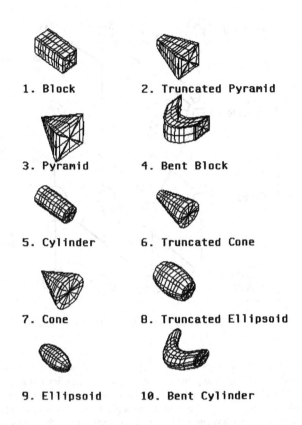

1. Block   2. Truncated Pyramid

3. Pyramid   4. Bent Block

5. Cylinder   6. Truncated Cone

7. Cone   8. Truncated Ellipsoid

9. Ellipsoid   10. Bent Cylinder

Fig. 18.1. The GEONS used in the system.

aspects. This means that only the aspects of a given GEON are represented that are qualitatively different in the sense that they have different compositions of visible faces. The faces are also represented in the same fashion. Faces that differ qualitatively have distinct representations. The features used for representing these faces are grouped edges. All edge groups differ in qualities like parallelism, symmetry and intersection of edges. Edges can be either straight or curved (convex/concave), and no measures of curvature are represented. Figure 18.2 illustrates this hierarchy.

18.3   Contour Processing

As presented in the previous section, the basic features used in the system are straight and curved contour segments. What we typically can get from image processing are linked edge points or contours. These need to be segmented at proper locations and further classified into straight and curved segments such that the segmented parts correspond (at least partially) to the face boundary groups in the aspect hierarchy. We notice that edge processing usually does not detect object boundaries perfectly. The edges can be broken due to various imaging problems. We would like to be able to group those broken segments into longer ones as much as possible.

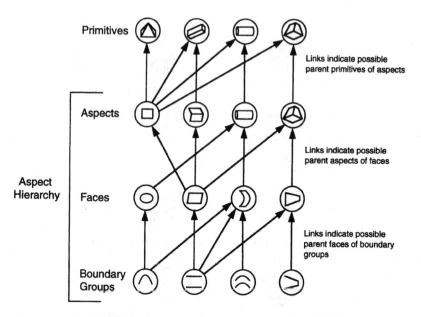

Fig. 18.2. From the grouped features faces are inferred...

To meet these requirements we have developed an algorithm for curve partitioning, colinear and cocurvilinear structure grouping, as well as for contour classification in terms of straight and curved [Li92]. The algorithm is based on the information theoretic *minimum description length* (MDL) principle and curve fitting techniques. For a given set of data (a contour), we first detect some candidate points, e.g. curvature extrema and zero crossings or the points from some polygon approximation. For each candidate point, we generate different hypothesis of the contour around the point, e.g. a straight segment, a curved segment, two connected straight segments, or two connected curved segments. For each of these hypotheses we compute a coding length of the data according to the MDL principle. The shortest coding length explains the data best and indicates whether to break up the segment or not. This principle is used both for colinear as well as cocurvilinear segment grouping.

In principle the same idea could be applied for contour classification, but from experience we know that this does not give reasonable results which can easily be explained as the MDL is purely data/model driven. So far we have only used the models line, circle and ellipse. Furthermore the data contains noise and the objects are rarely exactly like any of these models. This complicates the decision process of classification. To account for these problems we have also developed a more sophisticated classification procedure which combines MDL with other information. We could use for example the distance between the segments end points or the corresponding area bounded by the segment and the line between its end points. In the Figure 18.3 we show an example of the results produced by this method.

In principle the problem of curve partitioning is a problem of perceptual organization which has been studied by many researchers, e.g. [Asada86; Lowe85; Fischler86; Mokhtarian86; Wuescher91]. Our approach has a solid theoretical ground, the MDL principle. It is also *robust* to outliers and noise, and *invariant* under rotation, translation, uniform scaling of planar shape as well as projection.

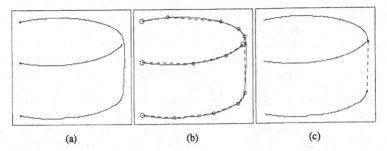

Fig. 18.3. An Example showing our contour processing approach. (a) the input data which is linked edge points, forming contours. (b) the result of a polygon approximation to produce candidate points. (c) the result of our approach: contours segmented at proper locations, the two curves belonging to the ellipse have been grouped into one and finally the grouped segments have been classified as straight (dashed line) and curved (solid) segments.

18.4   Recognition and Description

Many 3-D object recognition systems have been developed the last 30 years since Roberts historical system was presented. To simplify the problem and sometimes to expand the model domain, researchers has taken various different paths. To our knowledge no one has yet used any idea similar to our concept of *Vision as Process*.

The process of recognition is traditionally thought of as a single step matching of an object model with the available extracted image information. The matching principles can vary from strictly top-down verification of a model to full bottom-up hypotheses comparison of competing interpretations, using no model prediction. The models used are even richer in variation. The most common one is to use polyhedral models only. Systems based on matching polyhedrons can in general not be generalised also to a recognise objects with non-planar faces.

The *Recognition-by-Components* approach suggests a methodology which differs from earlier attempts in many respects. Biederman has suggested RBC as a theory for human object recognition. In the computer vision field Pentland and Dickinson recently has been most successful in implementing these ideas [Pentland86; Dickinson91]. The underlying assumption is that the human visual system has no explicit representation of the full geometry of all the objects that can be recognised. That would be unreasonable considering the enormous amount of information that would be involved. Instead the suggestion is that humans segment objects qualitatively into well-known basic geometric primitives, *Geometric Ions*, and "recognise" model objects as some joint configurations of these.

The work of Dickinson indicates the possibility to use geometric qualitative segmentation further to construct hierarchical part graphs of the GEONS. Each GEON primitive would then be partitioned into a number of distinct aspects, faces and face contour groups (see the Geometric Representation section above). Using a hierarchy of such geometric segmentations the matching can be performed at any level of abstraction, by top-down predictions or fully data-driven.

Mainly because of the potential of this last quality of Dickinsons system, called OPTICA, we have used it as a starting point for our experiments on 3-D object recognition in active vision.

In such a framework we can experiment with the matching procedure, we can chose to make partial interpretations and mix verifications of top-down predictions of missing information with bottom-up hypotheses "brain storming". Our wish to be able to stepwise verify predicted objects can be met. We can use the level of representations we find feasible for the particular state a certain hypothesis is in.

For this we have taken OPTICA as a starting point in our recognition tests. To fit our needs in the active environment we use more predictive processes and also represent partial matches as possible candidates to be further verified by predictions into the coming stream of information.

Presently our system has no contextual knowledge of the scene contents but this is being included in ongoing work. The way this can be done is by planting object hypotheses into the internal scene interpretation memory. If a specific object has previously been recovered it will immediately be retrieved from a scene object memory. If on the other hand the object instance is unknown to the system it will generate a geon prediction corresponding to this. The scene object memory also generates predictions of the same kind to verify and maintain its contents over time.

Now these predictions are used to impose top-down interpretations of the generated structured geon hypotheses, aspect graphs and face graphs that could be generated from the low level input of classified edges (curved/straight) guided by interest points. For more details on the geon/aspect/face decomposition see Dickinson [Dickinson90; Dickinson91; Dickinson92a; Dickinson92b].

To draw the attention to possible edges intersections and junctions in forming face we use traditional techniques based on interest points [Förstner87]. Presently the face composition rely on the extraction of closed contours from the classified edges. (refer also to earlier sections for details on these data)

The actual recognition is performed using all partially matching candidates and recording the best candidates in the data base. Partial matches will be kept, since they generate predictions about more objects. A schematic description is given in Figures 18.4 and 18.5

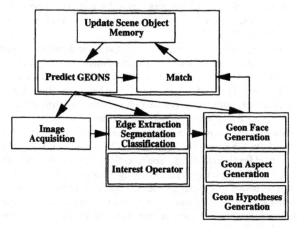

Fig. 18.4. The system is cycling in a Predict-Match-Update loop. The predict step use *Focus of Attention*, FOA, towards Image Aquisition, Edge Processing and Interest Operator, mainly by the use of *Region of Interest*, ROI. The GEON hypothesising step, including associated Aspects and Faces, are also tapered to prefer predicted structures. Succesful matches updates the Scene Interpretations with increased belief values while match failures decrease the predicted structures belief values.

18.5   Dynamic Representations and Feedback Reasoning

In an active vision system the activity and processing are exercised under a continuous flow of information. As time goes by more and more information is acquired. The system can take advantage of this flood but most not drown in it. The representation of the acquired information and interpretations

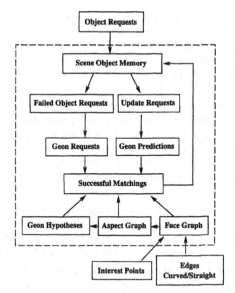

Fig. 18.5. The Scene Object Memory provides predictions of Geons in two ways, through full top-down object predictions and simle update predictions of previously found objects. The predictions are used in the top-down Matching with GEON structures, their Aspects and Faces. Faces are hypothesised from groupings of curved and straight edges where the intersections are predicted by the use of interest points.

derived should be provisions for maintaining already obtained knowledge. Furthermore, only objects under attention and in the field of view should be represented in some detail. The level of detail is determined by the task at hand, here interpretation. New hypotheses should hence only be generated when an object is seen and can be understood.

This ideal situation can seldom be reached because generally partial competing hypotheses are instantiated for unverified objects. We therefore maintain active hypotheses through matching with new supporting information - either upon request or automatically when such information is available.

Given that a hypothesis is maintained over time we can take advantage of the continous flow of information for verification. By using a predictive reasoning process new information can be aquired in a focused manner to resolve current problems in the interpretation process. This constitutes the active attentional mechanism at the high level. By the use of *active control of perception* the reasoning could even control the flow of low-level information.

Generally the *active vision* paradigm assumes the notion of a changing environment, changing viewing conditions and varying tasks for the visual system emanating from the needs of the seeing agent. This does not only imply that the system must act over continous time, but also that it has to adapt to such kind of external changes. There should be active and reactive behaviors. The system is supposed to interpret the scene in the terms of the primitives used, react on their dynamical changes, changes in appearances and disappearances and changes in views. The representations must be adapted correspondingly.

If the scene contents changes, that will directly affect the contents of the scene object memory but if only the viewing conditions vary, the system just need to compensate for that in the representation of its *view* of the scene, no object representation will then be affected.

Moving objects could in principle be tracked using parameterised representation of the primitives.

A major portion of the contents of the scene object memory will contain partial interpretations and tentative hypotheses. To be able to track the contents of the scene object memory, in all its diversity, the hypotheses must be related to their corresponding low level features and then tracked. That is the only way to ensure that the tracking will work for these types of representations in all their stages of verification. A drawback with such an approach is that there would be a large, complicated and hardly maintainable structure to maintain.

Let us instead assume the scene events to have moderate velocities and accelations compared to the cycle time of our predict-match-update loop (see next section on Maintenance of Interpretations). We then can expect the changes in the scene to be small in relation to what we represent. Following this idea we have experimented with a combination of reasoning and representation that is robust to minor changes in the input. The geometric representation will have a physical tolerance in pose and position directly related to a *region of interest* used in the reasoning process steps of prediction and matching.

The primary objective for the dynamics in our system is naturally to maintain the present knowledge in the scene object memory. For this we have included temporal information into the representations of the hypotheses and objects in the scene object memory. Most importantly we record the creation time and the time for latest update of each hypothesis, verified or partial.

To also be able to account for the changes in the the scene structure every hypothesis will have a *match cost* associated with it. This is a measure of the distance between the predicted structures within the associated *region of interest* and the factually extracted scene structures. We can thus indicate how well the latest match of the associated structure actually fit the features in the predicted *region of interest* for every hypothesis. The measure can be used to improve the prediction in the following cycles, but this has not yet been implemented. Furthermore the cost can be used to discard inferior competing hypotheses that are further away than the "best match" in relation to the last prediction. Using this cost principle we can treat the cases with slowly moving objects, egomotion compensation and other changes of viewing conditions.

Finally objects may enter or disappear from the field of view. That is treated by the maintenance function using slowly decaying memories of the hypotheses (see the section Maintenance of Interpretations). Presently we do not represent the dynamics of any individual object, i.e. motion is not calculated. Moreover the reasoning process has been constructed to work actively in relation to the dynamic state of the system or to adapt to the changes of goals that may follow.

The object hypotheses in memory need to be continuously updated and verified regardless of any events. If an object has no missing components we only need partial matches to verify the continous presence of the object. These objects have no concrete components related to them. Only the necessary structural information, scene connections and component relations are maintained. If no new objects are requested by the outside world, more processing can be used for verifying and maintaining the scene object memory. The instances in the memory that have been updated latest will be then have to wait for the "older" candidates to be updated first. As long as the object is not fully verified the maintenance of the interpretations previously stored in the memory is very similar to the recognition of completely unknown object instances. I.e. as long as we have not identified all geon components of the object or the object request has not been canceled due to information from another level we try to recognise the object along the same principles as described in the Recognition section above.

In principle all predictions are treated with equal priority. If however the seeing agent wants to look for an object and the request can not be met by the scene object memory, a top-down object prediction with high priority is generated. Then the updating of the memory contents will executed in the subsequent cycles.

18.6 Maintenance of Interpretations

Some of the qualities of an *active vision system* has a direct connection to the control of the flow information used by the system. In the interpretation maintenance we try to form stable picture of the environment as to be able to discard most of the information that cannot be used for either detection, verification or tracking of dynamic scene structures.

The system shall furthermore not be overwhelmed by the rich variations in the world. To avoid an explosion of the internal memory requirements interpretations shall not be discarded for newer ones until they are definitely out of date or proved erronous. A tolerance to minor changes in the world and viewing conditions will make this even more effective.

In principle, we may say that by reduced reactivity and tolerance each hypothesis tries to survive at the expense of newer hypotheses. In this the scene will be perceived more stably by the system than what is reflected in the flow of information. The exact relation between the reactivity of the system and the scene changes can be tuned within the different processes.

Interpretations already reached by the system need to be updated over time using the continous flow of information that can be aquired in a controlled manner. Furthermore, as mentioned in the previous section, the reasoning process needs to adapt to external as well as internal events. Such events include cene or view changes and change of *focus of attention* or can occur as a response to needs by the seeing agent. Finally it must be possible to delete interpretations that are not valid anymore. There may be no support for them in the information that can be aquired or they may just be inferior to some other physically overlapping hypothesis. Hence the system must be capable of *forgetting* parts of what it previously "thought" was the true scene structure.

For both these kinds of demands on the system a maintenance principle is required. The *predict-match-update* loop is a familiar approach dealing with dynamical systems. We think it is an obvious model also here. We want to keep a model interpretation maintained in a changing environment. So to be able to update our knowledge of the model, we *predict* its presence and nature so that the next time instant we can verify it against our perceptual information. If that prediction is verified it is called a *match* and the new knowledge is recorded in the model by an *update* procedure (see fig). There are advanced principles for such of predictive systems. For example a Kalman filtering approach is a possible alternative to our straight forward use of the *predict-match-update* principle.

A *Scene Object Memory* is realized by a simple database. Every memory/data base item is continously updated using the latest supporting information. The support is measured by the degree of match that can be met between how the item is predicted to be and how it came out in the acquired information, a *match cost*.

The information used in the interpretation process will vary over time even if the world or the viewer does not change. The key to the interpretation maintenance in our system is to attach to every hypothesis two *time stamps*, the creation time and the time for latest update (see also the previous section). Partly they are used to calculate a *recency factor* for each object hypothesis in the scene object memory, which indicates the time that has passed since the latest update or verification attempt was made on this particular memory item.

A *Region of Interest, ROI* of the present view is associated with each update request. If any of the components of an object can be fully or partially verified in its ROI that object will be updated. This ROI could be active selected by the seeing agent to account for egomotions and alternative prediction methods. Without such extra information the system can tolerate moderate motions by the observer or in the scene . Updating a memory object means that the update time will be changed to the present time and the belief factor of the object will be increased.

The belief factor is used in combination with a support value from the matching procedure to compare competing hypotheses. We use a simple approach with 5 integer values. Any object newly added to the memory will have a belief value of 3 associated with it. Each successful updating of verification will increase this value by one up to 5. If the updating or verification request was unseccessful this value will be decreased down to 0 when the object will be removed from the memory, i.e. forgotten.

18.7   Experimental Results

We now give a brief description of experiments with a system for geon based object recognition including maintenance of interpretations over time and under various changes of the scene.

In this environment we used a static camera for grabbing images on request. The object models used are BOXES and CUPS. They are modeled in terms of GEONS. A box as a single block and a CUP as a cylinder with handle modeled as a bent cylinder. Any CUP or BOX can have specific metrics attached to it to make it *a special kind of* CUP or BOX.

From a set of 19 images and 19 cycles in the maintenance process we have picked 7 to illustrate the functionalities of this system. The images between these seven contain no major scene or view changes.

From every image in the complete sequence we have extracted interest points and edges classified as straight or curved. (The extraction of these features are described in previous chapter).

In the 19 frames the objects of one box, and three cups have performed minor translations and rotations on the table top. The images 1, 3, 5, 8, 10, 13 and 15 with corresponding points, edges, closed contours, geon hypotheses and object memory are shown in fig. 18.6 thru fig. 18.12, at the end of the chapter. We have illustrated the interest points and edges. The closed contours in the next part of the figures are used for hypothesising geons. The approach here follow Dickinson [Dickinson92b].

We next show the geons found in this image and finally we present the status of the data base of recognised objects in the scene. The items in the data base are named and numbered and have an initial support of 3 when they are invoked. This supporrt or belief is incremented stepwise upto 5 if the hypothesis can be verified in the following cycles. Otherwise it will be decreased down to 1 and thereafter removed from the data base, or object memory.

The results show that when closed contours can be extracted and classified as faces or aspects of geons the maintenance is quite tolerant to major changes of the objects position and pose.

In the system there is a facility for controlling the region of interest, ROI. It is partly used for the updating of the scene object memory. Each previously recognized object only tries to verify itself in the vicinity of where it was last found. This has shown to be both good and bad. The good side of this is that as long as we can extract the supporting features close to a bounding box of the old hypothesis we can also maintain it with fairly high tolerance to changes. The drawback of using only closeness as criterion is that prediction of missing parts can only be performed in this ROI directly realted to the bounding box. These predicted missing parts have not often been possible to verify within such areas.

References

[Asada86] H. Asada, W. Brady: "The Curvature Primal Sketch", IEEE PAMI, Vol. PAMI-8, No.1, 1986, pp.2-14.

[Biederman85] I. Biederman: "Human Image Understanding: Recent Research and a Theory", *Human and Machine Vision II*, 1985, pp.13-57.

[Brunnström88] Brunnström K., Olofsson G., Eklundh J-O. (1988) "Hypotheses of Geometric Structure Generated in an Active Vision System", *Proc. IAPR Inter. Workshop on Computer Vision*, Tokyo, Japan

[Dickinson90] S. Dickinson, A. Pentland and A. Rosenfeld, "A representation for qualitative 3-D object recognitin integrating object-centered and viewer-centered models" In K. Leibovitc, editor, *Vision: A Convergence of Disciplines*. Springer Verlag, New York, 1990.

[Dickinson91] S. Dickinson, A. Pentland and A. Rosenfeld, "The recovery and recognition of three-dimensional objects using part-based aspect matching", Technical Report CAR-TR-572, Center for Automation Research, University of Maryland, 1991.

[Dickinson92a] S. Dickinson, A. Pentland and A. Rosenfeld, "From volumes to views: An approach to 3-D object recognition. *Computer Vision, Graphics and Image Processing: Image Understanding*, 55(2), 1992.

[Dickinson92b] S. Dickinson, A. Pentland and A. Rosenfeld, "3-D shape recovery using distributed aspect matching". *IEEE Transactions on Pattern Analysis and Machine Intelligence*, 14(2):174-198, 1992.

[Fischler86] M. Fischler, R. Bolles: "Perceptual Organization and Curve Partitioning", IEEE PAMI, Vol. PAMI-8, No.1, 1986, pp.100-105.

[Förstner87] W. Förstner, E. Gülch "A Fast Operator for Detection and Precise Location of Distinct Points, Corners and Centers of Circular Features", *Proceedings of Intercommission conference of ISPRS on Fast Processing of Photogrammetric Data*, Interlaken, Switzerland, June, 1987.

[Li92] M.X. Li: "Minimum Description Length Based 2D Shape Description", Tech. Report CVAP114, Dept. of Numerical Analysis and Computing Science, Royal Institute of Technology, 1992.

[Lowe85] D. G. Lowe: *Perceptual Organization and Visual Recognition*, Kluwer Academic Publishers, 1985.

[Mokhtarian86] F. Mokhtarian, A. Mackworth: "Scale-Based Description and Recognition of Planar Curves and Two-Dimensional Shapes", IEEE PAMI, Vol. PAMI-8, No.1, 1986, pp.34-43.

[Pentland86] A. Pentland, "Perceptual organization and the representation of natural form", *Artificial Intelligence*, 28:293-331, 1986.

[Olofsson86] Olofsson G. (1986) "Experiments with an Algorithm for Line Extraction", *Technical Report*, TRITA-NA-E8682, Dept. of Numerical Analysis and Computing Science, Royal Institute of Technology, S-100 44 Stockholm, Sweden

[Olofsson89] Olofsson G., Brunnström K., Eklundh J-O. (1989) "Hypotheses of Geometric Structure Generated in an Active Vision System", *Proc. SSAB-symposium*, Göteborg, Sweden

[VAP-ir.c.1.1] Jan-Olof Eklundh, Mengxiang Li, Göran Olofsson "Representations for 3-D geometric modeling" *ESPRIT Basic Research Action 3038 "Vision As Process, VAP" Internal Report IR.C.1.1* May 29, 1990. Dept. of Numerical Analysis and Computing Science, Royal Institute of Technology, S-100 44 Stockholm, Sweden

[Wuescher91] D.M. Wuescher, K.L. Boyer: "Robust Contour Decomposition Using a Constant Curvature Criterion", IEEE, PAMI, Vol. 13, No. 1, Jan. 1991, pp.41-51.

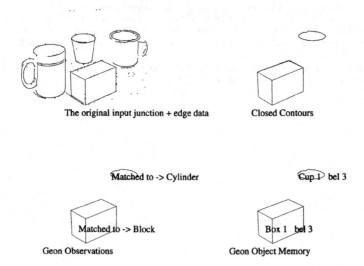

The original input junction + edge data   Closed Contours

Matched to -> Cylinder   Cup 1 bel 3

Matched to -> Block   Box 1   bel 3

Geon Observations   Geon Object Memory

Fig. 18.6. The initial setup. One cup and the box are recovered and assigned the initial belief value of 3. (Corresponding images 0, 1, 2).

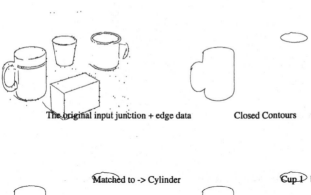

The original input junction + edge data   Closed Contours

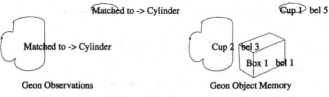

Matched to -> Cylinder   Cup 1 bel 5

Matched to -> Cylinder   Cup 2 bel 3
                                                      Box 1 bel 1

Geon Observations   Geon Object Memory

Fig. 18.7. One more object recovered, Cup 2. Cup 1 has managed to verify its closed top contour the last frames so that the belief value has increased to its maximum of 5. No supporting closed contour for the block geon of the Box 1 thus its belief value decresed accordingly. (Corresponding image 3).

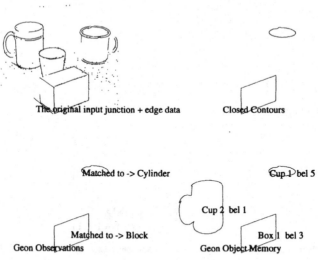

Fig. 18.8. Box 1 has been recovered in a new position close to its original, thus verified and increased belief. Cup 1 still does not fail. The Cup 2 has been moved out of the vicinity of its original position and cannot at all be recovered. (Corresponding images 4, 5, 6).

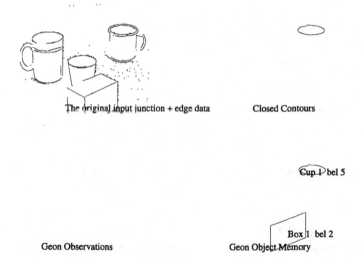

The original input junction + edge data   Closed Contours

Cup 1 bel 5

Box 1 bel 2

Geon Observations   Geon Object Memory

Fig. 18.9. Box 1 cannot be found but still lives on old belief values. Cup 2 has been deleted due to lack of support. (Corresponding images 7, 8, 9).

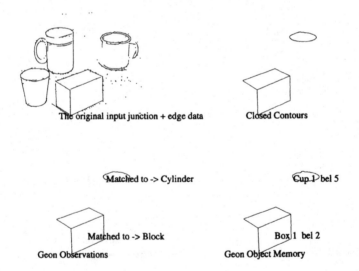

The original input junction + edge data   Closed Contours

Matched to -> Cylinder   Cup 1 bel 5

Matched to -> Block   Box 1 bel 2

Geon Observations   Geon Object Memory

Fig. 18.10. Box 1 has been moved but due to that, recovered and improved belief value. (Corresponding images 10, 11).

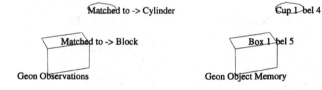

The original input junction + edge data   Closed Contours

Matched to -> Cylinder   Cup 1 bel 4

Matched to -> Block   Box 1 bel 5

Geon Observations   Geon Object Memory

Fig. 18.11. Box 1 has again moved but can still be strongly recovered with resulting high belief. Cup 1 has now been moved and turned with a little loss in belief due to problems of contour recovering in a frames 12 and 13. In frame 14 a new contour could be found and thus the final result of belief 4. (Corresponding images 12, 13, 14).

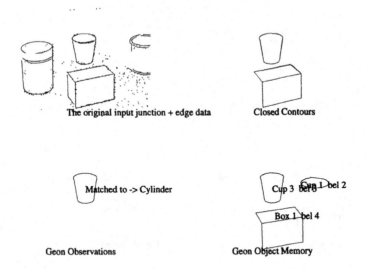

Fig. 18.12. Box 1 has problems with strange closed contour; doubled edges. Cup 1 has move out of image so no supporting data, but we remember it for a few frames with decreasing belief. New Cup 3 extracted from outer countour of a cylinder. (Corresponding image 15).

Part V

Active Scene Interpretation

From the beginning of the project, the VAP consortium agreed that contextual information (both spatial and temporal) and goal information are necessary to build a fully integrated and continuously operating vision system. To achieve continuous operation it is necessary to introduce control mechanisms which allow selection of:

- Where to look (gaze selection and control)
- What data to process (selection of Focus of Attentions) and
- Which operators to apply to such data.

In a continuously operating system the amount of information received is a problem in its own right. To ensure that internal models in the system does not grow out of bound, mechanisms for "intelligent forgetting" must be developed to ensure that the indexing of object models would not turn into a major obstacle. Initially the consortium decided to focus on goal driven processing, expecting that event driven processing would by added to the system as it evolved. This decision implied that both gaze selection and selection of visual cues is driven by either a centralised controller termed "the supervisor" or the scene interpretation module. In practise both methods were tested in the context of the skeleton system.

Most earlier research in scene interpretation and control has been carried out in the context of static images. Thus, well known systems such as ACRONYM [Brooks 1981], 3D Mosaic [Herman-Kanade 1986], 3DPO [Horaud-Bolles 1986] VISIONS [Hanson-Riseman 1978] are all based on the interpretation of a single images. These systems have in principle had access to an infinite amount of resources in the sense that response time was not a major issue in the design. In recent years there has been an effort to define an architecture that would allow continuous operation within the VISIONS system [Weems et al. 1989]. It is evident that the elimination of temporal context has simplified these systems, but the inability to obtain information from multiple view-points has at the same time constrained the interpretation of the images. In a number of situations it was necessary to develop detailed and highly specific methods to cope with ambiguous situations.

As few system which are able to interpret dynamic scenes are available, the VAP effort has had to build from scratch. This implies that there were free from constraints imposed by previous methods but also that there were few cues to guide the research. At an early stage in the project it was decided to divide the high level part of the system into an interpretation module and a centralised controller (the supervisor). The supervisor is responsible for transformation of user goal into actions that may be executed by one of the system modules. In the course of doing this the supervisor takes contextual information about the specific scene into account. Information available in the temporal context is used to direct attention towards areas in the scene where it is expected that objects of interest may be found. Once a visual goal has been defined, it is forwarded to the scene interpretation module or to any of the other system modules. In the initial design it was decided to test a hierarchical control scheme. In such a system, information to or from the supervisor goes through the scene interpretation module. However, the skeleton system was constructed to permit the supervisor (or any module) to communicate with any other module, so that experiments in distributed control are also possible.

The first chapter in part V discusses the problem of control from a general point of view and addresse architectural issues. This work has lead to the development of a general module architecture which is replicated at all levels of the skeleton system. The architecture is based on the hypothesise-match-update cycle (described in parts I and II) onto which a set of controllable access and utility functions have been added to facilitate both strict top-down/bottom-up communication and a black-board type of inter process communication. Based on general considerations, a centralised controller is then defined. This controller has been implemented as a rule based system which allows both centralised and distributed control of the entire system.

In chapter 20, the control architecture for the scene interpretation module is described in detail. Based on requests from the supervisor, the interpretation module uses its local scene database to define a gaze direction, a region of interest and to select a set of knowledge sources (recognition procedures) to be used in resolution of the goal command. For temporal maintenance, the control architecture has facilities for first order prediction of the new position of scene objects which may then be verified by the knowledge sources.

Chapters 21 and 22 describe the knowledge sources used for recognition of polyhedral and cylindrical objects. Both kinds of recognition procedures are model-driven in the sense that they exploit an encoded description of the set of possible objects to drive the recognition procedure. In the recognition of polyhedral objects, a hierarchical approach based on matching of groupings of straight line segments is used. Verification procedures are invoked to eliminate false hypotheses within the set of hypothesised objects For recognition of cylinders, a combined strategy which analyses both sets of straight line segments and their relation to ellipses is exploited. Initially, pairs of line segments are found. These segments may correspond to the occluding contours. A procedure for estimation of ellipses is invoked close to the end of these line segments. For both types of knowledge sources, experimental results that demonstrate the performance of the procedures on natural images are provided.

In chapter 23, a strategy for selection of view point is presented. This strategy is based on the information in a scene database and the system goal. Methods are described for handling of uncertainty associated with the projective transformation.The technique by which this information may be used in the selection of a gaze direction and a distance to the objects of interest is then presented. Experimental results are provided which demonstrate the utility of this strategy for an artificial scene.

Bibliography

[Brooks 1981] Brooks, R.: Symbolic Reasoning among 3-D models and 2-D images, *Artificial Intelligence*, **17**, pp. 285-348, 1981.

[Hanson_Riseman 1978] Hanson, A., & E. Riseman, Visions: A computer vision system for interpretating scenes, in *Computer Vision Systems*, A.R. Hanson & E. Riseman (Eds), Academic Press, New York, N.Y., pp. 303-334, 1978.

[Herman-Kanade 1986] Herman, M. & T. Kanade, The 3D Mosaic Scene Understanding system, In *From Pixels to Predicates*, S. Pentland (Ed.), MIT Press, Boston, Mass., pp. 322-358, 1986.

[Horaud-Bolles 1986] Horaud P. & B. Bolles, 3DPO: A system for matching 3-D Objects in Range Data, In *From Pixels to Predicates*, S. Pentland (Ed.), MIT Press, Boston, Mass., pp. 359-370, 1986.

[Weems et al. 1989] Weems, C., E. Riseman and A. Hanson, The Image Understanding Architecture, *Intl. Jour.of Computer Vision*, Kluwer, **2**(3), Jan 1989.

19. Control of Perception

Henrik I.Christensen and Erik Granum
LIA, AUC

19.1. Introduction

Control in general is aimed at achievement of a pre-specified performance for the system being controlled. The control is performed through change of process parameters based on knowledge about some 'performance index', information about present and past behaviour of the system, and a model of the system being controlled.

In the VAP project the aim of control is two fold: a) achievement of continuous operation and b) achievement of user specified goals (tasks). Goals may be divided into two categories according to their origin and duration. One set of goals is those specified by an external user, i.e., an industrial robot. The other category is implicit goals, which are internally defined to ensure proper operation and maintenance of information already obtained.

In this chapter the system level control issues addressed in the VAP project are outlined and discussed from a general point of view. Initially a general set of considerations related to controllable parameters, system organisation, and control strategies will be discussed. This discussion will form the basis for a description of the control methods which were adopted in the project. A production system for reasoning about goals and actions is outlined and its relation to other components in the system is discussed. Finally a set of lessons learnt, in term of control, from construction of the initial skeleton system is considered.

19.1.1. The context of vision

Recent research in computer vision has argued that vision should not be studied in isolation but rather as an advanced sensory modality for an (autonomous) agent. This implies that the, potentially time varying, task for the vision system is to provide information needed by the agent to carry out its task. In this context the vision system may be considered a single node in a closed loop control cycle, as shown in figure 19.1.

In this cycle the 'sensing' system received input to the planning module (goal specifications), and the environment (external events). In addition it produces world model information to the planner while it services may be utilised by the action module in tasks such as visual servoing. The vision system must thus have facilities for handling/satisfaction of not only user specified goals but it must also have facilities for detection and handling of unexpected event in the external world.

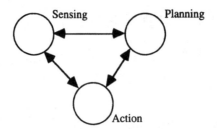

Fig. 19.1. The system context for use of a vision system.

19.1.2. The responsibility of control of perception

Control of perception may be divided into two categories according to its influence on the system operation: a) system level control, and b) module/algorithm level control. In category a) the aim of the control is coordination of processing and analysis to ensure continuous operation and satisfaction of externally specified goals. The task in this category is thus break-down of goal commands through planning into a set of sub-goals that may be accomplished by system modules. In addition monitoring of the execution by different modules allow change of the set of active modules and potentially change of module specific parameters. The main aim for the change of parameters is to allow handling of events and limitation of internal models to ensure the implicit goal related to continuous operation. the system level control is also responsible for scheduling of access and control of shared resources such as the camera head and the associated extrinsic and intrinsic parameters. In category b) (module control) the control task must ensure that processing of each new image (sample) is accomplished within a fixed time frame and that queries from other modules are honoured to allow such modules to complete their processing with their designated time-frame. The control at module level includes issues related to: What input data to request, to process, what data to produce for other modules, how to maintain local data to ensure 'optimal' performance. Much of the past research has been aimed at module level control in the form of selection of what regions in an image to process.

19.1.3. Past research

In computer vision very little work has been conducted on system level control. This may be due to the fact that very few fully integrated systems are available today. The most well known system capable of doing real-time analysis of images is the road-following system developed by [Dickmans et al 1988]. In this system the regions which are likely to contain the boundary of the road are predicted at three different locations with respect to the car: just in front of the car, 50 and 100 m in front of the car. The first region is used for immediate adjustment of driving parameters, while the other two regions are used for estimation of road curvature and thus used in a closed loop control loop which has a slower time response.

Another system which has been described extensively in the literature is the ALVEN system from Univ. of Toronto, which is aimed at analysis of ultra sonic images of the left ventricular. The control strategy used in the system is a predict-match-update cycle similar to the method used in this project. The control cycle for the ALVEN system is described in [Tsotsos 1987a; Tsotsos1987b].

Most of the remaining work on control is related to selection of the regions to be processed in the image. Examples of such work are [Califano et al 1991], [Rimey-Brown 1991], [Christensen et al 1990] [Christensen et al. 1991], [Jensen et al. 1992], and [Culhane-Tsotsos 1992]. Most of this work in based on a two phase processing model. The first phase is related to model invocation where cueing features are extracted using a pre-attentive strategy. Based on the set of hypotheses generated from the cueing features an attentive strategy, which facilitates verification/rejection of some of the asserted hypotheses, is applied. A good review of techniques for selection of attention and control applied in a number of active vision systems may be found in [Abbott 1992]

19.1.4. Outline of chapter

Initially the subjects of control, which may be changed to achieve a particular performance by the system, are outlined and discussed. Based on a categorisation of resources various strategies to control are presented and discussed in the context of system architectures. Based on rather simple control considerations the system architecture used in the project is outlined. The system architecture contains a centralised controller (a supervisor) which is responsible for coordintion of control, and communication towards an external user. The planning contained in the supervisor is subsequently described based on a description of a set of generic goal commands. Finally a sample implementation, which has been used to demonstrated the feasibility of the adopted approach, is described. At the end a number of lessons learnt throughout the project are outlined and a number of observations are summarised.

19.2. The subjects of control

In dynamic computer vision systems the use of resources is an important problem which must be controlled in order to achieve real time performance. Resources, which always will be too limited in the general case, may be categorised into three groups:

- Acquisition resources
    1. where to look (selection of view position and fixation point)
    2. selection of dynamic sensory parameters.
- Processing resources
    1. which data should be analysed?
    2. which algorithm should be applied?

- Storage resources
    1. amount of data in internal models.
    2. detail of data in internal models.
    3. complexity of internal models.

The acquisition resources are associated with the binocular camera heads used in this project (see chapter 9). The introduction of camera heads allow selection of view points for the scene/object being analysed, and it also allow dynamic selection of extrinsic and intrinsic parameters for the sensory system. Through control of these parameters it has been shown that in particular calculation of intrinsic images becomes much simpler (see [Aloimonos et al 1987])

Control of processing resources (i.e., CPU usage) determines the amount, selection, and detail of data to be analysed / interpreted, and it may also specify the algorithm to use, if several options are available. The choice of algorithm depends on the quality of data required and the necessary response time. This can formulated as control of the type of data manipulation to be applied for the current and controlled "Focus of Attention" (FOA) of the system.

Control of storage resources is directed at selection / maintenance of a limited amount of data in the internal models of the system. The control of the amount of storage resources determines the total amount of data available. For visual perception the world may be described at a continuous range of scales along the spatial and temporal dimensions. The control of scale of interest will also influence the amount of data, as it specifies the detail of description associated with objects.

Note that various combinations of the goals and scene dynamics will call for different and dynamically changing requirements for the balance between spatial and temporal resolution in the models.

In addition to the amount and detail of descriptors in internal models it is also possible to control the complexity of models. This includes control of the type and number of inter-object relations, and it might also be aimed at control of the complexity of descriptors associated with single objects.

All of the issues outlined above must be combined in suitable strategies to achieve an acceptable system performance in the context of external goals and a dynamically changing environment.

19.3. Control strategies

19.3.1. The system architecture reviewed

The breakdown of the vision system presented in chapter 2 and the associated modules describes in subsequent chapters provides the basis for the system architecture and the associated set of controllable modules. The architecture is shown in figure 19.2.

A calculating agent in figure 19.2 is defined as a thing that acts. A more elaborate definition of such agents may be found in [Crowley 1989]. The agents does all the data processing/analysis/interpretation required based on prior knowledge available in the "local knowledge base" and input data. The processing is performed in reaction to goal commands received from the perceptual controller (also called the 'supervisor') or a higher level agent.

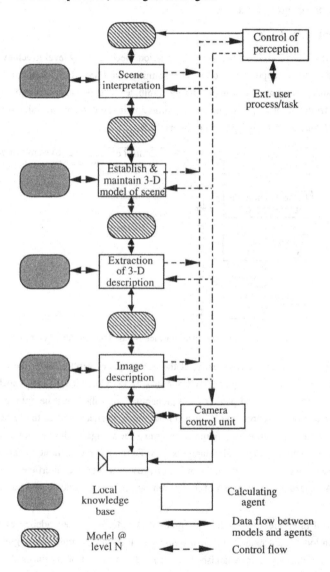

Fig 19.2. Outline of VAP system structure defined in chapter 2.

The "models" which connects neighbouring agents defines two-way asynchronous communication channels, which are used for data and/or control communication. The channels must be asynchronous as some of the agents operate at different cycle rates. I.e., it is not necessary for all agents to operate at the image acquisition rate. The dashed lines represent channels for communication of status and control information.

19.3.2. Centralised control

In a centralised control structure the planning of actions needed at any level to achieve system goals is performed by the perceptual controller. The planning is not necessarily static, in the sense that a complete plan is not required before processing is initiated. The perceptual controller receive status information from the modules and use such information in the dynamic planning of future actions. A centralised control structure is shown in figure 19.3.

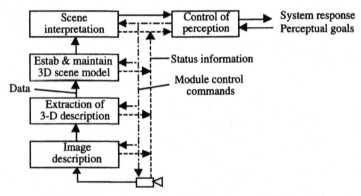

Fig 19.3.   A system structure which is based on centralised control.

As all planning is performed globally, at the perceptual controller level, the modules have a minimum of autonomy, and the global controller has detailed knowledge of the capabilities of each module. To perform task level planning the perceptual controller must be able to interpret local knowledge/data representations, to establish current status. Such a design of the system is in general undesirable as the integration of module/algorithmic knowledge in the perceptual controller will make it cumbersome to change individual modules. To relax this constraint a control interface can be incorporated into each module, in order to provide a generic interface to the perceptual controller. An alternative structuring of the system, which uses generic control interfaces in shown in figure 19.4.

The "control interface" per module of figure 19.4 reduces the dependency of the perceptual controller on local representations, and it transforms local information into a global framework, which is in accordance with an established protocol. The local control interfaces also enables use of more abstract goal commands, which are transformed to module specific actions by the interface.

Having a "abstract" interface to each module allow definition of a standard protocol, which allow communication of both data queries and goal commands, as described in section 19.6.

Centralised control is characterised by fast communication and a limited hierarchy which also facilitates detailed status monitoring and interaction in the event of unexpected situations. These benefits are, however, at the expense of detailed task knowledge in the planning / supervisory module which makes the system static, in the sense that it is extremely difficult to change individual modules. The trade off in centralised control is thus between speed and flexibility.

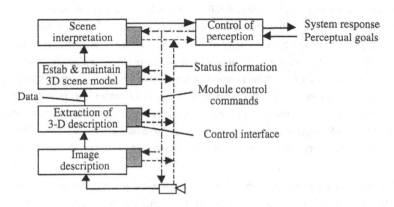

Fig 19.4. A system structure with centralised and *generic* control.

19.3.3. Decentralised control

In a distributed structure for local control the individual modules does the planning of actions required to arrive at a locally defined goals. The planning may involve use of data provided by lower level agents. For such situations the module can issue goal commands to the next lower agent (module). An elaborate description of distributed and hierarchical control may be found in [Lumia 1990]. A system structure which is based on distributed control is shown in figure 19.5.

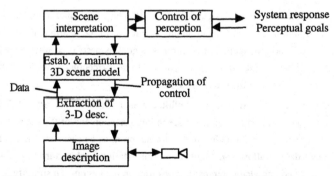

Fig 19.5. A system structure which is based entirely on distributed control.

In this structure the flow of data and control information is entirely local, using top-down or bottom-up propagation. The goal directed processing is achieved through successive refinement of goals top-down through the hierarchy, with data feedback bottom-up. Each of the local modules may use a perceptual cycle type of processing, where sub-goals are issued and evaluated based on data feedback from the lower level agents.

The control knowledge is represented locally in this architecture. A procedure may for example contain the set of required sub-actions needed to arrive at a solution. The use of distributed control knowledge provides a mean for use of very efficient control mechanisms, as it may be incorporated directly at the procedural level; i.e., the implementor of a particular algorithm may have specialised knowledge which allows for efficient control at the module level. It is, however, undesirable to incorporate domain knowledge of the procedural type, as it will make it more difficult to adapt a system for use in a new domain. To provide a "general" system, the control knowledge at the module level should be organised in a manner where more generic control functions are represented at the procedural level, while domain specific knowledge is represented using declarative techniques.

In the distributed control autonomy is the key-word but if it is combined with a hierarchical architecture as shown in figure 19.5 a problem is communication speed. Given a real-time environment modules are expected to produce information for the next higher level module at the end of each sampling period. The time for propagation of information for the image level to the upper most level will thus at least be equal to the sum of all the sampling periods in the system. This implies that responses to event in the external environment may be slow and the delay will typically deteriorate performance in term of control as interaction will be based on information which was acquired "some time ago". In the hierarchical control it is thus critical that the number of layers in the architecture is kept at a minimum or that distributed control is combined with other kinds of architectures such as the blackboard architecture suggested by [Riseman-Hanson 1978].

19.3.4. Hybrid control

The control architectures outlined in figures 19.4 and 19.5 have a number of undesirable properties. In the centralised architecture module autonomy is limited. In the distributed architecture system control is distributed and propagation of reactions and goal requests is correspondingly slow. If for example an event is detected it will not be propagated to the perceptual controller until a number of cycles later and notification of the user is thus delayed.

Another problem associated with the architectures in figure 19.4 and 19.5 is the purely hierarchical propagation of data. I.e., if the scene interpretation and/or the geometrical modelling module need image related information such information must be associated with 3D or semantic entities as tagged data to allow communication. The dataflow architecture has thus a significant impact on the data and knowledge representations used in the system. To simplify access to data it is consequently desirable to have facilities for direct access to data in any of the other modules. Such access may be provided using one of two different possibilities: a) the modules can be fully interconnected either through use of a blackboard architecture as used in the VISIONS system [Hanson-Riseman 1978] or through direct communication channels or b) a communication channel shared by all modules may be used for data requesting and replies. The fully connected architecture is only feasible if a blackboard architecture is adopted as use of direct communication links will imply too large an interdependency between different modules. To achieve real-time performance use of a blackboard architecture implies, however, construction of a new system as none of those available today support real-time processing (especially given the required data flow). The only general option left is thus the sharing of a common communication channel. Alternatively a purposive data flow architecture could be adopted. I.e., a communication channel is set up whenever there is an identified need for sharing of data. This option corresponds to a large extend to many of the architectures, which have been used, in systems build so far. This option does, however, not offer much new insight into system level control.

A hybrid system architecture which offers a simpler data representation, while maintaining possibilities for both distributed and centralised control is shown in figure 19.6.

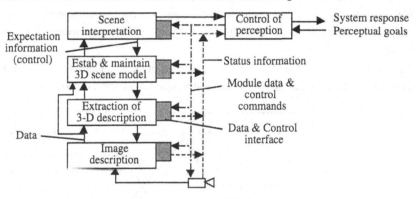

Fig 19.6. A hybrid system and control architecture.

19.4. System level control

19.4.1. Reasoning about goals

The user communicates with the vision system through an external interface. The interface receives goal commands, which specifies the needs of the user. Answers provided by the (individual) system (modules) are routed to the user in response to these goal commands or autonomous system actions, which have lead to detection of "significant" events in the scene. The commands the system may receive from a user are:

- Search   Determine if an object is present in the scene.
- Find   Recapture information related to a previously detected object.
- Describe   Provide information about location, pose, velocity for an object.
- Watch   Allocate resources for monitoring of an object and report changes.

- Track   Keep an object, which may be moving, in focus of attention.
- Relate   Relate two or more objects to each other. Skeleton relations are: above, on-
             top-of (and in contact with), beside (to the left & to the right), in-front-of.
- Classify   Identify a particular object or clique of objects.
- Explore   Maintenance and/or improvement of scene model. Specify where otherwise
             idle resources should be used.

The action that may be initiated in response to these commands are described in detail in the section 19.6 . It should be noted here that the planning of actions and the subsequent execution of such actions are organised in a hierarchical fashion, and only commands at high abstractions are described in detail. The completion of a goal will typically require specific actions by several modules, but here only the propagation of goal request to the scene interpretation module is considered. The variety of commands at low abstractions are still being developed and discussed. The relation between the different commands is shown in figure 19.7.

In figure 19.7 the uppermost box contains the commands which are related to description of spatial properties of the scene, while the lower right box contains the commands used for description of temporal phenomena and implicitly for belief maintenance in a dynamic environment.

The arrows in figure 19.7 indicate the relation between different commands; i.e. one may only describe an object if it already has been located and brought into an active focus of attention (FOA).

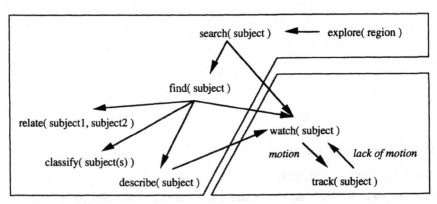

Fig 19.7. Relation between different user goal commands.

Commands which may be issued by the supervisor.

In order to control the vision system the supervisor must be able to issue commands to the modules which comprises the computational part of the system. Initially a hierarchical control approach has been adopted. This approach is to a certain extent complemented by heterarchical commands, as will be described in a later section.

For the hierarchical mode of control the supervisor communicates with the scene interpretation module, which is capable of identifying object and calculation of associated semantic and parametric information. For control of the scene interpretation module the following commands can be issued.

- si_search   Search for a particular object within the field of view.
- si_find   Recapture information about an object which has been detected earlier, provided the object is within the current field of view.
- si_describe Calculate attributes associated with an object in the current scene model. Attributes are restricted to position (centroid), size, pose, and velocity (path).
- si_track   Keep an object in focus of attention and report changes. If the object is approaching the border of the field of view or a defined sub-field notify the supervisor.
- si_watch   Monitor a particular object and report any spatial or temporal changes above a pre-specified threshold.
- si_relate   Provide information about the spatial and/or temporal relation between two objects.

Several of these commands can be issued in parallel to the scene interpretation module; i.e., several (sub)goals can be pursued in parallel. In order to do this it is necessary to associate some kind of priority with each goal command, as priorities are needed for conflict resolution. I.e., if two concurrent goals require actions that are incompatible the priority determines which of the goals that should override the other. In addition each goal command will have associated parameters which constraint the search for an answer. I.e., for the find command a object category or a specific object (the id) is provided to indicate "what object to look for?". If contextual information was used to derive the sub-goal this information will be used in a specification of "where to look?". The reasoning about goals in a temporal context is described in more detail in section 19.6.

19.4.2. Reasoning about system status

To make the planner reactive so that it may account for unexpected events in the scene, including action/task failure it is necessary to have access to feedback information that allow assessment of system status. Two kind of status information can be provided: module processing information and results. The module status information describes process parameters for a particular module. The status information may be related to:

- Number of items in the local data model?
- Number of items matched in the last cycle?
- Number of model invocations in the last cycle?
- Processing time for a sample

These parameters will allow introduction of simple resource control strategies. I.e., if the processing time is less that the expected too few data are analysed and lower level modules may be requested to provide more or more detailed information in a FOA. If too much data is provided it is of interest to determine if the data provided are of a reasonable quality. This may be determined through an analysis of the size parameters. I.e. if few item in the model are matched and there is a significant number of model invocations it may indicate that prediction is failing and data on a different temporal scale should be provided to facilitate stable tracking. On the other hand stable tracking may indicate that data fidelity is too high and more or coarser sampled data may be sufficient for the task at hand, to relieve resources for other and more urgent tasks.

In terms of results two kinds are available: positive and negative. The positive results reports that an item which has been requested has been found within the present set of FOAs. The object description has associated descriptive information such a position, type, confidence. The negative response indicates that a requested object has not been found or it indicates that a previously detected object no longer is in the FOA or the interpretation related to the original data has changed. Associated information indicates the type of semantic transition which has occurred.

In combination the two kinds of status information allow assessment of system status and communication of successes and failures to an external user.

19.5. Module level control

Local goal and data directed processing can be organised into a "hypothesise and test" structure. Such a structure provides also a simple means for incorporation of temporal context, as it generally is used for hypotheses generation (model invocation). At the module level it is therefore possible to adapt a structure as shown in figure 19.8.

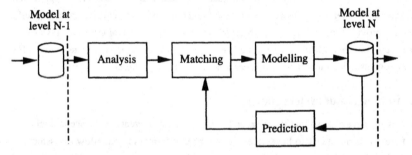

Fig 19.8. A hypothesise and test structure that may be suitable for processing at the module level N.

The structure in figure 19.8 is not capable of doing (external) goal directed processing. This may be achieved if a process, which is external to the module (N), is allowed to modify the content of the model at level N. Through introduction of expectations ((sub)goals) in the model base it might be possible to change the FOA etc. Sub-goal commands may in principle arrive from two sources, the next higher level or the perceptual controller. Goal request from the next higher level will be received implicitly as part of a top-down propagation of control information, while the goal command from the perceptual controller may be given explicitly. Goal commands are typically needed from the perceptual controller during event processing or initialisation.

To enable use of generic strategies in the perceptual controller a goal request from the controller is provided at a high abstraction level, which must be transformed locally at the module in order to be applicable for local control. For the transformation of goal command, propagation of control and local status monitoring it is desirable to introduce a "module control" block, which also handles all the local control actions required to ensure that the module has a degree of autonomy (re figure 19.5). An elaborated module structure which includes a "module control" block is shown in figure 19.9.

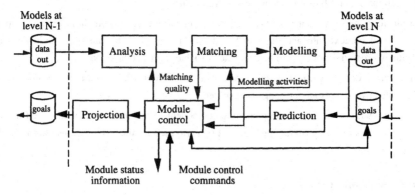

Fig 19.9. Structure for a module which has a local control block, that handles local control functions and provides a generic interface to the over-all control structure.

In figure 19.9 each of the models, which provide the medium for communication between adjacent modules, has been broken down into two for illustrative purposes. One of them contains the data derived information and the other contains in a similar format goal related information and requests produced by the next higher level. The mode of operation (bottom-up versus top-down) may then be controlled by assigning different weights to information in the two models when they are used for prediction.

To produce the "goal-model" for top-down control an additional "projection" block has been added in figure 19.9. This block perform the inverse action of the analysis block; i.e., in the "Extraction of 3D description" this module will perform a perspective transformation of goal requests in order to map data from 3D scene co-ordinates onto the 2D image domain. The transformation of data between different representations is performed at the level where the corresponding representation is inferred.

The module architecture outlined in figure 19.9 has been implemented as part of the skeleton system. The adaptations needed for implementation is described in chapter 2.

In terms of control only the perceptual controller is considered part of system control, while control of individual modules is covered as part of the description of those.

19.6. A rule based supervisor

19.6.1. Outline of supervisor & responsibilities

As outlined in earlier sections the supervisor is responsible for communication to external users, planning, and action monitoring. The communication to an external user will vary waith the user. I.e., for a robot questions and replies will be posed and interpreted using a robot specific protocol while a human user may use either a simple ASCII or graphical user interface. In term of the messages being communicated goal queries are formulated as described in section 19.4. The choice of a user interface is of little interest and will have little influence on the control theoretical considerations and it is thus not considered any further here.

For planning the supervisor must take the set of (concurrent) goals, domain information, spatial and temporal context into account and based on this the planning module should deliver a sequentially ordered list of tasks to be completed to achieve goal completion.

For status monitoring the strategies described in section 19.4 should be considered and implemented using a resource management strategy.

In the following in particular the plannning component of the supervisor is described in more detail. Throughout the description the breakfast scenario used for the VAP demonstrator is used as an example.

19.6.2. Goals and context

Given the set of available commands for control of the scene interpretation module each of the goal commands received from the user is described in detail, and related to the scene interpretation commands in the following sections.

Search command

When the *search* command is received from the user, it is the task of the system to determine if the specified object is present in the scene. In order to do this the following information is available to the planner:

- A-priori plans
- The current scene model
- An un-instantiated model of the objects including,
- Pre-recorded contextual information. (Associative information)

For determination of the existence of an instance of a particular object-class the present scene model is browsed, to determine if such an object already is available. If the object is present and up-to-date (i.e., with in the current FOA) the task is trivial, the system immediately reports success (an object id) and terminates. If the object is not present in the model it can be determined if some contextual objects are present. I.e., if a cup is not present in the scene model it can be determined if some of the objects which frequently occur in the vicinity of a cup are present. Typical contextual objects for a cup might be: a table, a spoon, a plate, or a coffee pot. If such objects are present in the scene they provide potential seed points (FOA indicators) for the subsequent search for the object. If no contextual information, which can constrain/guide a search for the object, is available the only option left is a search for the object initially within the field of view and subsequently within the entire scene. Such an unconstrained search is by nature very expensive and consequently such a command should have a relative low priority when several goals are pursued in parallel. In figure 19.10 an abstract representation of the *search* plan is shown.

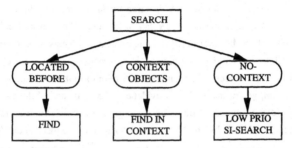

Fig 19.10. The plan for execution of the *search* command.

Find command

The *find* command is used for allocation of resources for updating of the model for a previously detected object. I.e., if a cup has been detected earlier and now more information about the cup is needed, consequently it must be brought into a local focus of attention. Again three different situations may be foreseen:

- object is already up-to-date (related to an active FOA),
- the object was last detected at a position which is inside the present field of view,
- the object was last detected at a location which is outside the present field of view.

To determine which situation is relevant for planning the supervisor can inquire the scene interpretation model and get information related to last known location, and the current setting of the camera system. This information is sufficient to determine the actions needed.

If the object is within the field of view and up-to-date no actions are needed, the result from the query is simply forwarded to the user, and the scene interpretation module is notified to indicate that there is an "interest" in this particular object (I.e. the list of present FOAs is updated with the region which contains the object of interest). The notification of the scene interpretation module may prevent that the object description is discarded as part of a model pruning (garbage collection) process, where "old" information is pruned in order to limit the size of dynamic models.

If the last known position of the object is within the field of view, but outside the current set of focus of attentions, the existence of the object at the last known location can be verified through a *si_find* command to the scene interpretation module.

In order to increase speed (reduce need for computational resources) it may be possible to perform model verification through direct access to a low level module. I.e., in a number situations the colour, texture, or some non-accidental geometric event associated with the object may be sufficient for verification of the existence of the object, as it for example has been described by [Biederman 1987]. Such primitives can be extracted by the image description module or the 3D geometric modelling system. Considerations related to which module should perform such actions has, however, not be pursued here. It is assumed that the scene interpretation module has control information encoded that will allow it to select the most efficient method for extraction of the information required to satisfy the si_find command.

If the *si_find* command fails, some kind of search for the object is required. For this, a strategy similar to the previously mentioned *search* can be adopted. I.e., through access to the scene interpretation module it can be determined if some contextual objects (*Cobjects*) are present in the current scene model. If such *Cobjects* are present the supervisor can determine close to which of the objects it should initiate a search. The planning of "where to look first" can be based on a simple cost-benefit algorithm, where the quality of evidence related to each of the *Cobjects* is evaluated together with their discriminatory abilities for the object of interest. Such a planning is simple to implement using a Causal Probabilistic Network (CPN) approach [Jensen et al. 1990].

If no *Cobjects* are present (a rare situation) the *find* command will be transformed into an unconstrained search for the object (as a consequence of the implied *search*). Given the test scenario, which has been chosen for the VAP system (the breakfast/table scenario), there should almost always be contextual information available. I.e., a primary internal task for the vision system is to determine the location of the ground plane and other kinds of landmarks. The location of the table is typically available and it can thus be used for constraining the search. The system must, however, have facilities which will prevent breakdown or failure even during initialisation.

If the last known location of the object is outside the current field of view, camera motion is required before an *si_find* or *si_search* command can be initiated. For this the supervisor can request camera motion. Given the system is attending several goals in parallel it is not given that the request for camera motion will be honoured. I.e., if the camera motion would conflict with another goal of higher priority the camera control system may reject the request. The request can either be rejected or it can be entered into a queue of pending request for camera motion. In order to have a responsive system the action chosen should be communicated to the supervisor and subsequently to the user, as such an action should lead to a re-evaluation of the current command priorities or cancellation of the goal command. In the initial system implementation only the command with the highest priority may request camera motion all other commands will have their request rejected and re-planning is consequently initiated immediately.

An abstract representation of the plan for the find command is shown in figure 19.11.

Watch command

When the *watch* command is issued, it is an indication from the user, that resources should be allocated to monitor changes related to an object. The detection of changes can be qualified, as not all kinds of changes will be of interest. I.e., it may be essential to have notification when/if an object changes temporal behaviour (the object stops, changes path, starts to move ...), while the continuous updating of spatial features might have a low priority. The watch commands does not need any explicit planning, but some implicit planning is needed, in order to ensure that the object of interest is within the current field of view. This can be achieved through use a *find* command followed by the *si_watch* command. Again, this assumes pure hierarchical control. This is not necessarily an optimal solution from a resource point of view. I.e., detection of changes related to an object can often be detected at levels below the semantic. For detection of dynamic changes related to a known object it may be possible to execute the watch command by analysing the optical flow for the 2D region that contains the object, or the motion of a non-accidental geometric entity.

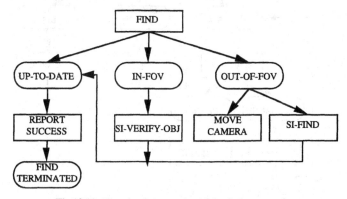

Fig 19.11. The plan for execution of the *find* command.

It should be noted that the watch command will constrain (large scale) camera motion, as the command will require the object to remain within the field of view.

Initially, the watch command is only used for detection of temporal changes, and it is assumed that the changes of interest have a semantic nature. The watch command recognises the following kinds of changes: initiation or termination of motion (temporal edges, violation of temporal smoothness assumptions), changes in path smoothness (lack of path coherence), and existence of objects (introduction or deletion of a particular class of objects).

The abstract plan for execution of the watch commands is shown in figure 19.12.

Track command

If the track command is issued, it is the task of the system to keep the specified object within the field of view and constantly update the description of the object. To perform this task, the object must be brought into the field of view (may be achieved using a find/search command), and the parameters of the camera system must be changed to focus/verge onto the object. If the object changes position the system must to the best of its abilities follow the object. A FOA should always, when possible, be centred on the object, and it must have associated mechanisms that will allow detection of other objects that potentially can occlude the object of interest. In the initial system such occlusions are detected but it will not perform actions to prevent such situations, as the planning of a path for the camera system for such situations is complicated, particularly in a domain which only is partially known. In the present form the track command is a 'gaze holding' operation and it may thus be carried out be image-level routines but at present it is carried out using the functionality scene interpretation module. The abstract plan for the track command is shown in figure 19.13.

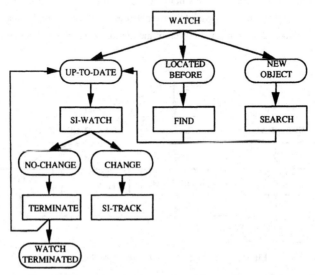

Fig 19.12. Abstract plan for execution of the watch command.

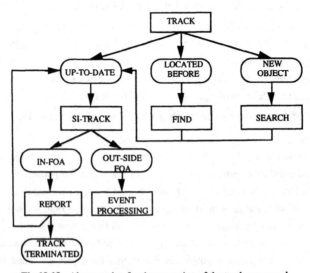

Fig 19.13. Abstract plan for the execution of the track command.

Classify command

This commands is used for labelling/identification of objects. When an object or a clique of objects has been recognised this commands can be used for adding a label to such objects. I.e., the system is capable of interpreting scenes with objects, but it may not have information available that will allow association of specific labels with individual objects. The recognition of objects such as a plate with fork and knife may identify a plate-setting, which has a large number of associated properties, which may be specified by the user through this command. In this command no planning is required. The information provided by the user is simply entered into the scene model (the "short term model"). This command can also be used for grouping of objects into explicit cliques, that may be references in future commands. I.e. plate, knife and fork may be labelled "plate-setting", and a subsequent command could specify *watch("plate-setting")*.

The introduction of cliques may, however, pose a problem as cliques typically will contain a number of separate objects, and consequently the clique constitutes a non-rigid "object" from a system point of view. I.e., if the *watch("plate-setting")* command is issued, how should this command be transformed into a track command if for example the knife is picked up by some one? I.e., should the system simply notify the user, or automatically decompose the commands into a track/watch command for each of the individual objects in a clique? This issue has not be settled in the present system. The initial system will always decompose commands for cliques into commands for each of the participating object.

Explore command

If idle resources are available they should be used for maintenance of the extracted information, including improvement of data. The strategy for improvement of data can be specified through use of this command. The explore command allow specification of the spatial/temporal regions where the main part of resources should be applied. It also provides information for the camera system, as it indicates which (3D) region that preferably should be kept within the field of view.

The strategy also specifies if the resources should be directed at improvement of spatial and/or temporal information. If the system is used in robot control, some tasks will require good spatial descriptions (bin picking or assembly tasks), while others will need temporal descriptions (navigation in a cluttered environment, where several other agents (robots or humans) operate in parallel).

The explore commands is used for explicit allocation of otherwise idle resources, but it should be noted that the explore command is also an implicit goal command, as the system is aimed at continuous operation it should be autonomous and consequently use idle resources for improvement of knowledge and exploration of the scenario to facilitate efficient execution of later commands. During initialisation a primary task for the exploratory resources is to establish the location of the ground plane, and other significant landmarks (listed as part of the domain knowledge). Such tasks will be executed autonomously even if the user has not yet specified any goals.

Describe command

The *describe* command is the most versatile as it is used for all user queries related to extraction of parametric descriptors for scene objects or phenomena. Such descriptors can be related to both spatial and temporal properties of the scene. This command is typically used for extraction of features related to objects which are in a current FOA due to a previous *find* command; i.e., a find command is typically issued just before a describe.

Phenomena in a scene may be described using a variety of different methods. I.e., the user can ask for a detailed description of the spatial properties of an object, or information such as "time-to-collision" for a moving objects (or a moving camera system).

The description of this commands is divided into two parts, one related to spatial phenomena and another related to temporal phenomena. In general the description command is used for fusion of information which is made available by some of the system modules; i.e., for certain task it may be appropriate to use data from the image description level (e.g., the image flow), while certain kind of information only will be available at the semantic level (say the specification of the kind of object being analysed; i.e., this is an instance *tea-cup* from the class *cup*).

Description of spatial phenomena is related to measurement processes, where it is of interest to provide a description of a particular object at a given level of detail. The specification of the required scale and accuracy of the description is essential for this command, as it influences the amount and type of processing required. The description of spatial information may be divided into the following categories, according to the item of interest:

- Objects or cliques of objects
- Sub-parts (volumes)
  • Surfaces
  • 3D primitives (vertices, lines, arcs, intersections)
- 2D primitives (vertices, edges, lines, arcs, edge-groupings)

Each of these categories will require different types of processing by the modules. For the object descriptors a low spatial resolution is usually sufficient, but a wide field of view is required to ensure that all or most of the objects of interest are considered, thus the scale of spatial description will also influence on the camera parameters.

For tasks such a navigation in a cluttered environment coarse spatial descriptors such as super-quadrics describing the convex hull of the object may be sufficient, at least for the planning of the overall path. If the system is used for manipulation tasks, a more detailed description is required, to ensure that an object actually is grasped by a gripper, and to determine the grasping strategy, which depend on the type and nature of the surface of the object. For use in such situations subpart descriptions are required to allow for selection/identification of the grasping point.

Relate command

The *relate* command is used for extraction of explicit relation information between objects in a scene. It is assumed that both objects already have been recognised by the system, and the parameters to this command is thus the object which should be related. In a dynamically changing world it is probably unrealistic to assume that relations between all objects in the scene may be maintained using a limited amount of resources. The relation between objects are thus produced in two ways. The interpretation within the system may generate such relation as a bi-product of other calculations. I.e. in order to recognise a plate it may be necessary to determine that it is located on-top-of a table. If such relations are produced autonomously be the system they are made available to the user, but the user is otherwise required to ask for calculation of such relations. In the process supervisor the planning is related to determination of the existence of the two objects and subsequently control of the camera system to ensure that both objects (if possible) are visible in the same view. The *si-relate* command is subsequently issued to the scene interpretation module.

19.6.3. An implementation

For implementation of reasoning about goals it was decided to use a rule based system for initial experiments. The aim of this work has been a fast implementation rather than a comprehensive performance.

In an early stage of this project it was decided that software when possible should be implemented in C or C++ in order to impose as few as possible constraints on use of available software and techniques. It is likewise desirable to have reasoning mechanisms which facilitates real-time operation. The C-Language Interfaced Production System (CLIPS), which has been developed by the US National Space Agency (NASA) is such a system. The CLIPS system is available from the COSMIC software distribution center at University of Georgia. Based on prior experience with CLIPS and reported work on real-time robotics control, CLIPS was chosen for the initial implementation of the process supervisor. The most recent version of CLIPS (4.3 & 5.1) includes facilities for representation of knowledge in terms of frame-structures. CLIPS consists of 2 databases, a *fact* database and an *agenda*. The facts database contain the knowledge presently available for reasoning, while the agenda contains a specification of the rules which are ready for activation.

The format of a rule in CLIPS is

```
(name "<comment string>"
        (facts needed for activation)
        •••••
=>
        (actions in response to facts match)
        •••••
)
```

The syntax is similar to that of the well known OPS5 system [Brownston 1985] which is available in Common LISP.

The activation of rules and resolution of conflicts is carried out through use of the RETE algorithm described by [Brownston 1985].

The CLIPS system is available in source format for a variety of computers. Given the availability of source code and a manual it is simple to embed CLIPS into other applications or provide interface functions within the CLIPS framework. Interface functions may be used for introduction of data or new actions for use on the right hand side of a rule (left hand use is also possible but difficult). To allow user interaction with the system a simple X/MOTIF interface was added to the CLIPS system. Through use of simple socket interfaces communication between an interface agent and the reasoning engine was established.

All the data needed for the supervisor has been implemented by means of frame representations, as such a structure facilitates inheritance where slots, for which no information currently is available, simply is assigned default values.

In the initial implementation only the commands *search, find, watch,* and *track* were implemented. For the specific communication to the scene interpretation module a 4 parameter set wa used. Each request to the scene interpretation module would contain the following information:

- Object of interest (i.e., cup)
- Expection location (derived form context information and the present scene model)
- Specification of camera parameters (where to look in term of an object-id: i.e look at cup-45)
- Size of Focus of Attention (determined from confidence in context information and concurrent set of goals)

In response the scene interpretation module would return either an object descriptoin specifying th presence of an object and it location and a unique id. Alternatively it might return failure, either "not-seen" or "moved outside field of view". For the first case an id corresponding to the goal query is returned or otherwise the id of the object which was lost is returned.

For the implementation of the supervisor the rule base is devided in a set of rules and some associated declarative knowledge. The rulebase contain a general set of rules, a described above, while the declarative knowledge encodes information specific to the domain of application (i.e. cup are usually found in the vicinity of tables, plates, cutlery, and pots). It is thus simple to change the rule base so that it may be used in other domains. The experiments reported a part of the scene interpretation chapter (16) used the prototype supervisor and no results will thus be reported here.

The prototype supervisor was implemented so hat it had full access to all controlable functions in the entire system through simple rule actions with a syntax like (send-message <module name> <command> [<parameters>]). This allow for a simple change to centralised control, which has been tested for using in relation to simple groupings. Such primitives may for example be used for maintenance over time. In such experiments it is evident that the reasoning in the supervisor at leat is an order of magnitude faster than the image processing which indicates that a least a certain amount of centralised control may be used before the centralised controller becomes the bottleneck.

19.7. Summary

In term of control of perception three issues have been addressed: system- and module architectures, and a supervisor for centralised control. Each of the issues are briefly summarised below.

For the system architecture a hierarchical structure very much in the spirit of the Marr paradigm [Marr 1982] was suggested. The structure was complemented by a common control channel to facilitate both distributed and centralised control. In initial experiment only the hierarchical structure was employed. Such an architecture simplifies module design but at the expense of speed and robustness. A problem in the hierarchical model is that dirct access to data is only available for neighbouring modules and semantic modules does consequently not have direct acces to image level features, which is considered a major disadvantage. It is thus considered essential that the developed structure is complemented by purposive facilities which exploits the access which may be achieved through the common communication channel. Another issue is the propagation of information along single communications channels, based on findings in purposive vision it does appear that more efficient structures may be implemented using simpler system architectures as for example suggested by [Aloimonos 1990] and [Christensen-Madsen 1993].

For the module architecture, which is described in more detail in chapter 2 and chapters 5 it is evident that the standardised structure which has been developed has been a major help in the construction of a fulle integrated system. The adoptation of such a structure has facilitated definition of a common protocol for both data and control requests and the computational procedures in each module have been "simple" to implement due to this standardisation. Preliminary expriments have suggested that the temporal maintenance to a large extend may be modelled as a recursive estimation problem (i.e. Kalman filtering for low level processes). The suggested and implemented structure has thus been successful in capturing the functionalities present in each of the modules. At the symbolic/semantic level it is however more difficult to apply standard recursive structures as maintenance here becomes a more diverse issue.

For centralised control a rule based approach was adopted both for planning and action monitoring. The implemented structure has a clear partition between generic functionalities and domain knowledge which allow for simple adoptation to other domains and the provided facilities may be used both for centralised and distributed control. In terms of processng speed 8-10 Hz may be obtained. A major weakness of the adopted approach is the inability to handle uncertainty. Any planning system using real-world information should preferably have facilities for dealing with uncertainty. One possible approach has been described by [Jensen et al. 1990] and [Rimey-Brown 1990]. This approach uses Belief network for representation and reasoning about domains. This method allow simple implementation of planning under uncertainty, but the domain knowledge is hardcoded into the system and the estimation of probabilities remains an open issue. Other paradigm such as Dempster-Schaefer should probably be considered before a paradigm is chosen for further experiments.

Bibliography

[Abbott 1992] L. Abbott, A survey of selective attention, IEEE Control Systems, Vol. 10, No.2, 1992

[Aloimonos 1987] J.Y. Aloimonos, I. Weiss & A. Bandopadhay, Active Vision, Int'l Journal on Computer Vision, pp. 333-356, 1987.

[Aloimonos 1990] J.Y. Aloimonos, Purposive and Qualitative Vision, DARPA Image Understanding Workshop 1990, Philadelphia, Penn, September 1990.

[Biederman 1987] I. Biederman, Matching Image Edges to Object Memory, International Conference on Computer Vision, London, June 1987

[Brownston et. al. 1985] Brownston,L. , R. Farrell, E. Kant and N. Martin, Programming Expert Sustems in OPS-5, Addison Wesley, 1985.

[Califano 1990] A. Califano, R. Kjeldsen, & R.M. Bolle, Data and Model Driven Foviation, 10 ICPR, Atlantic City, June 1990.

[Christensen-Granum 1990] H.I. Christensen & E. Granum, Initial Control Specification, VAP Internal report IR.E.1.1, February 1990.

[Christensen et al. 1991] H.I. Christensen, C.S. Andersen, & E. Granum, Control of Perception in Dynamic Computer Vision, SPIE Topical Workshop on Control Paradigms and Integration, Boston, Mass, Nov 1991.

[Christensen-Madsen 1993] H.I. Christensen & C.B. Madsen, "Purposive Reconstruction", CVGIP Image Understanding, Vol. 60, No. 1, pp. 103-108, July 1994.

[Crowley 1989] J.L. Crowley, Knowledge, Symbolic Reasoning, and Perception, In: Proc. Intelligent Autonomous Systems, Amsterdam, Dec. 1989.

[Culhane-Tsotsos 1992] S. Culhane & J. Tsotsos, A Prototype for Low Level Attention, ECCV-92. Sandini (Ed.), Santa Margarita, May 1992.

[Dickmanns 1988] E.D. Dickmanns. 4d-dynamic scene analysis with integral spatio-temporal models. In R. Bolles and B. Roth, editors, Proc. 5th Int. Symposium on Robotics Research, Tokyo (Japan), 1988. MIT Press.

[Hanson-Riseman 1978] Hanson, A.R. & Riseman, E.M., VISIONS: A Computer Vision System for Interpreting Scenes, in Computer Vision Systems, A.R. Hanson & E.M. Riseman, Academic Press, New York, N.Y., pp. 303-334, 1978.

[Jensen et al 1990] F.V. Jensen, J. Nielsen, H.I. Christensen, Use of Causal Probabilistic Networks as High Level Models in Computer Vision, Tech Report R-90-39, Aalborg University, Institute of Electronic Systems, November 1990.

[Lumia 1990] R. Lumia, The NASREM Architecture, In NATO Traditional and Non-traditional Robotic Sensors, T. Henderson (Ed.), Springer Verlag, 1990.

[Marr 1982] Marr, D., Vision, W. H. Freeman, San Francisco, 1982.

[Rimey-Brown 1991] R.D. Rimey & C. Brown, Controlling Eye Movements with Hidden Markov Models, Int'l Journal on Computer Vision, Submitted April 1991.

[Tsotsos 1987a] J.K. Tsotsos, Image Understanding, In: Encyclopedia of Artificial Intelligence, (Eds.) S.C. Shapiro & D. Eckroth, John Wiley & Sons, NY, Vol. II, pp. 389-409, 1987.

[Tsotsos 1987b] J.K. Tsotsos, Representational Axes and Temporal Cooperative Processes, In: Vision, Brain and Cooperative Computation, (Eds.) M.A. Arbib & A.R. Hanson, MIT Press, Cambridge, Mass, pp. 361-418, 1987.

20. Control of Scene Interpretation

Jiri Matas, Paolo Remagnino, Josef Kittler, and John Illingworth

UOS

20.1 Introduction

One of the main goals of visual sensing is to interpret the perceived visual data. By interpretation we understand the process of recovering information relevant to the goals of the autonomous system for which the visual sensor acts as one of its intelligent agents. This definition allows us to approach interpretation as a dynamic control problem of optimal resource allocation with respect to a given objective function.

At the level of the symbolic scene interpretation module of the of the Vision as Process (VAP) system, the surrounding environment is modelled as an organized collection of objects. A system goal therefore typically requests information about objects present in the viewed scene, their position and orientation, dynamics, attributes etc.

The main thesis behind the approach to symbolic scene interpretation in the Vision as Process (VAP) system is that spatio-temporal context plays a crucial role in the symbolic scene model prediction and maintenance. Another essential and distinctive feature of the novel approach is the active control of the visual sensor (mobile stereo camera head) based on the given visual goal and the current symbolic description of the scene. At any stage of processing, the spatio-temporal context is used to select the most suitable representation of objects permitting as efficient matching of image-derived data as possible.

The architecture of the scene interpretation module is based on the hypothesis that control actions implied by any visual task fall into three independent categories: active sensor (camera) control, control of the focus of attention (region of interest definition) and selection of the appropriate recognition strategy. The complex dynamic control problem can therefore be decomposed into a sequence of primitive visual behaviours. From the implementational point of view these primitive behaviours can be effected by issuing parameterised canonical control commands to the basic controllable entities of the module. These comprise i)camera next look direction, ii) camera position or zoom (only camera position control is currently available in the VAP skeleton system), iii) region of interest, and iv) knowledge source selection. The commands are implicitly encoded by the system supervisor in terms of the system goal and perceptual intentions.

A second distinctive feature of our approach to scene interpretation is the use of temporal context. Past experience in the form of information about recognised object is organized in a hierarchical database. In continuous interpretation this information is exploited to implement the focus of attention mechanism. Several 'forgetting' schemes are adopted to reflect dynamism of objects.

With all the building blocks of the VAP system in place [DD.G.1.3], it has been possible to demonstrate the merit of spatio-temporal context in scene understanding to validate the cornerstone of the VAP philosophy. This paper gives an account of the initial testing of the hypothesis in the setting of a simple table top scene of limited object dynamics. Typical experiments performed involve the verification of the presence of or the pose of a known object using resources commensurate to the information content of the spatio-temporal context established to date.

The presented work draws upon results in a number of research areas: active sensor control [Wilcox 90] [Rimey 92], selective perception [Brown 92], knowledge representation [Strat-Smith 87], learning [Draper 92], integration of knowledge sources [Garvey 88b]. The main novel feature of our approach is the close interaction of the system goal, sensor control and the visual task. More sophisticated individual components of high-level vision systems have been described in literature, eg. the uncertainty calculus used in the VISIONS system [Hanson-Riseman 78], the exploitation of geometrical constraints in ACRONYM [Brooks 83] or the integration of information from multiple views in the system proposed for terrestrial robots by Lawton et al. in [Lawton 88] [Lewitt 87].

The report is structured as follows. In Section 20.2 a brief overview of the VAP system is presented. The discussion is centered around the symbolic scene interpretation module; attention is given to the interface to supervisor and image description modules. The primitive visual behaviours which define the system capability are listed in Section 20.3.5. Section 20.3 describes the architecture of the Symbolic Scene Interpretation module, together with the main object recognition knowledge sources. Section 20.5 introduces the scenario adopted and describes the experiments conducted. Conclusions are drawn in Section 20.5.3.

20.2   VAP system overview

The concepts which are central to VAP have been outlined in the previous section. Their realisation was tested within the system architecture illustrated in Figure 20.1. From the point of scene interpretation the VAP system can be divided into four functional blocks. These blocks encapsulate the processes that transform lower level descriptions, into more abstract descriptions and correspond to fairly conventional ideas concerning levels of representation in a vision system (i.e.images $\rightarrow$ 2D primitives (lines, ellipses, curves, perceptual groupings, etc) $\rightarrow$ objects). Top-down flow of control information, also depicted in Figure 20.1, implements the mechanism of focus of attention. In addition, individual modules maintain temporally evolving models (either implicitly or explicitly) of the aspects of the world that they understand. These models are part of the mechanism for exploitation of context.

The lowest level input to the system is provided by the sensor. This is a limited resource with several controllable parameters including position, look direction, aperture, focus etc. Its parameters can be controlled by any other module; the system supervisor acts as an arbitrator of conflicting requests.

The image description module transforms image data into 2D percepts such as edges, lines, elliptical arcs, perceptual groupings etc. . The processing is carried out over several scales on all data currently available to the sensor and the internal model maintained by the module is constituted by the intrinsic parameters of the processes at this level. These parameters can be adjusted by top-down signals from higher levels if the module is consistently found to produce output which is not useful to the higher level modules.

The function of the scene interpretation module is discussed thoroughly in section 20.3; basically, the module accesses the 2D description and produces an attributed symbolic model which describes the type and pose of identified objects. The model is maintained over a larger spatial and temporal field than that considered by the sensor and the existing partial world model is the context used to select among the possible solutions available for goal satisfaction.

The top level module is the interface and supervisor which connects the system to the external world and which arbitrates requests from other modules for selective attention to be given to parts of that external world. It has reasoning capabilities to transform tasks into the visual goals that other modules can understand.

Overall, the system organisation is hierarchical and functions in a perceptual cycle where goals and parameters are given to a module and the results of this processing are compared to expectations or previous results with the differences being used to update models and guide subsequent actions.

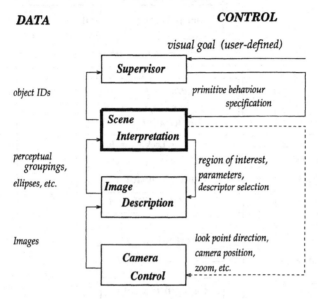

Fig. 20.1. Simplified VAP system architecture. Details of the Scene Interpretation module are depicted in figure 20.2.

20.3   Symbolic Scene Interpretation Module Architecture

An efficient use of resources for supervisor goal satisfaction poses a difficult dynamic control problem. In the process of the interpretation module design we observed that organizing resources into three basic operational units, *camera strategy unit, region of interest unit* and *the recognition unit* communicating via *the scene model database* significantly simplified the mapping of supervisor goals into control strategies.

The functionalities of the individual units are independent and complementary; it is therefore possible to factorise the system goal into a combination of a limited set of canonical parameterised control commands to the camera, region-of-interest and recognition units. Besides that, the appeal of the proposed structure stems for the fact that each of the modules corresponds to a well established high level concept. Based on the system goal and current knowledge about the surrounding environment, the camera strategy unit attempts to position and direct the sensor to simplify recognition (in accord with the paradigm of *active vision*); the region of interest unit selects for processing only relevant parts of acquired data (implementing the *focus of attention* mechanism). Information about recognised objects (the system 'history' or 'experience') is maintained in the scene model database enabling the other units to exploit *temporal context* in their operation.

Within this *distributed* framework, the central controller of the interpretation module is virtually nonexistent, its functionality degenerating into:

– passing the appropriate part of the supervisor goal in the form of a control command to individual units

– synchronisation of operation of the continuously operating units

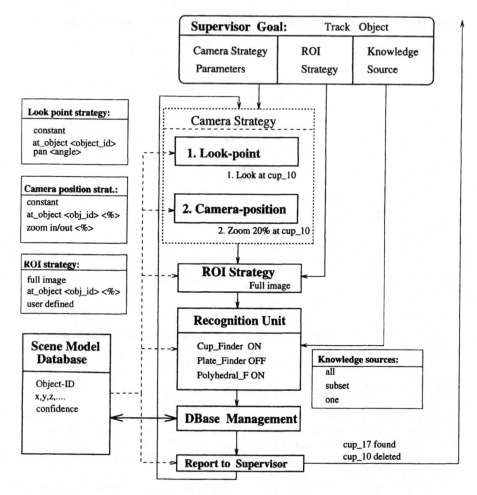

Fig. 20.2. Interpretation module architecture

The interpretation module structure is depicted in figure 20.2. The interpretation process is continuous; the loop in the centre of figure 20.2 presents all actions repeatedly taken to process consequent images. The camera strategy, region-of-interest and recognition unit change their mode of operation according to the respective part of a supervisor goal. Possible modes of operation are listed in boxes next to the main loop. An example of an operational mode for each of the basic units can be found inside the block representing the unit.

The in-depth description will proceed as follows. First, the basic operational units - the scene model database, camera strategy unit, region of interest unit and the recognition unit, will be described. Second, modification of the module operation as a response to various supervisor goals will be discussed. Finally, the continuous interpretation cycle, *the loop of perception*, will be presented.

20.3.1   The Scene Model Database

Exploitation of temporal context in scene interpretation through accumulation of information about objects recognised during the lifetime of a continuously operating vision system allows for a gradual improvement of performance. As the amount of information about the surrounding environment increases, more and more supervisor goals can be satisfied by database queries and/or spatially focused (and therefore efficient) visual processing.

In the scene model, objects are characterised by their recognition class (type), pose (ie. position and orientation), spatial extent and mobility. A confidence value is attached to each attribute. The choice of object attributes reflects the needs that the scene model database serves. The pose and spatial extent attributes (modelled by a minimal bounding parallelopiped) form a minimum object description sufficient for geometric reasoning used for verification of basic physical constraints (eg. no two objects occupy the same space). The pose and extent attributes are passed to the camera strategy unit when sensor attention is centred at a known object. The class information is related to the organisation of the recognition unit. All available recognition knowledge sources form a hierarchical tree (the following sequence can serve as an example of a branch: rotationally symmetric objects - cylindrical objects - cylindrical objects with similar diameter and hight). If a verification or update of parameters for a particular object is requested by the supervisor then the class attribute is used to invoke the appropriate (least general) recognition knowledge source; in conjunction with a unique object ID this enables fast indexing and retrieval of a detailed, object specific model with instantiated parameters, and, consequently, a more efficient and robust results are obtained.

It is important to stress that all objects, regardless of the recognition class, are represented in an identical way. All units within the module 'understand' the generic object representation and can access or store the information in a uniform way. Moreover, further refinement of the recognition class hierarchy or an introduction of a new knowledge source require changes only in the recognition unit as all class specific information is managed by the appropriate recognition knowledge source.

The building of a symbolic scene model in a system with an active mobile sensor is possible only when a coordinate transformation between the scene model reference coordinate systems and the camera coordinate system can be established and maintained. In order to avoid the accumulation and amplification of errors a hierarchy of local reference coordinate systems is used. The registration of the camera coordinate system and a local reference frame is established by means of recognition of an object whose internal coordinate system defines the reference frame.

The environment in which the interpretation module operates is constantly changing. Information about the pose of moving objects is out of date even before it is inserted in the scene model. The database manager takes this fact into consideration when updating confidence values of object attributes. When no new evidence about an object is available, eg. when the object is not in the field of view, the gradual aging of the pose estimate is modelled by an exponential decrease of the confidence level; the object mobility (durability) attribute defines the decay constant and hence the speed of the exponential forgetting process. For objects in the field of view, the evidence for object existence is temporally integrated using a simple counting scheme that updates confidence levels. Initially, an object hypothesis is assigned a confidence value equal to the likelihood of a correct match (the likelihood is part of the information output by the recognition unit). If a corresponding object is found in subsequent frames the confidence level in the hypothesis is increased (proportionally to the match likelihood) until a maximum value indicating absolute belief in the presence of an object is reached. Two hypotheses are assumed to correspond to a single object if the pose and extent parameters suggest volumetric overlap. Non-observation is taken as negative evidence and the confidence in the presence of the object is decreased by a constant. A velocity model for the motion and tracking of moving objects is not explicitly included in the current implementation.

20.3.2   Camera strategy unit

Goal driven sensor control is an indispensable part of an active vision system. Unlike a conventional camera controller, the camera strategy unit exploits the information stored in the scene module. Thus, instead of specifying the desired camera position and look point in terms of n-tuples of (Cartesian) coordinates, the supervisor goal defines camera movement indirectly by reference to objects stored in the database; eg. 'look_at cup_17 position 70%' (meaning: rotate the camera so that the center of cup_17 lies on the optical axis, select a viewpoint so that the cup projection occupies 70% of the field of view). This camera strategy abstraction simplifies the mapping between a high-level, user-defined goal (represented in a form close to natural language) from the mechanics of sensor movement. Moreover, increased modularity is achieved by insulating the supervisor from details of the database organization. The determination of camera parameters is not just a simple traversal of the reference frame tree and computation of camera-to-object coordinate system transformation; errors in object positions (the object of interest and those defining relevant local reference frames) must be taken into account (see [Remagnino 92a] and [Remagnino 92b] for details).

The following minimum capabilities are required of the underlying sensor controller effecting the camera strategy:

- (Cartesian) camera position control
- independent camera direction (look point) control

Other camera parameters, eg. focus, aperture, vergence (for a stereo pair), are assumed to be adjusted automatically by low level, reflexive as opposed to goal-driven, purposive processes. Speaking in terms of an analogy to the human body the interpretation module attempts to obtain a suitable viewpoint by controlling the head and neck rather than the eye movements.

The camera unit comprises two separate components, the look point strategy and the camera position strategy; table 20.1 summarizes their operational modes. The mapping of a supervisor goal into the operational modes is discussed in Section 20.3.5.

Table 20.1. Operational modes of the camera strategy unit.

command	parameters	description
Camera position strategy		
constant	$\emptyset$	No change in camera position
at_object	object ID, %	Move the camera so that the object fills a \<percentage\> of the field of view.
zoom_in/out	%	Move forward/backward along the line of sight (ie. look direction) stretching/shrinking the current field of view according to the specified percentage
user_defined	x, y, z	Move the camera to a specified position
Look point strategy		
constant	$\emptyset$	No change in look direction
at_object	object ID	Keep the look point at the center of the specified object.
pan	angle	Rotate the camera in the horizontal plane by \<angle\> degrees.
user_defined	x, y, z	Direct the camera towards the specified point

20.3.3   Region-of-Interest Unit

Selection of a restricted subset of incoming data for low-level processing, the main task of the ROI unit, provides a focus-of-attention mechanism complementary to active camera movements. The operational modes of the unit that determine the strategy applied for limiting the processed region, are listed in table 20.2. An interesting example of a symbolically defined ROI interest strategy, ie. using a reference to an object in the scene model, is demonstrated in the second experimental run described in section 20.5.

Table 20.2. Operational modes of the Region-of-Interest Unit.

Region-of-interest strategy		
command	parameters	description
full	$\emptyset$	Process the full image(s)
at_object	object ID, %	Frame the 2D projection of the object bounding box, stretch/shrink the frame by '%' percent, pass the 2D limits to the low-level modules
user_defined	xmin, ymin, xmax, ymax	Process the specified area only

Besides the expected beneficial impact on processing speed and data complexity, the use of ROI proved to improve the accuracy and reliability of low level vision processes that base the setting of control parameters on automatic estimation of image noise level.

20.3.4   The Recognition Unit

Bootstrapping operations of the interpretation module as well as supervisor goal satisfaction at least initially rely completely on the successful bottom-up performance of the recognition knowledge sources. The recognition unit serves the following purposes:

- provide an interface between the recognition knowledge sources and the scene model database, ie. pass information about
  - matched objects to the scene model manager
  - instantiated parameters of object hypotheses to individual knowledge sources to enable rapid indexing of relevant internal models
  - provide synchronisation for the recognition processes running in parallel.
- select and launch recognition KSs according to the types of objects which are the subject of the supervisor query

Currently three general KSs are running in the test environment; a polyhedral object matcher, a cylinder (cup) finder, and a plate finder (detector of rotationally symmetric planar objects). A detailed discussion of the knowledge sources is beyond the scope of this paper; see [Wong 93], [Hoad 92b], and [Hoad 92a] for description. The KSs draw on results of a complex set of low and intermediate level developments.

- a novel generalized Hough transform algorithm [Lee 92]
- a geometric modelling package [Wong 91a]
- perceptual grouping algorithms providing an intermediate image description in terms of collinear and parallel lines and various types of junctions in 2D [Matas-Kittler 92], [Etamadi 91], [Etamadi 92]
- a robust polygon extraction method [Wong 91b], [Cheng 92]

20.3.5   Visual Behaviour

As briefly indicated in section 20.2 the interpretation module operation is modified in response to goals specified by the system supervisor. The following set of user goals has been defined in the early stages of the VAP project [DR.E.1.2]:

- Search: Determine if a particular object (class) is present in the scene.
- Find: Re-find an object which has been recognised earlier.
- Watch: Allocate resources for maintenance of the description of a particular object.
- Track: Maintain description of a particular object and report continuously to the user.
- Explore: perform bottom-up driven exploration of the scene.

Table 20.3. Primitive behaviours applicable in the initialisation (bootstrap) phase, ie. before any particular objects are recognised. Behaviours marked with a '•' require a movable camera head.

Look point	Camera position	Region of Interest	Recognition Knowledge Sources	Description
constant	constant	full image	all	**explore** a static, predefined area
constant	constant	full image	some	**watch** for a certain type of object(s) in a predefined area (ie. selective explore)
constant	constant	full image	one	**search** for an object of a specific type in a predefined area
pan <*angle*>	constant	full image	all, some, one	• wide area **explore/watch/search**; the camera is panning <*angle*> degrees after every frame
constant	zoom in/out <%>	full image	all, some, one	• **explore/watch/search** with zoom-in or zoom-out (ie. camera moves along the line of sight)
constant	constant	user defined	all, some, one	**explore/watch/search** a user-defined area of the image

To achieve external goals, the supervisor performs detailed planning that results in a sequence of goals defining the operational modes of the functional units of the interpreter. From the point of view of an external observer, the operational mode of the interpretation module manifests itself as a primitive visual behaviour (the word 'primitive' is used to distinguish between the supervisor response to a user goal, called visual behaviour, and its components).

The operational modes of the camera strategy, region-of-interest and recognition units has been presented in previous subsections (20.3.2,20.3.3, 20.3.4). Originally, the operational modes, and indeed the whole structure of the interpretation module, were designed so as to facilitate satisfaction of the set of user goals listed in the beginning of the section. Then, in an attempt to justify the proposed module structure, the inverse problem was investigated: 'would all combinations of operational modes, ie. all primitive behaviours, result in a sensible, intuitively compelling behaviour?'. In fact, a number of previously unforeseen, reasonable primitive behaviours were discovered; see tables 20.3 and 20.4 for details. The tables list just the most compelling primitive behaviours as the number of possible combinations of operational modes is large. The definition of primitive behaviour in

Table 20.4. Primitive behaviours implementing various forms of focus of attention. Behaviours marked with a '•' require a movable camera head.

Look point	Camera position	Region of Interest	KS	Description
at_object <object_id>	constant	full image	one	**look at** an object. If the object moves, it is automatically **tracked** as the look point controller tries to keep it in the center of the field of view.
at_object <object_id>	at_object <object_id> <%>	full image	one	**•zoom on** an object. Keep the camera look point at the center of the object bounding box. Camera position ensures that the object covers <%> of the image area. To achieve the 'constant' object projection the camera must follow the object.
at_object <object_id>	constant	at_object <object_id> <%>	one	**•focus on** an object. Keep the camera look point at the center of the object bounding box. Process only a selected region of interest around the object. Note that this mechanism is similar to **zoom on** - the object of interest fills a constant portion of the processed part of the image. Here, efficiency is gained as smaller part of the image is processed; **zoom on** effectively increases resolution.
pan <angle>	at_object <object_id> <%>	full image	one	**•multiple views** of an object. As the look point mechanism pans the camera and the camera positioning module keeps the object in the center of the field of view, image of a single object from different views are acquired.
constant	constant	at_object <object_id> <%>	one	**dynamic region of interest.** The region of interest follows a moving object. Can be used for simple experiments with focus of attention without a movable camera head.

terms of modes of operation allowed for transformation of the ad-hoc group of supervisor goals into a much wider and yet consistent set (no combination of operational modes results in a senseless behaviour).

20.3.6   Loop of Perception

The complexity of the control strategy necessary for efficient operation of the interpretation module is greatly reduced by two factors. First, all planning for user goal satisfaction is performed at the supervisor level. Second, the resulting sequence of supervisor goal commands directly modifies the operational modes of the camera strategy, region-of-interest and recognition units. In this distributed control framework the module controller's main responsibility is to synchronise the cooperation of the individual independent units effecting the required *perceptual behaviour*. As perceptual behaviours are structurally identical, the interpretation process can be accomplished within a continuous, fixed cycle of operation - *the loop of perception*. The following four stages are repeated in the loop (see fig. 20.2):

1. Operational modes of all units are set according to the current supervisor goal.

2. Initiate the appropriate processes to determine the camera next look direction and position, and the region of interest in the image, using temporal context as required.
3. The relevant recognition knowledge sources are enabled and their launching is triggered by the low level image description as soon as it becomes available. The output of the knowledge source(s) is stored in the symbolic scene model database together with any confidence factors computed during the matching process.
4. The scene interpretation module controller reports the status of goal achievement to the supervisor.

General database management (confidence updates, garbage collection etc.) is performed in parallel within all phases. Data are passed between units indirectly through the scene model as described in section 20.3.1. The phases of the loop of perception are not stages of the execution flow; they should be rather viewed as sequence points when a certain set of parallel operations, necessary for the following phase, is completed.

20.4   Implementation Notes

The core of the interpretation module - the controller, camera strategy unit, region of interest unit and the recognition units are written in a public-domain production language CLIPS version 4.2 [Girratano 88]. CLIPS, "... a type of computer language designed for writing applications called Expert Systems"([Girratano 88], p. 1) facilitated fast prototyping and module implementation without compromising on efficiency (which was not of great concern anyway as the interpretation process spends most of the time waiting for the low-level image processing to be completed) . Small, specialized parts of the code were written in C (matrix and vector packages, transformations between coordinate frames). Interfacing C and CLIPS is seamless (CLIPS stands for "C" Language Integrated Production System).

The interpretation module can run as a part of the VAP system or in a stand-alone mode. In a stand-alone mode, useful especially for debugging and testing, communication with external modules (camera head, supervisor) is implemented with the help of control files. Within the VAP system, communication facilities are provided by the system skeleton SAVA.

20.5   Experiments in high level vision

The interpretation module presented in previous sections has successfully processed several image sequences. During the final review of the VAP I project, the module was in operation for several hours, interpreting scenes of breakfast scenario type, containing a table, plates, cups, and boxes. The objects in the scene were moved around as one would expect in such a scenario.

A sequence of images (figures 20.4-20.9) will be used to demonstrate the operation of the module. The objects involved are: 2 cups (one moving), a box, a plate (moving) and a table. The first experiment (figures 20.4-20.6) focuses on operation of individual recognition knowledge sources and on scene model database maintenance. In the second experiment, run on the same sequence, temporal context is exploited using the region-of-interest mechanism. Behaviours with active camera control are not presented in this paper; experiments including camera motion were only in preliminary stages at the time of writing of the report.

It is assumed in the experiments that object 'table' has already been identified; the tabletop plane to camera transformation is therefore known. The previous stage of interpretation concerned with establishing table-to-world coordinate transformation, a particular case of a reference frame tree transversal, is described elsewhere [Remagnino 92b].

The two experiments will be presented as a commented log output by the system. Information about the interpretation process is formatted as in the following template:

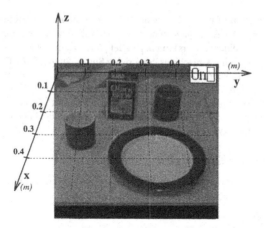

Fig. 20.3. The 3D coordinate system, the table-top reference frame, used in the two experiments

```
----------- Frame frame_number ----------
perceptual behaviour: supervisor goal
   look point          : command (mode of operation)
   camera position     : command (mode of operation)
   region of interest: command (mode of operation)
   knowledge sources : list of knowledges sources


+ object-type_ID [confidence_level] 3D_position
+ object-type_ID [confidence_level] 3D_position
  object-type_ID [confidence_level] 3D_position
- object-type_ID [confidence_level] 3D_position
```

The log entry starts with the processed 'frame number'. The 'frame number' should help readers to relate the numerical information about object position presented in the log file to the graphical results shown in figures 20.4-20.9. Next, the supervisor goal defining module's perceptual behaviour is displayed. To save space, this information is shown only when a new goal is received. The operational modes of the look point, camera position, region of interest, and knowledge source units follow. The complete list of commands (operational modes) can be found in tables 20.1 and 20.2. The last section of the log entry contains information about object stored in the scene database. Object related data is structured as follows:

1.   – '+' marks newly detected objects
     – '−' marks objects to be removed from the scene model
2. object_type and a unique ID
3. confidence level of object hypotheses [in square brackets]
4. x,y and z triplet defining the 3D position of the object centroids. The 3D coordinate frame used in the experiment, the local reference frame of the tabletop, is depicted in figure 20.3.

Unfortunately, information about time was not included in the log. It is difficult to give general estimates of the time needed to complete one cycle of perception as numerous factors, mainly outside the interpretation module itself, determine the overall performance. One of the most significant factors, the execution time of low-level image processing, can be eliminating by use of special-purpose hardware developed within the VAP project. The timing is also significantly influenced by the speed and configuration of the local network as units of the interpretation module (region-of-interest, camera control, database management) as well as individual recognition knowledge sources (plate, cup and polyhedral object finders) run in parallel ; (the system has been tested on a number of heterogeneous networks of UNIX machines comprising Sun3s, Sun4s, SPARC Is and IIs). Last but not least, the execution time is greatly reduced by applying any of the focus-of-attention strategies. In the worst case scenario (all KS running, all image processed, no special purpose HW, a single SPARC 2) the time of one cycle is roughly 1min; on average (multiple SPARC IIs, no image processing HW, region of interest focused on a object) a loop of perception is completed in less then 5 seconds .

In the following sections two experiments are presented.

20.5.1 Experiment 1: Explore

This section discusses an experiment which demonstrates operation of the interpretation module in the data-driven, *explore* mode.

```
----------- Frame 1 ----------
perceptual behaviour: explore
   look point         : constant
   camera position    : constant
   region of interest : full image
   knowledge sources  : plate_finder cup_finder poly_finder

+ poly_1   [1.00] 0.01 0.24 -0.07
+ plate_9  [1.00] 0.42 0.29 0.00
+ cup_11   [1.00] 0.34 0.12 0.04
```

The scene model database is empty in the beginning of the interpretation process; so far no objects have been recognised. The supervisor can therefore issue only one of the bootstrapping goals listed in table 20.3 (ie. a goal not requiring object ID as a parameter). The log listing shows that the interpretation process is operating in the *explore* mode. The selection of the *explore* behaviour need not be a consequence of receiving a supervisor goal; the interpretation module enters the default *explore* mode automatically if no supervisor goal is issued (all other changes in perceptual behaviour are induced by supervisor goals).

The perceptual goal description confirm that all available knowledge sources are launched (plate, cup and polyhedral object finder). The recognition knowledge sources detected three objects labelled poly_1, plate_9 and cup_11.

Comparison of the database contents and image 1a of figure 20.4 indicates that the estimate of plate_9 and cup_11 pose is good. The pose of tea box poly_1 is estimated at 7cm below the tabletop plane - an obvious error. Moreover, the dark cup in the upper-left corner was not found. The imperfect performance of the recognition procedures is not of great concern; our approach is based on the assumption that temporal integration of (inherently noisy) recognition results is robust enough to lead to stable scene interpretation.

```
----------- Frame 2  ----------
- poly_1   [0.00] 0.01 0.24 -0.07
  plate_9  [3.00] 0.42 0.29 0.00
  cup_11   [2.00] 0.34 0.12 0.04

----------- Frame 3 ----------

  plate_9  [4.00] 0.42 0.29 0.00
  cup_11   [3.00] 0.35 0.12 0.04
+ poly_22  [1.00] 0.37 0.24 0.14
```

The presence of plate_9 and cup_11 in the scene has been confirmed by recognition knowledge sources in frames 2 and 3. Confidence levels for both objects have been increased using the updating scene discussed in section 20.3.1. The weak poly_1 hypothesis was removed when its confidence level dropped to 0. A new (and once again incorrect) polyhedral object hypothesis was instantiated in frame 3.

```
----------- Frame 4 ----------

   plate_9 [5.00] 0.42 0.29 0.00
    cup_11 [4.00] 0.36 0.11 0.04
-  poly_22 [0.00] 0.37 0.24 0.14

----------- Frame 5 ----------

   plate_9 [5.00] 0.42 0.29 0.00
    cup_11 [4.50] 0.39 0.10 0.04
+  poly_40 [1.00] 0.31 0.22 0.13
```

In frame 4 the confidence level of the plate_9 hypothesis has reached the maximum, 'absolute certainty' level. The increase in confidence in the cup_11 hypothesis is smaller as the cup, contrary to the stationary plate, moves. The change in the cup position forces the confidence updating scheme to consider various options, eg. 'Am I getting very noisy measurements of a static cup pose?', 'Is the cup moving?', 'Perhaps the old cup was taken away and a new one put close to the original cup?'. Although the right interpretation (ie. motion) is chosen, the confidence level is updated with more restraint to cater for the other options.

Note the good agreement between the cup trajectory in images 1-5 of figure 20.4 and the recorded data. The poly_22 hypothesis was removed from the database; the new box hypothesis, poly_40, is very close to reality.

```
----------- Frame 6 ----------
perceptual behaviour: watch (cup, plate)
   look point        : constant
   camera position   : constant
   region of interest: full image
   knowledge sources : plate_finder cup_finder


   plate_9 [5.00] 0.42 0.29 0.00
    cup_11 [5.00] 0.43 0.12 0.04
   poly_40 [0.98] 0.31 0.22 0.13
```

In frame 6 a new supervisor goal, *watch (cup,plate)*, is issued. The operational modes of the basic units are modified accordingly. In this case only the polyhedral knowledge source is turned of; the camera and region of interest strategy remains unchanged. The new supervisor goal could be issued as a consequence of :

- a user request
- detection of a triggering event in the data passed to the supervisor
- entering a new stage of a dynamic plan

However, reasoning about perceptual behaviour is not in the scope of the interpretation module.

```
----------- Frame 7 ----------
```

```
plate_9 [5.00] 0.42 0.29 0.00
cup_11  [5.00] 0.46 0.12 0.04
poly_40 [0.96] 0.31 0.22 0.13

----------- Frame 8 ----------

plate_9 [5.00] 0.42 0.29 0.00
cup_11  [5.00] 0.48 0.12 0.04
poly_40 [0.94] 0.31 0.22 0.13

----------- Frame 9 ----------

plate_9 [5.00] 0.42 0.29 0.00
cup_11  [5.00] 0.48 0.12 0.04
poly_40 [0.92] 0.31 0.22 0.13

----------- Frame 10 ----------

plate_9 [5.00] 0.41 0.28 0.00
cup_11  [3.00] 0.48 0.12 0.04
poly_40 [0.90] 0.31 0.22 0.13
```

No new objects are detected in frames 7-10. The pose of object plate_9 is very stable, changing within the system precision of 1cm. Images 7a-9a show that movement of cup_11 was correctly followed. At image 10 the cup was removed from the scene and consequently the cup recognition source didn't detect it. The confidence level was decreased, but the object hypothesis will remain in the scene model for another 2-3 frames until the confidence falls to 0.

The confidence level of the poly_40 hypothesis decreases slowly from frame 6 when the polyhedral knowledge source was switched off. The updates of confidence in poly_40 and cup_11 model two completely different phenomena. In the case of cup_11 strong evidence is produced by the recognition knowledge source that the object has disappeared. No such information is available about poly_40; the module is effectively 'blind' to polyhedral objects as the appropriate knowledge source is not running. However, the scene model must take into account the *aging* of the information related to poly_40. The speed of confidence change is governed by the expected average durability of object pose of an 'unviewed' tea-box (the same strategy is used for objects not in the field of view).

```
----------- Frame 11 ----------

plate_9 [5.00] 0.40 0.27 0.00
cup_11  [2.00] 0.48 0.12 0.04
poly_40 [0.88] 0.31 0.22 0.13

----------- Frame 12 ----------

plate_9 [5.00] 0.38 0.25 0.00
cup_11  [1.00] 0.48 0.12 0.04
poly_40 [0.86] 0.31 0.22 0.13

----------- Frame 13 ----------

plate_9  [5.00] 0.38 0.23 0.00
- cup_11 [0.00] 0.48 0.12 0.04
poly_40  [0.84] 0.31 0.22 0.13
```

The interpretation process progresses as expected in frames 10-13. The cup_11 hypothesis is finally removed. The left-to-right motion of plate_9 is correctly tracked.

```
---------- Frame 14 ----------

   plate_9 [5.00] 0.37 0.21 0.00
   poly_40 [0.82] 0.31 0.22 0.13

---------- Frame 15 ----------
perceptual behaviour: explore
   look point          : constant
   camera position     : constant
   region of interest: full image
   knowledge sources : plate_finder cup_finder poly_finder

   plate_9 [5.00] 0.38 0.20 0.00
   poly_40 [0.80] 0.31 0.22 0.13
+  cup_116 [1.00] 0.24 0.35 0.04
```

The supervisor switches the mode back to *explore*. A new cup hypothesis is created.

```
---------- Frame 16 ----------
   plate_9  [5.00] 0.38 0.19 0.00
   poly_117 [1.00] 0.11 0.23 0.01
-  cup_116  [0.00] 0.24 0.35 0.04
```

In the last frame a new polyhedral object, poly_117, is detected. It is not recognised that it is actually a noise measurement induced by the same tea-box as poly_40; the hypothesis are not merged, the old is replaced by the new one.

20.5.2   Experiment 2: Focus of Attention

The first experiment presented the basic functionalities of the scene interpretation module: the temporal integration of recognition results, confidence maintenance, change of perceptual behaviour in response to a supervisor goal. In the second experiment an additional mechanism, region of interest control, is used to implement focus of attention.

```
---------- Frame 1 ----------
perceptual behaviour: explore
   look point          : constant
   camera position     : constant
   region of interest: full image
   knowledge sources : plate_finder cup_finder poly_finder


+ poly_1 [1.00] 0.01 0.24 -0.07
+ plate_9 [1.00] 0.42 0.29 0.00
+ cup_11  [1.00] 0.34 0.12 0.04
```

The focus-of-attention and explore experiments process the same sequence; consequently, the results will be identical as long as the perceptual behaviour (supervisor goal) remain the same.

```
---------- Frame 2  ----------

- poly_1 [0.00] 0.01 0.24 -0.07
  plate_9 [3.00] 0.42 0.29 0.00
```

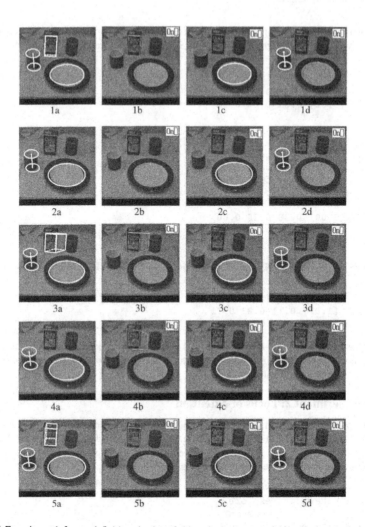

Fig. 20.4. Experiment 1, frames 1-5. (a) projection of objects in the scene model into the image plane. (b)(c)(d) object hypotheses generated in the frame by polyhedral, plate, and cup recognition knowledge sources

```
cup_11  [2.00] 0.34 0.12 0.04

---------- Frame 3 ----------
perceptual behaviour: track (cup_11 150)
  look point        : constant
  camera position   : constant
  region of interest: at_object cup_11 150%
  knowledge sources : plate_finder cup_finder

  plate_1 [2.98] 0.42 0.29 0.00
  cup_11  [3.00] 0.35 0.12 0.04
```

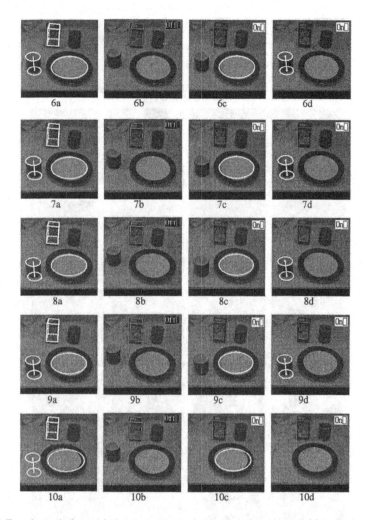

Fig. 20.5. Experiment 1, frames 6-10. (a) projection of objects in the scene model into the image plane. (b)(c)(d) object hypotheses generated in the frame by polyhedral, plate, and cup recognition knowledge sources

At frame 3 a new supervisor goal, *track (cup_11 150)* is issued. The region of interest unit operational mode is modified to ensure that only a restricted area around cup_11 is processed by the recognition knowledge sources. The 2D region-of-interest is computed as follows. First, all visible corners of the object's 3D bounding box are projected on the image plane. The minimum and maximum coordinate values in horizontal and vertical directions define the 2D bounding rectangle. The 2D region-of-interest (ROI) is obtained by stretching the bounding rectangle by a factor defined by the second parameter of the track command (clipping is performed if a part of the region-of-interest is outside the image). The above definition guarantees that the symbolic, object-centered ROI dynamically follows the object of interest.

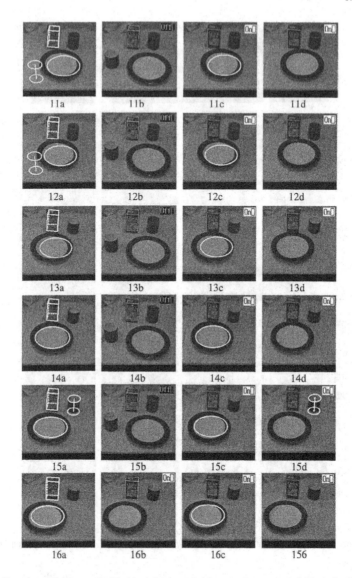

Fig. 20.6. Experiment 1, frames 11-16. (a) projection of objects in the scene model into the image plane. (b)(c)(d) object hypotheses generated in the frame by polyhedral, plate, and cup recognition knowledge sources

The cooperation of the ROI mechanism with the recognition knowledge sources is carried out through information stored in the symbolic scene model. After every frame, the 3D bounding box information necessary for ROI computation is updated using the recognition results. The benefits of ROI are twofold: First, a significant speed up (approx. 10x in the presented experiment) is achieved as the recognition time is roughly proportional to the number of pixels processed. Second, low-level image processing routines that use image statistics for self-tuning (eg. automatic threshold selection for hysteresis linking in edge detection) improve performance because only information in the vicinity of the object is taken into consideration. In the presented experiment, only the second type of benefit can be observed; the speed up would normally result in more frequent acquisition of images (compared with the *explore* mode). Consequently, a larger number of images (per unit time) with smaller object movements would be processed. However, both experiment were performed on a pre-recorded sequence of 15 frames.

The 'stretch' parameter of the track command enables the supervisor to control the area of processing. The parameter setting, based on previous object mobility and loop of perception execution time, must reflect the following relationships. On the one hand, a smaller area of processing means faster execution of the loop of perception while on the other hand the object of interest can (partially) move out of a small ROI and thus render detection impossible.

```
---------- Frame 4 ----------

  plate_9 [2.96] 0.42 0.29 0.00
  cup_11  [3.50] 0.38 0.10 0.04

---------- Frame 5 ----------

  plate_9 [2.94] 0.42 0.29 0.00
  cup_11  [4.50] 0.38 0.10 0.04

---------- Frame 6 ----------

  plate_9 [2.92] 0.42 0.29 0.00
  cup_11  [5.00] 0.42 0.12 0.04

---------- Frame 7 ----------

  plate_9 [2.90] 0.42 0.29 0.00
  cup_11  [5.00] 0.46 0.12 0.04
```

The cup_11 object is successfully tracked in frames 4, 5, 6 and 7. Comparison of the cup_11 pose with results obtained in the *explore* experiment shows that pose estimates in the two sequences are close (within 2cm) but not identical. This observation can be explained by adaptation of the low-level modules (edge detection) to noise specific to the ROI. The plate_9 object is effectively out of the field of view; its confidence level is updated according to the forgetting scheme described in the presentation of the 'explore' experiment.

```
---------- Frame 8 ----------

  plate_9 [2.88] 0.42 0.29 0.00
  cup_11  [5.00] 0.46 0.12 0.04

---------- Frame 9 ----------

  plate_9 [2.86] 0.42 0.29 0.00
```

```
cup_11   [4.00] 0.46 0.12 0.04
----------- Frame 10 ----------
perceptual behaviour: track (plate_9 150)
  look point          : constant
  camera position     : constant
  region of interest: at_object plate_9 150%
  knowledge sources : plate_finder cup_finder

  plate_9 [3.86] 0.41 0.28 0.00
  cup_11   [3.98] 0.46 0.12 0.04
```

Cup_11 disappears after frame 7. The supervisor responds to the decrease of confidence in cup_11 by issuing a new command *track plate_9 150*. The decrease of confidence in cup_11 proceeds according to the slow, out-of-field-of-view scheme as cup_11 is not fully inside the new ROI.

```
----------- Frame 11 ----------

  plate_9 [4.86] 0.40 0.27 0.00
  cup_11   [3.96] 0.46 0.12 0.04

----------- Frame 12 ----------

  plate_9 [5.00] 0.38 0.25 0.00
  cup_11   [3.94] 0.46 0.12 0.04

----------- Frame 13 ----------

  plate_9 [5.00] 0.38 0.23 0.00
  cup_11   [2.94] 0.46 0.12 0.04

----------- Frame 14 ----------
perceptual behaviour: watch (plate, cup)
  look point          : constant
  camera position     : constant
  region of interest: full image
  knowledge sources : plate_finder cup_finder

  plate_9 [5.00] 0.37 0.21 0.00
  cup_11   [1.94] 0.46 0.12 0.04
```

The processing proceeds as expected. The plate_9 was successfully tracked. Confidence in cup_11 starts to decrease rapidly when a new goal, *watch (plate, cup)* is received.

```
----------- Frame 15 ----------

  plate_9 [5.00] 0.38 0.20 0.00
  cup_11   [0.94] 0.46 0.12 0.04
+ cup_72 [1.00] 0.24 0.35 0.04

----------- Frame 16 ----------

  plate_9 [5.00] 0.38 0.19 0.00
- cup_11   [-0.06] 0.46 0.12 0.04
  cup_72 [0.00] 0.24 0.35 0.04
```

Due to the selected behaviour and original high confidence value the cup_11 hypothesis has not been discarded before frame 16 as no evidence against its presence was produced in the recognition process. However, if a more sophisticated geometric reasoning scheme was implemented, the cup hypothesis could have been disposed of when the volume of plate_9 intersected the space supposedly occupied by cup_11 thus creating an internal inconsistency in the scene model database.

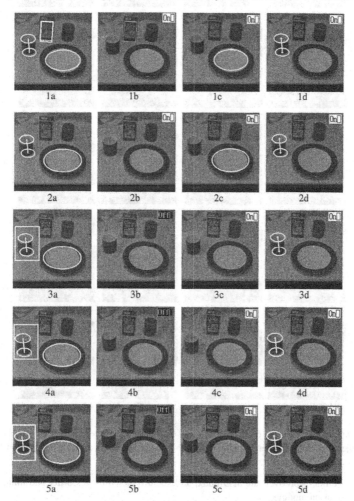

Fig. 20.7. Experiment 2, frames 1-5. (a) projection of objects in the scene model into the image plane. (b)(c)(d) object hypotheses generated in the frame by polyhedral, plate, and cup recognition knowledge sources

20.5.3 Conclusion

A novel framework for control of scene interpretation has been proposed. It has be shown that decomposition of the interpreter into camera, region of interest and recognition units allows the module to respond to a broad class of visual goals by a simple mapping of the goal into operational modes

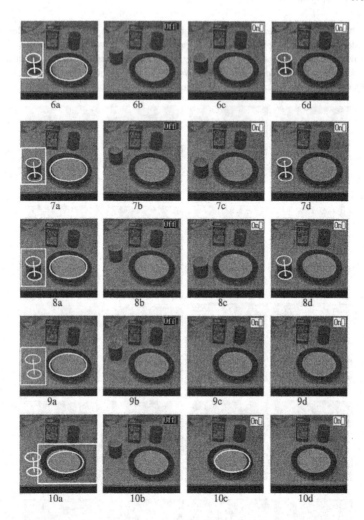

Fig. 20.8. Experiment 2, frames 6-10. (a) projection of objects in the scene model into the image plane. (b)(c)(d) object hypotheses generated in the frame by polyhedral, plate, and cup recognition knowledge sources

of individual units. Moreover, the distributed architecture increases flexibility and maintainability of the interpretation module.

The two experiments described in section 20.5 clearly demonstrate the merits of the spatio-temporal context for scene interpretation. Information about objects in the scene is stored in a hierarchical scene model. Several 'forgetting' schemes are adopted to reflect dynamism of objects. The database is used to guide future spatially focus interpretation.

In the context of a prototypal indoor scene, the breakfast table-top, the interpretation module was capable to recognise, track and focus on objects with reasonable robustness and acceptable speed.

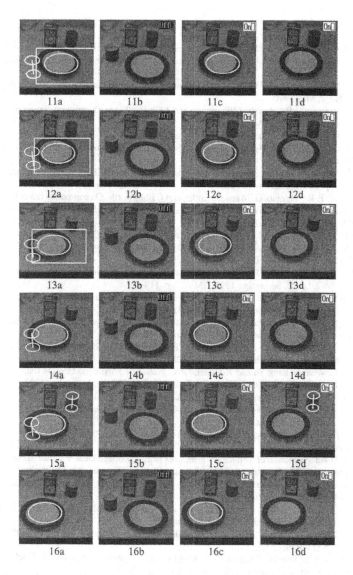

Fig. 20.9. Experiment 2, frames 11-16. (a) projection of objects in the scene model into the image plane. (b)(c)(d) object hypotheses generated in the frame by polyhedral, plate, and cup recognition knowledge sources

References

[DD.G.1.3] J. Kittler, J. Illingworth, G. Matas, P. Remagnino, K C Wong, H I Christensen, J.O. Eklundh, G. Olofsson, M.Li. Sybolic Scene Interpretation and Control of Perception VAP Project Document DD.G.1.3, March 1992.

[DR.E.1.2] H I Christensen and E Granum: "Specification of Skeleton Control Structure", VAP Deliverable DR.E.1.2, May 1990.

[DR.D.2.3] A Etemadi, J-P Schmidt, J Illingworth and J Kittler "Scene Description Representation", VAP Deliverable DR.D.2.3, August 1991.

[Etamadi 91] A. Etemadi, G. Matas, J. Illingworth, and J. Kittler. Low-level Grouping of Straight Line Segments. In *British Machine Vision Conference 1991*, Glasgow, September 1991. BMVC. pp. 118-126

[Etamadi 92] A Etemadi "Robust Segmentation of Edge Data", Proc 4th IEE Intern Conf Image Processing and Applications, Maastricht, 1992.

[Remagnino 92a] P. Remagnino, G. Matas, J. Kittler, and J. Illingworth. On computing the next look camera parameters in active vision. In *10th European Conference on Artificial Intelligence*, Vienna, August 1992. ECAI. pp. 806-807

[Remagnino 92b] P. Remagnino, G. Matas, J. Kittler, and J. Illingworth. Control in the bootstrap phase of a computer vision system. In *4th International Conference on Image Processing and its applications*, Maastricht, April 1992. IEE. pp. 85-88

[Matas-Kittler 92] G. Matas, and J. Kittler. Contextual Junction Finder. In *British Machine Vision Conference 1992*, Leeds, September 1992. BMVC. pp. 119-128

[Lee 92] H M Lee, J Kittler and K C Wong"Generalised Hough Transform in Object Recognition", Proc 11 Intern Conference on Pattern Recognition, The Hague, 1992.

[Wong 91a] K C Wong, J Kittler and J Illingworth "Analysis of Straight Homogeneous Generalized Cylinders Under Perspective Projection", Proc International Workshop on Visual Form, Capri 1991.

[Wong 91b] K C Wong, J Kittler and J Illingworth:"Heuristically Guided Polygon Finding", Proc British Machine Vision Conference, pp 400-407, Glagow, 1991.

[Wong 93] K. C. Wong, Cheng Yu and J. Kittler, "Recognition of Polyhedral Objects Using Triangle-pair Features", The Special Issue of IEE Processings Part I on Image Processing, 1993. To appear.

[Cheng 92] Y. Cheng, K. C. Wong and J Kittler"The Recognition of Triangle-pairs and Quadrilaterals from a Single Perspective View", Proc IEE 4th Internat. Conference on Image Processing and Its Applications, Maastricht, 1992.

[Hoad 92a] P. Hoad and J. Illingworth"Detection of Flat topped cylinders from a single view", submitted for publication

[Hoad 92b] P. Hoad"Extraction of Volumetric Primitives from 2D Image Data",MPhil/PhD transfer report, September 1992, Department of Electronic and Electrical Engineering, University of Surrey

[Girratano 88] Joseph C. Giarratano" CLIPS User's Guide ", Artificial Intelligence Section, Lyndon B. Johnson Space Center, June 3, 1988

[Garvey 88a] Thomas D. Garvey, John D. Lowrance and Martin A. Fishler " An Inference Technique for Integrating knowldedge from Disparate Sources"

[Garvey 88b] Thomas D. Garvey"An Experiment with a System for Locating Objects in Multisensory Images", SRI International,

[Brooks 88] Rodney A. Brooks "Visual Map Making for a Mobile Robot", In Martin A. Fischler and Oscar Firschein, editors, *Readings in Computer Vision*, pp. 438-443

[Brooks 83] Rodney A. Brooks " Symbolic reasoning among 3-dimensional and 2 dimensional images", *Artifcial Intellingence*, 17:349-385, 1983.

[Hanson-Riseman 78] A. R. Hanson and E M. Riseman " VISIONS: a computer system for interpreting scenes. In Allen R. Hanson and Edward M. Riseman, editors, *Computer Vision Systems.*, Academic Press, New York, 1978.

[Draper 92] Bruce A. Draper, Allen R. Hanson and Edward M. Riseman "Learning Knowledge-Directed Visual Strategies", In *Proc. of the DARPA Image Understandin Workshop*, 1992, pp. 933-940

[Lewitt 87] T. Lewitt, D. Lawton, D. Chelberg, P. Nelson and J. Due "Visual Memory for a mobile robot", In *Proc. of the AAAI Workshop on Spatial Reasoning and Multisensor Fusion, Morgan Kaufman Publishers, Los Altos, California, 1987*

[Lawton 88] D.T. Lawton, T.S. Levitt, and P. Gelband " Knowledge based vision for terrestrial robots", In *Proc. of the DARPA Image Understandin Workshop*, 1988, pp. 933-940

[Rimey 92] Raymond D. Rimey "Where to Look Next using a Bayesin Net: An Overview", In *Proc. of the DARPA Image Understandin Workshop*, 1988, pp. 927-932

[Strat-Smith 87] Thomas M. Strat and Grahame B. Smith "The Core Knowledge System", SRI International, Technical Note No. 426, October 1987

[Wilcox 90] Lambert E. Wixson "Real-time Qualitative Detection of Multicolored Objects", In *Proc. of the DARPA Image Understandin Workshop*, 1990, pp. 631-638

[Brown 92] Christopher M. Brown " Issues in Selective Perception ", In *Proc. of the DARPA Image Understandin Workshop*, 1992, pp. 21-30

21. Recognition of Polyhedral Objects Using Triangle-pair Features

K. C. Wong, Cheng Yu and Josef Kittler

UOS

Abstract

This paper is concerned with the problem of model based recognition of polyhedral objects from a single perspective view. A hypothesize-verify paradigm based on the use of high level knowledge constraints derived from local shape properties is presented. In the recognition system, two inter-mediate features, namely triangle-pair and quadrilateral are employed as key features for model invocation and hypothesis generation. A verification process for performing a detailed check on the model-to-scene correspondences is developed. To reduce the number of implausible hypotheses generated from scene-to-model intermediate feature assignments, two geometrical constraints, namely distance and angle constraints are employed. A list of closed polygons and C-triple pairs extracted from a 2D intensity image by means of edge and intermediate feature detection process is used as an input to the matching system. The intermediate feature grouping process starts by identifying junctions created by pairs of line segments and then forms triples by combining pairs of junctions which share a common line. These triples are then scanned by a procedure which connects them into meaningful geometric structures. As a by-product of the recognition method the relative pose of the 3D polyhedral objects with respect to the camera is recovered. Extensive experimental results are reported to confirm the feasibility of the proposed method.

21.1 Introduction

Model-based recognition of 3-D polyhedron under the perspective projection is an interesting and practically important topic in computer vision. Many man-made objects are polyhedral, especially industrial parts, buildings, furniture, etc. In the absence of 3-D sensors which in any case are either expensive, provide slow 3D-data acquisition or are difficult to calibrate, the interpretation of such objects may be based on the recognition, from their 2-D perspective projections, of the spatial planar polygons which constitute their surfaces.

In spite of the relatively simple form of polyhedral objects, their recognition has proved to be a very difficult problem. This situation is a consequence of the elusiveness of the solution to the problem of recognizing spatial planar polygons under perspective projection. Furthermore, an image usually contains data from many objects, as well as spurious and missing data caused by shadow, occlusion, surface markings and poor segmentation. Furthermore, the explicit depth information is not preserved during the process of projection. However, there are informative invariant 2D geometric features which can be extracted by initiating perceptual grouping operations that organise isolated low level primitives into larger scale structures conveying meaningful geometric cues. But in general such features are not powerful enough to resolve all the ambiguities inherent in a single perspective image. However, if complemented by high level, object specific constraints, both the robustness and efficiency of the matching process can be improved. Geometric models, in particular can provide highly constraining predictions for recognising well-defined object types such as polyhedrons.

In general, there are several distinct phases in model-base matching of rigid objects. Two off-line stages are model generation for constructing a CAD-like database of models, and model analysis for identifying and organising model features into structures for matching and for developing strategies for execution of the matching task. The two main run-time stages are hypothesis generation and verification. The former consists of extracting interesting 2D geometric features from an image and then generating possible poses of scene objects so that the subsequent object verification process is provided with tight constraints on where to search for confirmatory evidence of model existence. The latter, model verification process performs a detailed check of the projected 3D features against 2D image data, confirming feature presence and accounting for features which are not observed. Most of the existing recognition systems using the above mentioned approach rely on geometric cues derived from the geometrical relationships between model-scene feature correspondences [Lowe 87], [Horaud 87].

In this paper we present a model-based recognition system for identifying the scene-to-model correspondences from a single perspective image. A hypothesize-verify paradigm based on local shape descriptions, namely triple-pair and quadrilateral features, is described. Geometric constraints derived from these key features are exploited to complete the battery of tools required to recognise general polyhedral objects.

Recently Lei [Lei 90] has developed a method for the recognition of spatial planar polygons under perspective projection based on a view point invariant, namely, the cross ratio. However the approach is available for polygons of five sides or more only which restrict the type of polyhedron that can be recognised based on this features. Lowe [Lowe 87] experimented with groups of powerful 2D non-accidental viewpoint independent cues in his SCERPO system to reduce substantially the number of inconsistent matches that may be considered in the matching process. Having selected a subset of informative features from the image, he proposed an iterative method to estimate the poses of the scene objects by refining the chosen initial transform parameters using a progressively greater number of hypothesized model-scene correspondences. However, the assignment of the initial values is a non-trivial task. Horaud [Horaud 87] developed an effective hypothesis-verification scheme using triplet as a key feature. He provided a constructive method to recover the pose of a scene object using the geometric relationships between the corresponding model and scene vertices without any restriction on the angles between edges. Feasible solutions were searched for from a tessellated Gaussian sphere.

As mentioned earlier, our approach is based on triangle-pair features [Cheng et al. 92]. The main idea of proposing the use of triangle-pair features derives from the observation that spatial planar polygons are the constituent surfaces of polyhedral objects. Thus, they do not exist in isolation but rather are in specific geometric geometric relations with each other. A quadrilateral feature can be analysed as a special case of a triple-pair feature. There are several reasons for employing these features as key features : the number of projections of triangle-pairs (called C-triple pairs) and quadrilateral features are generally manageable; they are qualitative viewpoint invariant geometric primitives; The transformation between a model and a camera frame can be completely determined using these features. The robustness of these features can be easily enhanced using an interactive environment between the matching phase and low level and feature grouping process.

To reduce the number of implausible hypotheses generated from scene-model triangle-pair and quadrilateral feature assignments, two effective geometric constraints, namely distance and angle constraints, are derived and incorporated in the matching process. Only those hypothesized model-to-scene corresponding features which satisfy both the distance and angle constraints will be considered in the computational intensive verification process. As a by-product of the matching process, the transformation defining the pose of the model with respect to the camera can be easily obtained using one of the two methods described in the paper.

Our framework based on a hypothesize-verify technique using triangle-pair and quadrilateral features is very effective and intuitive. The derivation of the distance and angle constraints is simple. Furthermore, other interesting geometic primitives can be easily incorporated into our system to handle more complex viewing environment.

The paper is organized as follows. In the next section, the solution of perspective equations for a spatial triangle is derived. The triangle recognition problem is considered in the context of candidate solutions obtained from the perspective equations. This will introduce an effective way to use distance and angle constraints to prune out implausible solutions and to identify a unique object from the model base which explains the observed data. Third, the equations of distance and angle constraints are applied to the problem of detecting quadrilaterals. In Section 21.3, the methods of estimating the scene object pose using the recovered triangle-pair or quadrilateral is described. In Section 21.4, the modules integrated into our polyhedral object recognition system are presented. Extensive Experimental results obtained using the proposed method on three real images are presented in Section 21.5. Finally, conclusions are drawn about the proposed method.

21.2   The geometric constraints of intermediate features

In this section, the triangle-pair (quadrilateral) primitive which can be used as a key feature for generating hypotheses will be introduced. Two geometrical constraints imposed by this intermediate feature will be derived. The pruning power and reliability of the features will be studied in the experiment. First, the analysis of the perspective projection of a spatial triangle is briefly overviewed in the next subsection.

21.2.1   A Perspective Projection of a Triangle

The difficulty in finding the correct model for a given perspective projection of a triangle is that different 3D triangular surfaces under perspective projection will produce the same 2D projected triangle. That is , in general all triangles in the database can be projected onto a single image triangle. One can always find a pose and an appropriate distance between the centre of the triangle and the image plane to achieve this.

If the corresponding points between a triangle and its projected image are known, then the perspective relationships between the model and its image can be computed easily. The details of the derivation and analysis of the perspective relationship can be found in [Fischler-Bolles 81]. Here only a brief version of the derivation is outlined. Suppose $\triangle ABC$ in Figure 21.1 is an object model and $\triangle A'B'C'$ is its projected image. The point O is an origin with respect to a camera coordinate system and O' is the origin of an image plane. Letting $\alpha = \cos \angle A'OB', \beta = \cos \angle B'OC', \gamma = \cos \angle C'OA',$ $k_b = k_a x$ and $k_c = k_a y$, We have

$$a^2 = k_a^2(1 + x^2 - 2x\alpha) \tag{1}$$
$$b^2 = k_a^2(x^2 + y^2 - 2xy\beta) \tag{2}$$
$$c^2 = k_a^2(1 + y^2 - 2y\gamma) \tag{3}$$

where a, b and c are the lengths of the segments AB, BC and AC respectively, and k_a, k_b and k_c are the distances of vertices A, B and C of the triangle from the origin. Denoting $n_1 = (\frac{b}{c})^2$ and $n_2 = (\frac{b}{a})^2$, after some manuplation, the following biquadratic polynomial equation in one unknown x is derived,

$$P_4 x^4 + P_3 x^3 + P_2 x^2 + P_1 x + P_0 = 0 \tag{4}$$

where the coefficients P_i are a function of n_1, n_2, α, γ and β. The solution of equation (4) can be determined in close form [Dehn 60] or by iterative techniques [Conte 65]. In theory there are

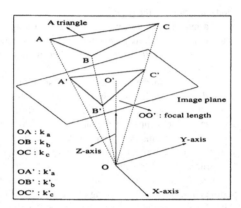

Fig. 21.1. A perspective projection of a spatial triangle.

eight solutions for the eqs. (1)-(3). Since for every real positive solution there is a real negative solution, in fact there are at most four possible solutions. For each positive real solution of equation (4), we can determine from equation (1) a single positive real value for the vertex distance $k_a = a / \sqrt{[x^2 - 2x\alpha + 1]}$ and consequently $k_b = k_a x$. From equation (3) on the other hand, we have $y = \gamma \pm \sqrt{\gamma^2 + (c^2 - k_a^2) / k_a^2}$. For each real positive value of y we obtain a value of k_c from $k_c = k_a y$. Thus, it has been seen that the solution of k_a, k_b and k_c of the perspective equation with respect to the camera coordinate frame can easily be computed.

As the point correspondences are not normally known, there are up to a maximum of twelve possible solutions for distances k_a, k_b and k_c for any arbitrarily selected triangle model and a given 2D projected triangle extracted from a perspective image. Furthermore, there are corresponding values k_a, k_b and k_c for every constituent triangle surface of the polyhedral objects stored in the model base. Note that, a generic vision system would normally contain models of many objects. In view of these factors a detailed matching of each polyhedral object model against all the triangle features in order to obtain the best possible scene interpretation is therefore generally infeasible. Methods have to be developed for grouping isolated image primitives such as triangles into larger scale structure conveying meaningful geometric cues. In our framework, quadrilateral and triangle-pair structures are exploited as key features for model invocation and hypothesis generation. The model-scene high-level feature assignments which satisfy further constraints then become feasible candidates for consideration for the pose determination. In aim is for most model-scene assignments to be rapidly dismissed from consideration following a cursory examination using the constraints. In the next section, we will introduce the triangle-pair feature and its two geometric constraints will be derived.

21.2.2   The constraints : Triangle-pair

The main idea behind the proposed method derives from the observation that spatial polygons are the constituent surfaces of polyhedral objects. Thus, they do not exist in isolation but rather are in specific geometric relations with each other. These relations can be used to resolve inherently ambiguous triangle-pairs which arises when triangular polyhedral object faces abut each other. We shall consider such a triangle-pair recognition problem first. Recall that the correspondence between the vertices of a model and of observed triangles is not known a priori. For example in Figure 21.2(a) there are two possible potential matches which have to be considered between a triangle-pair model

$\triangle ABC$ and $\triangle ACD$ and a triangle-pair image $\triangle A'B'C'$ and $\triangle A'C'D'$ which are listed in **Table A**. To get a handle on the problem we shall invoke various constraints to reduce the solution set.

Table A

Scene	A'B'C'	A'C'D'	A'B'C'	A'C'D'
Model	A B C	A C D	D A C	C A B

In particularly we shall use higher level knowledge constraints. This in no way compromises our approach because the ultimate aim of identifying the model triangle from its projection is to recognise the polyhedron in which the triangle-pair constitutes one or part of its abutting faces.

The first constraint we adopt is a distance constraint. The basic principle of this approach is as follows. Suppose A'B'C'D' is a projected line drawing of a polyhedron (see Figure 21.2(a)). Obviously it consists of $\triangle A'B'C'$ and $\triangle A'C'D'$ abutting each other at the common edge $A'C'$. Note that the cardinality of the set of all the solutions k_a^i, k_b^i and k_c^i of the perspective equation of the image $\triangle A'B'C'$ and k_a^j, k_c^j and k_d^j of the image $\triangle A'C'D'$ for each triangle-pair model in the database can be as many as 16. However, given that $A'C'$ is the projection of the common edge AC between the model triangle-pair, only those solutions which satisfy $k_a^i = k_a^j$ and $k_c^i = k_c^j$, will be considered as plausible solutions. If the number of plausible solution is one, a unique model which corresponds to the feasible solution will be declared as the object in the scene. Usually using the distance constraint, the correct triangle model can be identified. But when the number of pairs of model triangles satisfying the distance constraint is greater than one, we must use an additional constraint to resolve the residual ambiguity. One such option is an angle constraint.

In Figure 21.2(a), **P** and **Q** are the normals to $\triangle ABC$ and $\triangle ACD$, respectively. Here the positive direction of the normals is always pointed to the inside of the polyhedron. From Figure 21.2(a) we denote,

$$\frac{OA}{OA'} = \frac{k_a}{k_{a'}} = g; \qquad \frac{OB}{OB'} = \frac{k_b}{k_{b'}} = h; \qquad \frac{OC}{OC'} = \frac{k_c}{k_{c'}} = r; \qquad \frac{OD}{OD'} = \frac{k_d}{k_{d'}} = s;$$

then

$$\mathbf{P} = \mathbf{BC} \times \mathbf{BA} = \begin{vmatrix} \mathbf{i} & \mathbf{j} & \mathbf{k} \\ rx_c' - hx_b' & ry_c' - hy_b' & f(r-h) \\ gx_a' - hx_b' & gy_a' - hy_b' & f(g-h) \end{vmatrix} \tag{5}$$

$$\mathbf{Q} = \mathbf{DA} \times \mathbf{DC} = \begin{vmatrix} \mathbf{i} & \mathbf{j} & \mathbf{k} \\ gx_a' - sx_d' & gy_a' - sy_d' & f(g-s) \\ rx_c' - sx_d' & ry_c' - sy_d' & f(r-s) \end{vmatrix}$$

Where (x_i', y_i') is an image point with respect to the origin O' of the image plane and f denotes the focal length. The angle φ between $\triangle ABC$ and $\triangle ADC$ is $\varphi = \pi - cos^{-1} \frac{\mathbf{P} \cdot \mathbf{Q}}{|\mathbf{P}||\mathbf{Q}|}$ The measured angle φ should correspond to the actual angle between the faces of the hypothesized polyhedron. Since this information can be assumed to be known a prior, the angle can be used to prune out the remaining inconsistent solutions.

Having discussed the distance and angle constraints, the integration of these constraints into the recognition framework results in the following procedure. First, all the models satisfying the distance constraints are selected from the model base for further consideration. If there is one such object model, we can then conclude that the object in the scene has been identified. Otherwise, the pre-computed angles between surfaces of the hypothesized models are compared with the angle φ computed from an image. The object model satisfying the angle constraint will be associated with the scene object.

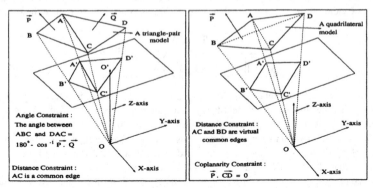

(a) Constraints of a triangle-pair feature   (b) Constraints of a quadrilateral feature

Fig. 21.2. A perspective projection of a triangle-pair and a quadrilateral feature.

21.2.3   The Constraints : Quadrilateral

In this section, the technique of solving the 3-point perspective problem is used as a tool for
identifying each quadrilateral in the scene. First note that each quadrilateral stored in the model base
can be decomposed into two sets of triangular surface pairs using the diagonal of the quadrilateral
as a common edge between them. For example, a quadrilateral ABCD shown in Figure 21.2(b)
is decomposed into two triangular surface pairs $\triangle ABC$ and $\triangle ACD$ taking the diagonal AC as
a common edge and $\triangle ABD$ and $\triangle DBC$ taking the diagonal BD as a common edge. The four
possible combinations between two sets of triangular surface pairs stored in the model and a triangle-
pair taken from an image are listed in **Table B**. After selecting a correspondence between the model

Table B

Scene	A'B'C'	A'C'D'	A'B'C'	A'C'D'	A'B'C'	A'C'D'	A'B'C'	A'C'D'
Model	A B C	A C D	C D A	C A B	D A B	D B C	B C D	B D A

and the image, for example the first group in **Table B**, that is, $\triangle ABC$ corresponding to $\triangle A'B'C'$ and
$\triangle ACD$ corresponding to $\triangle A'C'D'$, we solve the two perspective projection equations to obtain two
groups of solutions. If the solutions for k_{a1} and k_{c1} under model $\triangle ABC$ are equal to the solutions
found for k_{a2} and k_{c2} under model $\triangle ACD$ respectively, then we can conclude that quadrilateral
A'B'C'D' is possibly the projection of model ABCD.

If quadrilateral ABCD is the true corresponding model, it must be a planar polygonal surface.
Hence, if the triangular surface ABC is part of the quadrilateral ABCD, the point D of the triangle
BCD must lie on the plane of triangle ABC. This implies that the vector **CD** must be orthogonal
to the normal **P** of the plane defined by triangle ABC. The formula for computing the normal **P** is
given in equation (5) and the vector $\mathbf{CD} = (sx'_d - rx'_c)\mathbf{i} + (sy'_d - ry'_c)\mathbf{j} + f(s - r)\mathbf{k}$. If the point D
is co-planar with $\triangle ABC$, then $\mathbf{P} \cdot \mathbf{CD} = 0$. Therefore the decision formula is

$$\begin{vmatrix} sx'_d - rx'_c & sy'_d - ry'_c & f(s-r) \\ rx'_c - hx'_b & ry'_c - hy'_b & f(r-h) \\ gx'_a - hx'_b & gy'_a - hy'_b & f(g-h) \end{vmatrix} = 0 \qquad (6)$$

The values of g, h, r and s computed using a triangular surface pair between models and the image,
and the image points are substituted into equation (6) to check whether the quadrilateral model

and the projected quadrilateral satisfy the constraint of coplanarity. The quadrilateral model which agrees with the coplanarity constraint will be registered as a consistent match.

21.3 Pose Estimation

As a by-product of the recognition method the corresponding vertices of a quadrilateral or a triangle-pair with respect to a camera frame are recovered. Using this information, the relative rotational parameters of the mapping from the model to its instance in the image can be computed. The transformation defines the pose of the model with respect to the camera. It can be obtained using one of two methods described in this section.

Consider the perspective geometry of a camera model depicted in Figure 21.3, the image plane is assumed to be in front of the center of projection so as to acquire an upright image of the scene. The focal length, **foc** is the normal distance from the center of projection to the image plane. Based on the above configuration, the position of the scene vertex P_s can be expressed in a camera frame centered at the origin **E** as $P_s = R_{MC} P_w + T_{MC}$, where R_{MC} is the relative orientation between the model and camera frame and T_{MC} is a translation vector.

To determine the relative rotation, we decompose the rotation transform into model-to-vertex R_{MV} and camera-to-vertex R_{CV} transforms. The transformation R_{MV} of vertices P_w with respect to an object model coordinate system to vertices P_v with respect to a vertex-based coordinate system (see Fig. 21.3) can be expressed as $P_v = R_{MV} P_w$ where $R_{MV} = \begin{pmatrix} m_{1x} & m_{1y} & m_{1z} \\ m_{2x} & m_{2y} & m_{2z} \\ m_{3x} & m_{3y} & m_{3z} \end{pmatrix}$ and $M_i = (m_{ix}, m_{iy}, m_{iz})$ is a unit vector. Likewise, The transformation R_{CV} of vertices P_s' with respect to a camera coordinate system, which is centered at the origin of the world coordinate system **O**, to vertices P_v with respect to a vertex-based coordinate system can be expressed as $P_v = R_{CV} P_s'$ where $R_{CV} = \begin{pmatrix} c_{1x} & c_{1y} & c_{1z} \\ c_{2x} & c_{2y} & c_{2z} \\ c_{3x} & c_{3y} & c_{3z} \end{pmatrix}$ and $C_i = (c_{ix}, c_{iy}, c_{iz})$ is a unit vector. Having determined R_{MV} and R_{CV}, the rotation transform which maps an object model P_w to the scene feature point P_s' with respect to the camera coordinate system centered at the origin **O** of the model frame, can be written as $P_s' = R_{MC} P_w$ where $R_{MC} = R_{CV}^t \times R_{MV}$. In the following subsection, the meaning and methods of computing the unit vectors M_i and C_i for the case of triangle-pair and quadrilateral features will be discussed in detail. For this purpose, let us suppose the model-scene corresponding vertices $ABCD$ and $A'B'C'D'$ offer plausible candidate.

21.3.1 The triangle-pair feature case

Let E_1, E_2 and E_3 be AB, AC and AD respectively, described with respect to an object model frame (see Figure 21.2 (a)). An orthogonal vertex-based frame can be constructed using the Gram-Schmit orthogonalization process, $M_i' = E_i - \sum_{j=1}^{i-1} \frac{E_i \cdot M_j'}{\|M_j'\|^2} M_j'$, and then normalised to obtain unit vectors $M_i = \frac{M_i'}{\|M_i'\|}$. Likewise, the unit vectors C_i can be determined from $A'B'$, $A'C'$ and $A'D'$ using the same procedure.

21.3.2 The quadrilateral feature case

Consider the vectors **AB** and **AC** (see Figure 21.2(b)), described with respect to an object model coordinate system. The orthogonal x-, y- and z-axis of a coordinate system basis can be constructed by $M_1 = \frac{AC}{\|AC\|}$, $M_2 = \frac{(AB \cdot M_1)M_1 - AB}{\|(AB \cdot M_1)M_1 - AB\|}$ and $M_3 = \frac{M_1 \times M_2}{\|M_1 \times M_2\|}$ respectively. Likewise, the unit vectors C_i can be determined from $A'B'$ and $A'C'$ using the same procedure. This method can be used for the case of a triangle-pair feature for computing the parameters of the rotational transformation by considering the vectors of any two edges of triangle ABC or ACD. However, we find that using edges from two different surfaces to compute the relative rotation is more accurate and stable.

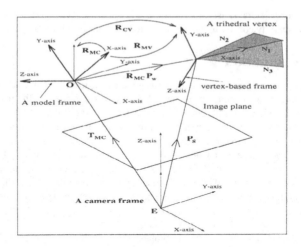

Fig. 21.3. A simplified pin-hole camera model

21.4 Matching Strategy

21.4.1 Feature Extraction

In this section, we briefy present a simple algorithm for grouping 2D line segments into closed polygons and triangle-pair 2D features that correspond to plausible physical 3D structures of polyhedrons. The algorithms starts by identifying junctions created by two line segments whose end points are mutually proximal to within a junction threshold, measured along the line of intersection [Etamadi et al. 91]. Having extracted the V-junctions from a scene, triples are formed by combining pairs of junctions which share a common line. These triples are then scanned by a procedure which connects them into polygon structures. Heuristic rules are used to control the combinatorial explosion associated with unconstrained association of junctions and triples. For additional details on the method of extracting closed polygons the reader is referred to [Wong et al. 91]. In here, we shall describe the proposed method of identifying triangle-pair 2D features. The algorithm proceeds as follows.

- First, a list of triples, namely *C-triple*, is generated by merging pairs of V-junctions which share a common line segment, namely *intermediate segment*. The resultant triples must have two end points located on the same half plane with respect to the intermediate segment.
- The aim of forming the triangle-pair 2D features is to find all *C-triple pairs*. To form a *C-triple pair*, the two merging *C-triples* must have a common segment. The intermediate segment of at least one of the two merging *C-triples* must be the common segment. Figure 21.4(a) illustrates the case where the intermediate segments of the two merging *C-triples* form the common segment. In Figure 21.4(b), the intermediate segment of the *C-triple* (**A**) shares a common segment with one of the end segments of *C-triple* (**B**). Such cases will be identified as valid structures for further consideration in grouping triangle-pair 2D features.
- The result of the merging process is a list of *C-triple pairs* which are likely to be the projections of 3D triangle-pair features. These feasible candidates will be employed to provide a tight constraint on where to search for complementary V-junctions which yield plausible triangle-pair 2D features. In searching for significant triangle-pair 2D features, the algorithm attempts

to identify V-junctions at the end points of the segments radiating from the common line segment of a *C-triple pair*. A triangle-pair 2D feature is then identified if at least one V-junction is extracted from each of the two merging *C-triples*. For example, two plausible triangle-pair 2D features $\{ e_1, e_2, e_3 \}$ - $\{ e_1, e_6, e_7 \}$ and $\{ e_1, e_4, e_5 \}$ - $\{ e_1, e_6, e_7 \}$ shown in Figure 21.4(c) are interpreted as 2D projections of a 3D triangle-pair structure.

Clearly, the robustness and reliability of the feature extraction depends entirely on the extraction of V-junctions and *C-triples*. In order to cope with inadequate low-level processing and partial occlusion, the poorer quality 2D junctions and *C-triples* can be accommodated by varying the threshold on the junction region size. However, extraneous and spurious closed loops and triangle-pair 2D features which do not arise from the projections of 3D local geometric shape may be extracted. Fortunately, the number of these features is generelly manageable in the hypothesis generation and verification process and hence the computational cost is kept low. Many false hypotheses which are generated from these spurious features can be pruned away using the geometric constraints derived in this paper. Furthermore, the confidence measures computed by mapping the hypothesized models to the scene features using the transformation determined from the false model-scene assignments are relatively small and hence can be rejected. Having selected plausible geometric features from the line map, the method of generating feasible hypotheses by matching the extracted scene features against the model description will be described in the next section.

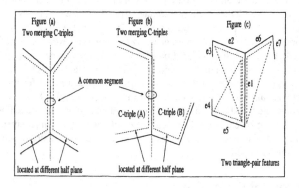

Fig. 21.4. (a), (b) and (c) *see text*

21.4.2   Hypothesis Generation

The role of this runtime stage is to identify model-scene feature assignments which satisfy the geometric constraints. The feasible model-scene feature assignments are used to estimate the pose of the target objects in the scene so that the subsequent verification process is provided with tight spatial constraints on where to search for confirmatory evidence of model instance. The procedures of this stage can be described as follows.

- The combinatorial assignments of the corresponding vertices between the triangle-pair and quadrilateral geometric features and the model description are considered. Using the model-scene triangle assignments, a list of possible solutions can be determined from the perspective equation (4). For each possible solution, the position and surface normal of the scene triangle measured with respect to the camera frame is computed.
- The model with at least one quadrilateral or triangle-pair satisfying the geometric constraints will be registered as a consistent interpretation. The relative rotation and translation embedded

in the pose defining transformation are then estimated using the geometric relationships derived from the admissible model-scene assignment candidates.

Having determined the pose of each hypothesis, the method of computing the confidence measure by correlating the hypothesized model with the 2D scene features such as 2D junctions and straight line segments will be described in the next section.

21.4.3 Verification Process

The role of the verification process is to perform a detailed check of the correspondence between hypothesized object models and image data, confirming features present and accounting for features which are not observed. In this process, infeasible hypotheses are pruned away and the most plausible candidate is selected as an instance of the target object. The procedures exploited in the verification process can be described as follows.

- The 2D description of each hypothesized object is generated by backprojecting the model using the transfomation computed in the hypothesis generation module. First, the hypothesized objects containing hidden surface which is interpreted as a quadrilateral or a *C-triple* of a triangle-pair feature extracted from a scene are removed from the candidate list.
- For the remaining candidates we first count the number of 2D junctions of the hypothesized object that coincide with the junctions extracted from the scene image. Two junctions are said to coincide if they are within a proximity threshold value and their angles and orientation must be within pre-specified allowable tolerances. After comparing every projected junction of the hypothesized object with the 2D junction extracted from the scene, the hypotheses with the greatest number of matched junctions will be invoked for consideration in the next stage. The aim of this stage is to select the hypothesis of greatest coincident of features with the scene data. To achieve this, the nearest scene line from each projected 2D model line is identified within an allowable threshold and then each of the identified nearest scene lines is divided by the corresponding projected line. These computed quotients are then summed up and divided by the number of visible projected edges of the hypothesized model to yield a confidence measure for the hypothesis.

21.5   Experimental Results

In this section, some experimental results are presented to demonstrate the effectiveness and robustness of the matching strategy incorporated into our recognition system. All the test images are taken with a standard CCD camera and were processed on Sun-4 Sparcstation with code written in C language. Real images containing polyhedral objects were employed to test the efficacy of our method. An object-centred and viewer-centred right-handed coordinate systems were used to define the vertices of object models and scene objects respectively. The first experiment involved 4 polyhedral object models of different shape. The test image contained the target objects being placed in random orientation without occlusion. The aim of this experiment was to test the effectiveness and reliability of the two high-level geometric features employed as seed features for generating feasible hypotheses. The vertices of the polyhedral object models are shown in Figure 21.5. The test image shown in Figure 21.6 (a) was processed by a Canny edge detector, and straight line segments shown in Figure 21.6 (b) were extracted from its thresholded output using a Hough-Based line algorithm. It can be seen that edges do not always meet exactly at a junction. Also there are many lines extracted from the scene which participate in the high-level feature grouping process even though they do not correspond to physical 3D edge structures.

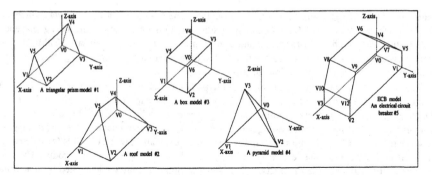

Fig. 21.5. Object Models

For this test image, 4 quadrilaterals (**C1**,...,**C4**) and 22 triangle-pair features (**F1**,...,**F22**) were found. These are tabulated in **Table C** along with the line segments forming these features. The C-triple notation (e1, e2, [e3, e4]) indicates a common junction has been identified as the intersection of the edges (e2, e3) or (e2, e4). Closed chains of more than 4 sides extracted by the grouping process were not considered. Five triangle-pair features **F4, F5, F9, F10 and F14** which were produced during hypothesis generation do not correspond to the projection of a 3D triangle-pair feature. It is worth noting that both the true and spurious features are employed as key features for generating feasible hypotheses in the matching process.

All the high-level features extracted from the scene were matched exhaustively against the preprocessed description of the 3D models. Column [t] of **Table D** lists the number of feasible model-to-scene feature assignments extracted before applying geometric constraints. As can be seen from the table, the number of solutions can very large. In all cases, there were feasible solutions found when establishing the perspective geometric relationships between individual triangle features of the models and the test scene. The total number of feasible solutions determined from matching the triangular prism #1, roof model #2 and box model #3 against all the quadrilateral features **C1**,..., **C4** and triangle-pair features **F1**, ..., **F22** was 3438, 3492 and 4862 respectively (see bottom row of **Table D**). As the pyramid model #4 does not contain surfaces made up from 4 sides, the matching of the quadrilateral features against the pyramid model was not necessary. In the case of matching the pyramid model # 4 against the triangle-pair features **F1**, ..., **F22**, the number of feasible solution was 1114. The admissible solutions were then checked using the distance constraint derived from the high-level features.

In the experiment, a reasonably large error in matching the endpoint positions of the common line segments shared by two triangle features was allowed. This was to prevent the 'pruning out' of true high-level features due to inadequate low level processing. After applying the distance constraint, the number of feasible candidates for the models #1, #2, #3 and #4 matching against the extracted scene features was reduced by 56.2 %, 65.8 %, 46.7 % and 64.8 % *i.e.* about half the hypotheses were rejected. Details of the matching of models against each extracted scene feature are shown in column [d] of **Table D**.

The admissible candidates satisfying the distance constraint were then tested with the angle constraint. In general, we chose a large allowable tolerance hwen comparing the corresponding angles between abutting model and scene surfaces. In this experiment, the tolerance was chosen to be 20^0. In allowing such a high tolerance, false hypotheses will necessarily be retained in the list and passed to the verification phase which will slightly reduce the overall efficiency of the recognition process. However, the probability of rejecting the correct interpretation is greatly reduced. After applying the angle constraint, the proportion of geometrically admissible candidates to feasible solutions without applying the geometric constraints for the models #1, #2, #3 and #4 reduces to 10.4 %, 7.4 %, 17.2 % and 8.6 %, respectively. A summary of the matching of models against each scene feature using both the distance and angle constraints is shown in column **[a]** of **Table D**. The number of implausible hypotheses for each model to high-level feature match using the geometric constraints has been pruned down dramatically.

Some examples of incorrect and poor quality hypotheses generated by matching the box model #3 against feature **F1, F6** and **F18** are labelled **H1, H2** and **H3** respectively (see Figure 21.6(c)). The perspective distortion of the hypothesized backprojected model **H2** shown in Figure 21.6(c) was quite severe. Representative hypotheses generated by matching the triangular prism model #1 against the scene features **F14** and **F6** are shown in Figure 21.6(d) and marked with **H1** and **H2** respectively. It is worth noting that the scene feature **F14** taking part in the matching process was produced by two extraneous line segments due to effects such as shadow. Some hypotheses generated by matching roof model #2 against the scene feature **F8, F9** and **F19** are labelled **H1, H2** and **H3** respectively (see Figure 21.6(e)). The poses of these hypothesized models were very unstable with respect to the supporting surface. Representative incorrect hypotheses arising from matching the pyramid model #4 against the high-level scene features **F6, F8, F4** and **F18** are shown in Figure 21.6(f) and labelled with **H1, H2, H3** and **H4** respectively.

The confidence measure of each admissible hypothesis was computed. The highest confidence measure in matching a model against a high-level feature is tabulated in column **[q]** of **Table D**. At this point, the highest confidence measure of each object model was identified. The hypotheses generated by matching the models to the correct scene features are shown in Figure 21.6(g) and (h). The hypotheses generated by matching the triangular prism #1 against the scene features **F12** and **F8** yielding the highest confidence measures are shown in Figure 21.6 (g) marked **S1**. All the hypotheses were very close to the correct one and their confidence measures are quite significant (all above 81 %). Another correct scene feature **C1** yielded a relatively high confidence measure (89.0 %) when matching against the triangular prism #1. The plausible hypotheses generated by matching the roof model #2 against the scene feature **F6**, and which yield the highest correspondence measure, are labelled **S2** and depicted in Figure 21.6 (g). The translation vectors of two of these hypotheses were very different from the true one. Fortunately, these hypotheses yield a very low confidence measure (about 32 %). The hypotheses **S3** and **S4** generated when matching the pyramid model #4 and box #3 against the feature **F1** and **C3** are shown in Figure 21.6 (i). Some of the hypothesized box models **S4** are suspended above the supporting surface.

In this experiment, the triangular prism #1, roof model #2, box model #3 and pyramid model #4 matched the scene object #1, #3, #4 and #2 correctly. The confidence measure in each of these cases is 90.9 %, 72.1 %, 85.7 % and 90.8 % respectively. After the extraction of high-level features was completed, the hypothesis generation and verification process took 2 minutes and 47 seconds for identifying the correct solution. The residual ambiguous hypotheses did not affect the reliability and computational efficiency of the recognition system. Most of these false hypotheses were associated with relatively low confidence measures and could easily be rejected on that basis.

The second experiment involves the same polyhedral object models as in the previous experiment. The test scene contained three target objects surrounded and occluded by other objects (see Figure 21.7(a)). The triangular prism and roof model were placed in an unstable position. The aim of this experiment was to study the computational feasibility and robustness of the proposed method of object recognition in the presence of clutter and subject to partial occlusion. The output of the low level processing contained many spurious line segments which were due to effects such as shadowing (see Figure 21.7(b)). Also there were some V-junctions in the image caused by self occlusion. In this example, 2 quadrilaterals (**C1,C2**) and 8 triangle-pair features (**F1,...,F8**) were identified and tabulated in Table E. The labels of the line segments forming these features are tabulated in **Table E**. Only three of all these extracted features correspond to any 3D physical structure of the object

	Table C	triangular prism #1				roof model #2				box model #3				pyramid model #4			
Index	Edge labels	t	d	a	q	t	d	a	q	t	d	a	q	t	d	a	q
C1	(28, 31, 32, 30)	54	9	6	89.0	54	20	4	62.3	118	88	22	50.0	-	-	-	-
C2	(1, 2, 8, 7)	48	48	12	63.9	52	16	12	65.3	116	96	66	83.6	-	-	-	-
C3	(2, 3, 4, 5)	56	46	25	63.9	54	30	16	53.7	108	96	70	85.7	-	-	-	-
C4	(1, 6, 9, 3)	48	48	0	-	56	16	6	54.6	104	96	39	80.6	-	-	-	-
F1	(10, 14, [15, 16]) (10, 13, [12, 18])	144	106	18	60.6	144	88	18	50.4	192	112	32	39.3	48	23	4	90.8
F2	(3, 2, [5, 8]) (3, 9, 6)	144	46	6	37.4	144	30	6	34.8	208	72	24	83.3	48	11	0	0.0
F3	(3, 2, [5, 8]) (3, 1, [6, 7])	144	26	10	32.2	148	46	14	36.6	192	104	48	84.0	48	15	5	52.2
F4	(12, 10, [13, 14, 17]) (12, 19, 18)	144	66	14	52.8	144	50	12	48.8	192	120	16	30.8	50	17	6	69.8
F5	(10, 15, [14, 16]) (10, 19, 18)	152	86	18	52.0	152	72	16	54.5	208	104	32	44.8	50	27	2	76.8
F6	(27, 20, 22) (27, 26, 24)	144	74	26	51.4	152	68	16	72.1	192	104	32	39.0	54	23	4	46.3
F7	(27, 20, 22) (27, 25, 24)	144	74	26	50.4	152	68	16	68.1	192	104	32	39.9	54	23	4	49.6
F8	(30, 33, 29) (30, 28, 31)	156	44	10	90.9	148	42	8	51.9	208	88	16	50.3	48	10	5	52.2
F9	(27, 23, 21) (27, 26, 24)	144	74	20	53.6	144	56	12	68.9	208	136	32	33.3	48	23	6	39.6
F10	(27, 23, 21) (27, 25, 24)	144	74	20	51.1	144	56	12	70.3	208	136	32	48.3	48	23	6	42.1
F11	(3, 4, 5) (3, 1, [6, 7])	144	56	10	69.4	152	28	4	53.1	192	56	24	84.0	50	9	3	49.1
F12	(30, 33, 29) (30, 32, 31)	156	44	10	90.9	148	36	10	32.3	224	96	0	-	48	6	3	52.2
F13	(3, 4, 5) (3, 9, 6)	144	78	16	66.5	156	66	6	20.6	208	88	32	81.5	56	34	6	33.6
F14	(12, 15, [14,16]) (12, 19, 18)	144	46	4	53.8	144	28	4	-	192	56	8	-	50	6	0	-
F15	(2, 5, 4) (2, 8, 7)	144	122	34	68.4	152	102	8	51.6	224	184	48	82.9	54	36	4	48.0
F16	(1, 6, 9) (1, 7, 8)	156	80	16	69.5	160	72	10	53.4	208	96	40	79.1	60	31	5	31.4
F17	(2, 8, 7) (2, 3, [4,9])	152	56	2	48.8	152	28	4	50.6	192	88	8	81.1	48	0	0	-
F18	(1, 2, [5,8]) (1, 3, [4,9])	144	28	10	15.5	148	46	16	49.3	192	104	72	82.6	48	23	14	35.9
F19	(1, 6, 9) (1, 2, [5,8])	156	46	6	23.6	160	32	8	34.9	208	88	24	79.1	54	11	0	-
F20	(2, 1, [6,7]) (2, 3, [4,9])	144	34	16	63.2	144	54	14	36.1	192	112	32	81.1	48	27	14	58.7
F21	(2, 1, [6,7]) (2, 5, 4)	144	44	10	60.0	144	20	4	50.6	192	96	32	82.6	48	0	0	-
F22	(1, 7, 8) (1, 3, [4,9])	144	52	12	67.9	144	24	2	53.3	192	72	24	84.0	54	14	5	36.1
	Total	3438	1507	357	-	3492	1194	258	-	4862	2592	837	-	1114	392	96	-

models. Thus this image is a difficult test for any interpretation procedure.

All the high-level features were exploited as seed features and were matched exhaustively against the 4 models shown in Figure 21.5. Based on the perspective projection analysis, the total number of feasible solutions determined for the triangular prism #1, roof model #2, box model #3 and pyramid model #4 was 1344, 1348, 1896 and 412, respectively. After applying the distance constraint, the number of geometrically admissible solutions in each case was reduced by 68.2 %, 75.8 %, 61.2 % and 79.1 %, respectively. The residual candidates were then checked with the angle constraint. After performing the test, the proportion of remaining admissible candidates dropped to 8.3 %, 7.1 %, 8.3 % and 2.9 %. These results illustrate the powerful degree of constraint available using the distance and angle constraints.

Representative hypotheses generated by matching the triangular prism #1, roof model #2, box model #3 and pyramid model #4 are shown in Figure 21.7 (c), (d), (e) and (f) respectively. Two hypotheses generated by matching the base of triangular prism #1 and roof model #2 against the projected planar face **C1** of the small cube are shown in Figure 21.7 (c) and (d) and labelled with **H**. Significantly, the shortest distance computed for the two hypothesized models, with respect to the camera frame, differed from the true value by a factor of five. The confidence measures in matching of the scene features against the pyramid model #4 were relatively weak (all below 40 %). Hence, we can declare that the pyramid model is not present in the scene. Some pyramid hypotheses of confidence measures above 30 % are shown in Figure 21.7(f).

The confidence measure of each geometrically admissible candidate was computed. The highest confidence measure obtained when matching an object model against each extracted feature is tabulated in column **[q]** of **Table F**. The hypothesized object model yielding the highest confidence measure was determined. The hypotheses generated by matching the object models to the correct high-level features are shown in Figure 21.7(g). The hypotheses generated by matching the triangular prism #1 and roof model #2 against the scene features **F3** and **F2** yielding the highest confidence measures is shown in Figure (g) marked **S2** and **S1** respectively. In this experiment, the triangular prism #1 and roof model #2 matched the scene object #2 and #1 correctly. The confidence measure in each of these cases is 76.8 % and 78.1 % respectively. When computing the confidence measures for hypotheses generated for the box model, the most plausible but incorrect candidate, among the hypotheses generated by matching against scene feature **C2**, gave a confidence measure 2.3 % lower than the best four hypotheses (55.3 %) generated from matching the model against the scene feature **C1**. This wrong interpretation (shown in Figure 21.7(e) labelled **H**) was due to the fact that the quality of the 2D line description of the target box model extracted from the scene was slightly poorer than the small cube. This was because the target box model was partially occluded by the pyramid model, hiding some of the significant local geometric 3D structures. Furthermore, the 2D description of the small cube and target box were very similar in terms of 2D shape under perspective projection. As a result, the assignments between the box model and the 2D features of the small cube will always be geometrically admissible. However, the correct hypothesis can be found if a bounded distance from the camera to the target object is known a priori. The computed shortest distance for the four hypotheses generated by the small cube at the top of the list differed from the edge of the supporting surface by a factor of four. Hence, these *invalid* hypotheses can be pruned away from the list. In practise, the bounded area of the supporting surface may not be generally known in advance. However, the box model could only be identified from the scene by making this weak assumption. The hypotheses generated by matching the box model #3 against the correct scene feature **C2** is shown in Figure 21.7(g) and labelled with **S3**. The models superimposed onto the scene using the computed transformation are shown in Figure 21.7(h).

Index	Table E Edge labels	Table F triangular prism #1				roof model #2				box model #3				pyramid model #4			
		t	d	a	q	t	d	a	q	t	d	a	q	t	d	a	q
C1	(1, 3, 24, 12)	52	46	13	38.3	56	16	2	38.9	104	96	23	55.3	-	-	-	-
C2	(20, 23, 22, 21)	56	42	16	61.7	48	32	16	51.4	96	96	55	53.0	-	-	-	-
F1	(4, 5, 30)- (4, 7, [8, 28])	156	36	6	17.0	156	12	0	-	224	8	0	-	48	0	0	-
F2	(11, 10, 27)- (11, 8, [7, 28])	152	58	8	53.3	152	46	8	78.1	208	80	40	32.7	54	28	4	35.3
F3	(17, 16, [18, 19])- -(17, 15, 9)	160	70	14	76.8	168	36	4	22.0	192	16	0	-	64	22	0	-
F4	(17, 16, [18, 19])- (17, 25, 13)	144	36	16	69.8	148	20	8	46.6	192	32	16	38.4	48	0	0	-
F5	(22, 21, [20, 29])- (22, 6, 25)	156	20	4	35.0	160	34	12	43.2	240	128	16	25.4	50	5	1	39.8
F6	(22, 23, 20)- (22, 6, 25)	168	38	12	41.7	156	34	16	43.9	240	120	0	-	50	4	0	-
F7	(2, 13, 25)- (2, 9, 15)	156	54	6	22.3	156	76	20	39.2	208	152	8	22.1	50	27	7	31.3
F8	(17, 16, [18, 19])- (17, 25, 6)	144	28	16	69.8	148	20	10	46.6	192	8	0	-	48	0	0	-
	Total	1344	428	111	-	1348	326	96	-	1896	736	158	-	412	86	12	-

Next, the proposed matching strategy was tested on a cluttered scene consisting of a **ECB** model (electrical circuit breaker) mounted on the wall in our vision laboratory (see Figure 21.8 (a)). The target ECB model was surrounded by several arbitrary shape features. A simplified version of the **ECB** model #5 is shown in Figure 21.5. In this experiment, all the object models shown in Figure 21.5 were involved in the matching process. The aim of this experiment was to study the effectiveness and capability of the system in recognising model objects in a complex environment. Figure 21.8 (b) shows the output of the Hough based line finding process.

In this test image, 4 quadrilaterals (**C1,...,C4**) and 6 triangle-pairs (**F1,...,F7**) were identified from the scene using the feature grouping process. The labels of the line segments forming these features are tabulated in **Table G** The quadrilateral features **C1** and **C2** was generated by the 'instructions face' of the **ECB** model and an adhesive memo paper stuck on the wall, respectively. It is worth noting that the scene feature **C4** was not extracted from a complete planar surface of the **ECB** model, as the edge index 13 was extracted from the edge of an upright surface of the **ECB** model. In fact, there was only one relevant quadrilateral feature extracted from the scene. Three of the triangle-pair features **F5,F6** and **F7** were generated by noise. One triangle-pair feature formed by edges (19, 20, 24) and (19, 9, 29) was removed from the list of candidates as the edge 9 was parallel to 19 which does not correspond to the physical 3D structure in general view.

The preprocessed 3D descriptions of the models were matched exhaustively against the high-level features extracted from the scene. The perspective analysis of quadrilateral and triangle-pair structures precomputed from the model and the scene features was performed. Before applying the distance and angle constraints, the number of feasible solutions determined for the triangular prism #1, roof model #2, box model #3 and pyramid model #4 and **ECB** models #5 was 1144, 1094, 1612, 288 and 1886, respectively. After applying the distance constraint, the number of admissible candidates in each cases was reduced by 68.0 %, 67.7 %, 56.9 %, 67.2 % and 72.6 %. The angle constraint was then applied on the residual candidates. After performing this test, the number of feasible model-to-scene assignments satisfying the geometrical constraints were reduced dramatically. The proportion of the plausible candidates in each case dropped to 6.9 %, 7.0 %, 13.1 %, 3.3 % and 10.2 %, respectively. These matching results show that a significance amount of the performance gain was obtained by exploiting the pruning power of the geometric constraints of the high-level features.

The confidence measure of each admissible hypothesis listed in column **[a]** of **Table H** was computed. The corresponding highest confidence measure is listed in column **[q]** of **Table H**. Representative hypotheses generated by matching the triangular prism #1, roof model #2, box model #3, pyramid model #4 and ECB model #5 are shown in Figure 21.8 (c), (d), (e), (f) and (g) respectively. It is interesting to note that the scene feature **F7** participating in the matching process was produced by spurious line segments due to effects such as shadow. Some hypotheses generated by matching triangular prism #1, pyramid model #4 and ECB model #5 are shown in Figure 21.8 (c), (f) and (g) marked **H**, respectively.

The hypothesized object model yielding the highest confidence measure of sufficient quality was interpreted as an instance of the object in the scene. Computing the confidence measures for the matches of the **ECB** model against all the scene features (see column **[q]** of **Table H**), the highest confidence measure (68.2 %) was observed for the correct scene feature **F2**. As the confidence measures in matching scene feature **F2** against the other four object models were relatively low (all below 50 %), the correct model for the scene object was identified.

The hypotheses generated by matching the ECB model #5 against the correct scene feature **F2**is shown in Figure 21.8(h) and labelled with **S**. One of the hypotheses was suspended in space. The models superimposed on the scene image using the computed transformations are shown in Figure 21.8(i).

It is important to note that the highest confidence measure computed when matching the roof model and the scene feature **C1** was very high (94.8 %). This is due to the fact that the rectangular base of the roof model under perspective analysis is similar to the scene feature **C1** of a rectangular shape (see Figure 21.8(d) marked **W**). Furthermore, the only feature involved in computing the confidence measure was the visible rectangular surface of the transformed roof model in an accidental view.

Index	Edge labels (Table G)	tri prism			roof model			box model			pyramid model			ECB model		
		t	a	q	t	a	q	t	a	q	t	a	q	t	a	q
C1	(1, 6, 5, 2)	54	20	61.6	60	8	94.8	120	44	62.9	-	-	-	136	64	45.4
C2	(11, 14, 15, 10)	54	6	38.6	58	3	60.8	116	11	36.9	-	-	-	126	18	31.1
C3	(7, 28, 3, 27)	56	18	52.0	48	10	52.0	96	44	58.3	-	-	-	96	42	63.4
C4	(7, 28, 13, 27)	56	14	51.9	56	13	70.9	98	47	53.9	-	-	-	98	51	60.5
F1	(28, 3, 27) - (28, 12, [16, 21])	156	0	-	148	10	-	192	32	35.1	48	1	48.1	240	8	68.0
F2	(28, 7, [27,26])- (28, 12, [16,21])	144	4	12.9	148	4	30.2	192	24	33.8	48	2	37.0	232	10	68.2
F3	(7, 26, [4,25])- (7, 27, [3, 13])	156	12	45.5	144	12	44.2	192	16	46.9	48	3	54.8	236	10	61.2
F4	(7, 26, [4, 25])- (7, 28, [3, 12])	156	2	19.0	144	12	44.2	192	8	46.9	48	0	-	236	6	67.1
F5	(7, 26, [4, 25])- (7, 28, 13)	156	0	-	148	16	43.95	192	8	46.9	48	0	-	236	0	-
F6	(19, 8, [23,22])- (19, 20, 24)	164	8	41.1	148	0	-	224	16	27.6	48	5	40.9	252	8	23.4
F7	(21, 18, 17)- (21, 12, [28, 3])	216	10	34.8	216	4	-	288	0	-	72	1	44.2	332	10	33.8
	Total	1368	94	-	1318	92	-	1902	250	-	360	12	-	2220	227	-

21.6   Conclusions

In this paper, we have presented a hypothesize-verify approach to polyhedral recognition based on the use of geometric constraints derived from local shape properties. In the framework, two intermediate features, namely triangle-pair and quadrilateral are employed as key shape descriptors for identifying a manageable number of geometrically feasible model-to-scene candidates. Two effective geometric constraints, namely the distance and angle constraint, have been derived and incorporated into our recognition system. Many infeasible hypotheses can be swiftly pruned away from the hypothesis list. Only those hypothesized model-to-scene correspondences which satisfy the distance and angle constraints are considered in the subsequent (computationally intensive) verification process. The integration of the feature extraction, hypothesis generation and verification process are described. Extensive experimental results using real images have been presented. They verify the effectiveness and reliability of our proposed method. As a by-product of the matching process, the transformation defining the pose of the scene objects with respect to the camera has been recovered. The experimental results show that the localization method has reasonable accuracy in estimating the pose of the target object. In general, the success rate for identifying and localising target objects using the proposed paradigm is relatively high.

Clearly, one of the prerequisites of the proposed method is the ability to extract triangle-pair and quadrilateral features from low level primitives such as junctions from image data. To accommodate the problems of noise, oversegmention or undersegmentation, the poorer quality junctions will participate in the feature extraction process by allowing large thresholds on proximity and orientation checks. As a result, both the number of plausible and implausible features included in the matching process will be increased. However, the probability of rejecting the correct model-to-scene feature correspondences will be greatly reduced. Moreover, the growth of hypotheses generated from the model-to-scene assignments will not affect the computational performance of the system, as a majority of these hypotheses which are geometrically infeasible will be pruned away.

Most of the existing verification processes are based on identifying correspondances among image features for hypothesised low level primitives whose positions are computed from the estimated 3D position of the target object. The performance of this approach will degraded significantly when verifying model hypotheses generated in a complex scene. To increase the robustness of the verification process, the use of spatial and temporal context will be considered in future. The spatial description of a scene can be constructed by aggregating matching results derived over a time of interest period. For example, a box model was interpreted as a small cube in the scene. In this instance, if the supporting surface such as a table is identified and maintained in the global scene description, the incorrect interpretation could then be rejected using the bounded space of the table.

References

[Cheng et al. 92] Cheng Yu, K. C. Wong and J. Kittler, "Using quadrilateral and triangle-pair features for recognising polyhedral object, The 4th International Conference on Image Processing and its Applications, pp 425-428, April 1992.

[Conte 65] S. D. Conte, "Elementary Numerical Alalysis" McGraw Hill, New York, 1965.

[Dehn 60] E. Dehn, "Algebraic Equations" Dover, New York, 1960.

[Etamadi et al. 91] A. Etemadi, J-P Schmidt, G. Matas, J. Illingworth and J. Kittler, "Low-level Grouping of Straight Line Segments", Proc. of 1991 British Machine Vision Conference, Glasgow, pp 118-126, 1991.

[Fischler-Bolles 81] M. A. Fischler and R. C. Bolles, "Random Sample Consensus: A paradigm for Model Fitting with Applications to Image Analysis and Automated Cartography", Communications of ACM, vol. 24, No. 6, pp 381-395, 1981.

[Horaud 87] R. Horaud, "New Methods for Matching 3-D Objects with Single Perspective Views", *IEEE Trans. on Pattern Anal. and Machine Intell.*, Vol. PAMI-9, No. 3, pp 401-412, May 1987.

[Lei 90] G. Lei, "Recognition of Planar Objects in 3-D Space from Single Perspective View Using Cross Ratio" IEEE Trans. on Robotics and Automation, Vol. 6, No. 4, pp 432-437, Aug. 1990.

[Lowe 87] D. G. Lowe, "Three-Dimensional Object Recognition from Single-Two Dimensional Images", *AI*, 31, pp 355-395, 1987.

[Wong et al. 91] K. C. Wong, J. Kittler and J. Illingworth, "Heuristically guided polygon finding", Proc. of 1991 British Machine Vision Conference, Glasgow, pp 400-407, 1991.

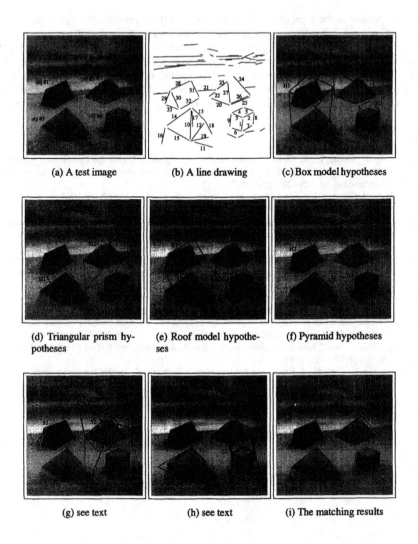

(a) A test image   (b) A line drawing   (c) Box model hypotheses

(d) Triangular prism hypotheses   (e) Roof model hypotheses   (f) Pyramid hypotheses

(g) see text   (h) see text   (i) The matching results

Fig. 21.6. The scene image and the experimental results

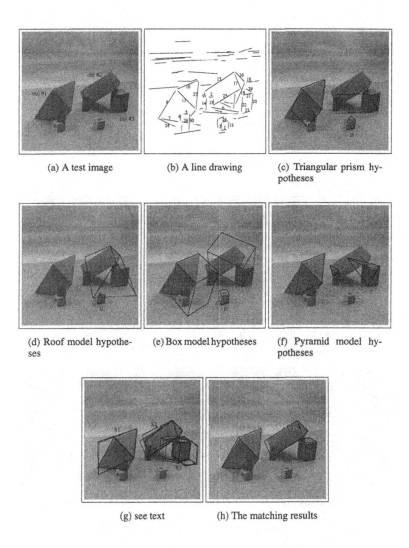

(a) A test image

(b) A line drawing

(c) Triangular prism hypotheses

(d) Roof model hypotheses

(e) Box model hypotheses

(f) Pyramid model hypotheses

(g) see text

(h) The matching results

Fig. 21.7. The scene image and the experimental results

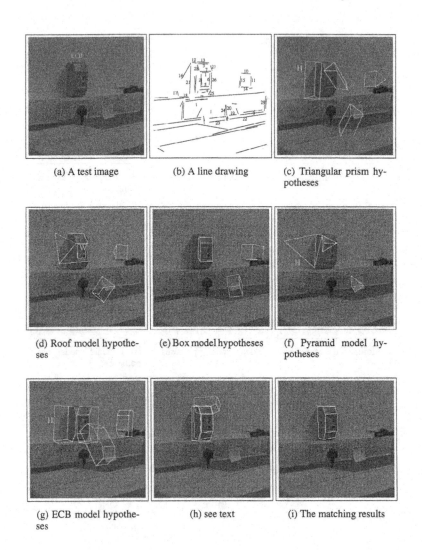

(a) A test image

(b) A line drawing

(c) Triangular prism hypotheses

(d) Roof model hypotheses

(e) Box model hypotheses

(f) Pyramid model hypotheses

(g) ECB model hypotheses

(h) see text

(i) The matching results

Fig. 21.8. The scene image and the experimental results

22. Recognition of 3D Cylinders in 2D Images by Top-Down Model Imposition

Paul Hoad and John Illingworth

UOS

22.1 Introduction

Object recognition is an important and basic goal of computer vision. Most research thus far has concentrated on recognition of objects from their shape. Successful object recognition has been shown for the matching of 2D shapes to 2D image data. However, the task of recognising 3D objects from their 2D projections in an image has proven less successful, even in the case of polyhedral objects. Methods have generally required the unreliable inference of intermediate 3D information or have been based on specific, metric models of the objects which are to be identified. Little success has been achieved in the effective recognition of curved or generic 3D objects from 2D image data[Murrey 88].

In this paper the recognition of a generic class of curved object shape is considered: that of flat topped cylinders. It is shown that 2D image data can be processed to find instances of 3D cylinders by combining bottom-up hypothesis generation with top-down generic model based matching of lines and ellipses. The method starts by detecting "quasi-parallel" pairs of straight lines to define a cylinder axis and extent. This information then constrains the possible shape and location of 2D ellipses that could be produced from a flat topped cylinder. The difficult task of unconstrained ellipse detection is thereby avoided and robust ellipse detection can be achieved using specialised methods. In this study a Hough based ellipse detector is adopted to find approximate ellipse parameters and reliably identify subsets of points which belong to a single ellipse. More accurate ellipse parameters are determined by standard fitting methods. The use of line information to constrain ellipse detection is a considerable improvement over a method which tries to find and combine the line and ellipse data independently in a purely bottom-up fashion.

The following section discusses the proposed method for cylinder detection while section 3 gives examples of its experimental performance on both synthetic and real image data. Finally, section 4 offers some concluding remarks on lessons learnt and directions in which it may be extended.

22.2 Description of method

A cylinder is a parameterised shape, instances of which can be distinguished by their height and their radius. The basic shape can be constructively defined by sweeping a circular plane of constant radius along a straight axis which is perpendicular to the circular plane. Ideally, the projection of a flat topped cylinder into a 2D image produces a pair of approximately parallel straight lines whose endpoints are bridged by elliptical curves. The top of the cylinder projects to a complete ellipse but self-occlusion means that only about half of an ellipse is visible at the bottom of the cylinder. However, in reality a more fragmented description of an ellipse is obtained when images are processed by current techniques. This is illustrated in Figure 22.1.

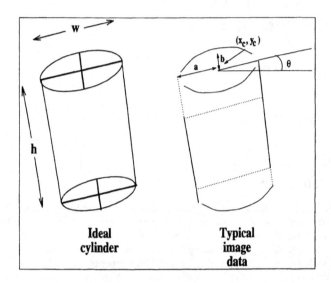

Fig. 22.1. Illustration of Idealised and Realistic Cylinder Image

Detection of individual image features which compose a cylinder can be difficult. Edge detection methods[Canny 86],[Petrou-Kittler 92] are somewhat unreliable and often produce fragmented curves which are not ideal input to subsequent procedures for straight line and ellipse detection. Numerous line detection methods exist and several are reasonably successful although none is perfect[Rosin-West 89; Princen et al. 89; Burns et al. 86]. Typical problems are the fragmentation of lines because methods rely on connectivity of edge data, or the inaccurate estimation of line parameters because of image noise processes.

Ellipse detection is even more difficult than straight line finding[Porrill 90]. The most common method of ellipse detection comprises fitting an ellipse to connected sets of edge pixels i.e. edge strings[Ellis et al. 91; Rothwell et al. 91]. The regression methods employed for fitting are typically non-robust to the presence of outlier data and therefore to obtain good estimates the raw connected edge data has to be segmented so that candidate strings which are passed to the fitting routine comprise edges from a single ellipse. This segmentation is achieved by measuring the curvature along the string and breaking strings at points of high curvature. However, several difficulties can arise. Image noise processes mean that curvature estimates are noisy. The distribution of curvature values created by noise overlap significantly with that due to joins of curved segments. This means that a simple threshold to identify break points is not foolproof. In order to make the method work it is necessary to select a low threshold on curvature values which produces an over fragmentation of the strings. This behaviour creates difficulties for the fitting stage of the method as the methods are ill-conditioned when given short strings of edge pixels. In order to get accurate ellipse parameters the fitting routine needs information from a large part of the elliptical curve i.e. $\frac{1}{3}$ of the ellipse boundary[Porrill 90]. In addition, not all parts of the ellipse contribute equally to the calculation of the fit parameters[Bookstein 79]. The high curvature ends of the ellipse provide stronger constraints on the fit than the low curvature parts but typically it is the lower curvature parts that are detectable most easily in images i.e. the bottom elliptical curve of the cylinders. All these factors means that the low level data used for cylinder detection is subject to error and general bottom-up driven methods of cylinder detection are prone to fail.

In this paper data-driven, bottom-up processes are used to suggest hypotheses for cylinders but these then are used in a top-down, predictive way to invoke robust and specific processing for the detection of confirmatory features. The image structures used for cylinder hypothesis generation are pairs of opposing and partially overlapping, parallel lines. These provide minimal evidence for determination of cylinder parameters, as shown in Figure 22.1. The image lines suggest a hypothesis for the sides of the cylinder which are of length equal to the maximum projected lengths of the pair onto each other. The separation of the line pair specifies or constrains the cylinder width. Having these estimates it is possible to predict the location and size of most of the parameters of the image ellipses at the top and bottom of the cylinder. The size of the major axis of the ellipse is equal to half of the width of the cylinder and its center should lie at the middle of the lines which connect corresponding endpoints of the cylinder sides. The orientation or slant of the ellipse will be the same as the angle of inclination of the parallel sides. In orthographic projection only the minor axis of the ellipse is not well determined although it is constrained to be smaller than the width of the cylinder. However, the size of the minor axis depends on the tilt of the cylinder with respect to the observer and in perspective projection this can be independently estimated by the amount of convergence between the detected image sides. Thus an approximate estimate of the minor axis can be calculated from the sides of the cylinder hypothesis.

The above discussion depends on finding the parallel lines which represent the cylinder sides correctly. Due to the vagaries of feature segmentation methods it is not guaranteed that the cylinder sides will be found correctly and therefore detected image features cannot constrain the ellipse parameters quite as well as might be first thought. It is therefore necessary to adopt methods which take account of this. We have therefore used the lines only to define a restricted region where the ellipse features should be found and used the same information to constrain the area of parameter space which needs to be considered in a Hough Transform method to find the ellipse center. The Hough Transform method is an interesting method to use as it provides coarse determination of parameters and at the same time provides the feature segmentation and outlier rejection capability that is necessary for accurate parameter determination via least square fitting techniques.

Now that the basic ideas and advantages of the method have been presented we will consider in more detail the algorithm actually implemented. Figure 22.2 shows a flowchart of the method starting from an image and ending in confirmed flat-topped cylinders. Standard edge detection, such as the Canny edge detector is applied to the image. Lines are detected using a Hough Transform based line detector[Princen et al. 89]. These lines are then searched for pairs which have the same orientation to within an angular tolerance of 10 degrees and when projected one onto the other have visible overlapping sections. Ideally the lines should overlap fully but because of feature segmentation difficulties this may not be the case. The amount of overlap as a ratio of the length of the cylinder hypothesised can be used as a pruning factor to reject some hypotheses. Other simple heuristics such as requiring the lines to be a minimum distance apart (typically 10 pixels) can also be used to reject improbable hypotheses.

The detected pairs of lines define regions of the image where ellipses should lie if they are part of a cylinder. The search for ellipses is restricted to the edge data which lies in a simple rectangular region centred at the midpoint of the line connecting corresponding line end-points. The size of rectangle is chosen as a fixed multiplying factor larger than the inferred major and minor axis sizes. In the current implementation this factor is set at 1.5.

Different methods have been used for the detection of ellipse parameters associated with the

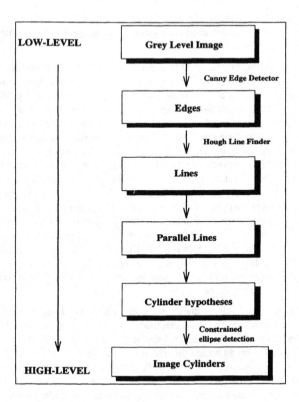

Fig. 22.2. Block Diagram

top and bottom of the cylinder. The top cylinder is unobscured and therefore a large proportion of it is often visible and good parameter estimates can be obtained. The bottom ellipse suffers from obscuration and the portion visible is not the most helpful for determining ellipse parameters. In fact we have adopted the strategy of finding a single ellipse as strong confirmatory evidence for the cylinder but do not place strong requirements on the second ellipse other than it is consistent with the other evidence.

The determination of ellipse parameters is based around the method of ellipse detection suggested by Tsuji[Tsuji-Matsumoto 78]. It uses the fact that pairs of points that lie on an ellipse diameter will have edge angles which differ by π. Therefore finding these pairs provides evidence for an ellipse with center coordinate at the midpoint of the line joining the pair. This evidence or vote for an ellipse can be combined with other evidence in a Hough transform method with an accumulator array which is congruent with possible image center positions. Ellipses for which there is much evidence will contribute many votes to the same element of the accumulator array and this peak of votes can be detected by simple analysis and thresholding of clusters in the accumulator array. The threshold value for peaks can be determined using loose knowledge of the expected size of the ellipse.

An often mentioned criticism of Hough methods is that they require large accumulator arrays and are computationally demanding. However in the case of this particular application this is not so. The region of image space where ellipse centers can occur is well constrained to a narrow strip of values around the central axis defined by the midpoint of the sides. Thus the accumulator is only 2D and its size is not particularly large. In addition the number of input edge points is much less than the total number of edges in the image as the sides have defined a restricted window of interest. It should also be noted that in our implementation we have removed pixels that form the sides of the cylinder from consideration prior to searching for ellipse centers.

A further advantage of the Hough method is that it simultaneously estimates ellipse center parameters and identifies the subset of points which are part of the ellipse. Only those points which form pairs that contribute to the peak bin in the Hough accumulator space are part of the ellipse. These can be isolated either by keeping a linked list of image-point to accumulator cell associations during accumulation or by first finding the center position using the Hough method and then making a second pass through the edge data looking for points which contribute to the peak.

Once a candidate center of sufficient size has been found and the contributing edge points have been isolated then the remaining parameters are determined by regression fitting using the method suggested by Bookstein[1][Bookstein 79]. The resulting ellipse parameters can then be compared for consistency with the information supplied by the lines and a goodness of fit measure for the whole cylinder can be defined.

22.3 Experimental Results

The proposed method has been tested on many images, both synthetic and real world data. It was programmed in C and run on a Sun-4 workstation. Figure 22.3 shows some of the results of various stages of the processing for a simple image consisting of a synthtically generated cylinder[2] (Image Name: Cylinder). Figure 22.3(a) shows the original image while 22.3(b) is the result of applying Canny edge detection. It can be seen that most of edges are successfully extracted although the nearer edge of the top ellipse is not connected to the far edge of the same ellipse. This would cause difficulties to a simple ellipse fitting routine. It can also be seen that the exact point of join of the sides and the ellipses is somewhat difficult to determine. They merge fairly smoothly one into another. This phenomena means that the break point of the connected string which includes the sides and the ellipses may result in some stray side edge points being included into the ellipse string. These outliers can introduce serious bias into the determination of the ellipse parameters. Figure 22.3(c) shows the cylinder hypotheses deduced from detection of parallel lines. Note that the upper and lower ellipses have been drawn as circles with radius equal to the separation of the lines (the circle is a bound on the size of the actually observed ellipse). The smaller (incorrect) cylinder hypotheses derives from a pair of short lines which have been found on the near curves which corrspond to the edges of the top of the cylinder. The cylinder hypotheses are used to define rectangular regions to search for the top ellipse of each cylinder. Figure 22.3(d) shows the windowed regions and highlights the edge pixels within them. Prior to ellipse finding the edges associated with the sides are removed. This both removes extraneous pixels to make the method more reliable process and slightly speeds up the center finding as fewer points are involved in the pairing process. Figure 22.3(e) shows a representation of the Hough peak which is detected at the center of the ellipse together with the edge points that contributed to it. It should be noted that both the near and far edges of the ellipse are identified and can simultaneously be included in the subsequent fitting procedure. The subset of identified data is extremely clean and therefore accurate, unbiased ellipse parameters are found. The estimated ellipse is shown overlayed on the edge data in Figure 22.3(f). Finally, a check is made that

[1] An implementation of the Bookstein algorithm was kindly supplied to us by C. Rothwell and A. Zissermann of the Robotics Group, University of Oxford, U.K.

[2] Synthetic images were generated using the public domain package Rayshade, written by Craig Kolb

some evidence is available for the bottom ellipse of the cylinder. It can be seen that in this case the method works very well.

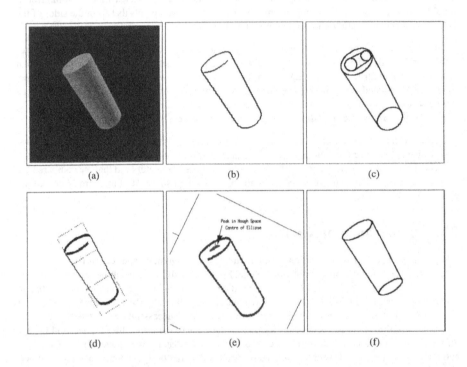

Fig. 22.3. Results on Synthetic Image of a Cylinder

An example of the use of the method on a real data image is shown Figure 22.4 (Image Name: Cup). The cup produces a complicated edge image as it includes surface markings and other noise processes. From the edge data many sets of parallel lines can be extracted and many cylinder hypotheses are made. Most are spurious, for example the pairing of the table edge and cup edge. In this particular example 22 cylinder hypothese were found. However, the number of cylinder hypotheses can be quickly pruned using small amounts of domain specific knowledge about factors such as the expected size of cylinders or their expected orientations. The final result of cylinder finding in this example is the single cylinder in Figure 22.4 which corresponds correctly to the cup.

A further example of the method is shown for the table top scene in Figure 22.5 (Image Name: Table). This shows two cups, a circular plate and a packet of tea. It can be seen that despite the precesnce of these several objects and the small image size of the two cups the method successfully finds the two cylindrical cups.

As with all image processing and vision techniques the cylinder detection method is not perfect and in some cases can fail. One such case is illustrated in Figure 22.6 (Image Name: TwoCups). The reason for this failure is poor edge and line detection caused by low local constrast of the rear cup against the background. The lines which form a cylinder hypothesis in the region near the rear cup are too short and as a consequence the windowed area for ellipse detection is misplaced and ellipses

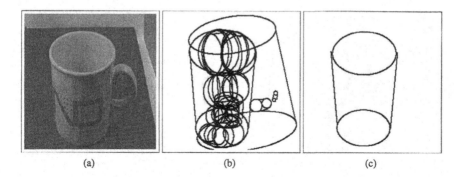

(a) (b) (c)

Fig. 22.4. Results on Real Image of a Cup

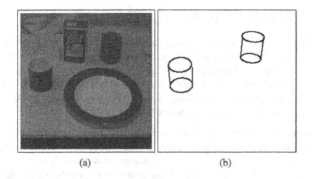

(a) (b)

Fig. 22.5. Results on Real Image of a Table Top Scene

cannot be found.

The computational speed of the proposed method depends on the complexity of the scene imaged but is not excessive. The invocation of cylinder hypotheses by looking for line pairs is extremely quick but the robust detection of the ellipses by the Tsuji method limits the speed of the method as it involves selecting pairs of edge points (for n edge points, a complexity of n^2). Table 1 shows, for each of the four examples presented , times in seconds for the edge detection, line detection, cylinder hypothesis invocation and ellipse finding stages of the process. The final column gives the total time for the method.

Table 22.1: Computation Times for Example Images

Image Name	Canny Edge Detection	Hough Line Detection	Cylinder Hypotheses	Ellipse Detection	Total
Cylinder	4.5	0.4	0.1	0.9	5.9
Cup	7.9	2.0	0.7	34.7	45.3
Table	8.0	2.1	0.8	40.8	51.7
TwoCups	9.1	2.8	0.4	15.5	27.8

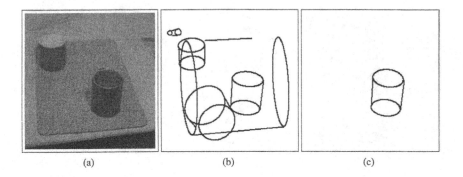

        (a)   (b)   (c)

Fig. 22.6. Results on Real Data Image of Two Cups

22.4   Discussion and Conclusions

In this paper we have shown how cylinder recognition can be achieved using low level hypothesis generation followed by predictive, top-down search for confirming image features. The use of prediction allows the adoption of specific, robust methods for detection of the image features, in this case Hough based methods of ellipse detection. The method works well on real image data and is reasonably fast. This speed benefit is also partially due to the predictive nature of the process which focusses attention to only those image parts where features are likely to occur.

There are several possible improvements to the method or avenues for future investigation. Firstly, one of the early problems in the detection process occurs if the edge detector fails to find all reasonable edges. The edge detector used in this paper automatically chooses a threshold on edge magnitude. It may be beneficial to the ellipse detection process to repeat edge detection on the windowed image region used to find ellipses as a more appropriate threshold which extracts additional image structure may be selected. Alternatively, edge detectors which do not adopt the step edge model used by the Canny detector might prove more adept at extracting roof edges that occur at the front edges of closed flat-topped cylinders. A second problem is that the ellipse detection stage requires the detection of pairs of diametrically opposed edge points. As in the case of the lower ellipse of a cylinder such points may not be present. However, alternative Hough based processes[Yuen et al. 89] have been suggested which do not have this requirement. Indeed these methods could be used in addition to the Tsuji ellipse detector (with suitable normalisation) as they share a common ellipse centre accumulator array. Other recent Hough based ellipse detection methods[Xu et al. 90] could also be considered. Finally, the process depends on finding the sides of the cylinder. This has been adopted as the sides are generally easier to find from an image analysis point of view and therefore are more reliable model invocation cues than ellipses. However, if some of the sides are obscured then it would be nice to drive the cylinder recognition process top-down from the few ellipse hypotheses than can be independently found. These suggestions are topics for current and future work.

References

[Murrey 88] Murray D., "Strategies in Object Recognition", GEC Journal of Research, Vol 6, No 2, pp 80-85, (1988).

[Canny 86] Canny J., "A Computational Approach to Edge Detection.", IEEE Trans. on Pattern Analysis and Machine Intelligence, Vol 9, pp 679-698, (1986).

[Petrou-Kittler 92] Petrou M. and Kittler J., "Optimal Detection of Ramp Edges", IEEE Trans. on Pattern Analysis and Machine Intelligence, (1992)

[Rosin-West 89] P.L. Rosin and G.A.W. West, Segmentation of Edges into Lines and Arcs, Image and Vision Computing, Vol 7, pp 109-114, (1989).

[Princen et al. 89] Princen J, Yuen H.K., Illingworth J. and Kittler J., "A Comparison of Hough Transform Methods.", Proc of IEE Conference on Image Processing and its Applications, Warwick, U.K., pp 73-77, (1989).

[Burns et al. 86] Burns J.B., Hanson A. and Riseman E., "Extracting Straight Lines", IEEE Trans. on Pattern Analysis and Machine Intelligence, Vol 8, pp 425-455, (1986).

[Porrill 90] Porrill J., "Fitting Ellipses and Predicting Confidence Envelopes using a Bias Corrected Kalman Filter.", Image and Vision Computing, Vol 8, pp 37-41, (1990).

[Ellis et al. 91] Ellis T., Abood A. and Brillault B, "Ellipse Detection and Matching with Uncertainty", Proc of 2nd British Machine Vision Conference, Glasgow 1991, pp 136-144, published by Springer-Verlag, ISBN 0-387-19715-X.

[Rothwell et al. 91] Rothwell C.A., Zisserman A., Forsyth D.A. and Mundy J.L., "Using Projective Invariants for Constant Time Library Indexing in Model Based Vision.", Proc of 2nd British Machine Vision Conference, Glasgow 1991, pp 62-70, published by Springer-Verlag, ISBN 0-387-19715-X.

[Tsuji-Matsumoto 78] Tsuji S. and Matsumoto, "Detection of Ellipses by a Modified Hough Transform", IEEE Trans. on Computers, Vol 27, pp 777-781, (1978).

[Davies 89] Davies E.R., "Finding Ellipses using the Generalised Hough Transform", Pattern Recognition Letters, Vol 9, pp 87-96, (1989).

[Bookstein 79] Bookstein F., "Fitting Conic Sections to Scattered Data", Computer Graphics and Image Processing, Vol 9, pp 56-71, (1979).

[Xu et al. 90] Xu L., Oja E. and Kultanen P., "A New Curve Detection Method: the Randomised Hough Transform.", Pattern Recognition Letters, Vol 11, pp 331-338, (1990).

[Yuen et al. 89] Yuen H.K., Princen J., Illingworth J. and Kittler J., "Detecting Partially Occluded Ellipses using the Hough Transform.", Image and Vision Computing, Vol 7, pp 31-37, (1989).

23. Camera Control for Establishing the Current and Next-Look Direction in an Active Vision Object Recognition System

Paolo Remagnino, Josef Kittler, Jiri Matas and John Illingworth

UOS

Abstract

This paper discusses aspects of the control of an active computer vision, object recognition system. A prototype control mechanism able to estimate and modify the extrinsic parameters of a single camera has been designed and implemented. The system is capable of continuous and active adjustments of the camera in response to a given visual goal. The implementation of the mechanism takes into account errors in the position of objects stored in the memory of the vision system, and errors due to image segmentation. The advantages and drawbacks of the methods used are discussed and its performance is demonstrated and quantified in experiments using synthetic and real image data.

23.1 Introduction

Computer vision is concerned with giving machines the capability to use image data to sense and interpret the world. Attainment of visual compotence would allow machines to flexibly interact with their environments and thereby autonomously perform complex tasks such as navigation and manipulation. In this paper representations and processes for control of a simple vision system consisting of a single mobile camera are discussed. Specifically, the paper considers and gives examples of implemented processes which:

1. provide a simple but efficient way to estimate the extrinsic parameters of a single camera in a closed environment, and
2. shows how a specified visual goal can be satisfied using an active sensor in an environment described by means of simple objects represented in a coarse manner.

Of particular interest is the development of methods that can effectively cope with problems arising from the propagation of errors and uncertainties in both the sensed data features and the prestored/created environmental models.

Conceptually a machine vision system consists of many interacting procedures and knowledge sources. The procedures and prestored knowledge are used to interpret and process the raw image data into more abstract descriptions. In the system discussed in this paper these abstract descriptions are in terms of objects and their relative spatial dispositions. Typical procedures used extract image edges, lines and curves and match them against pre-stored knowledge of objects shapes and expected spatial configurations of sets of objects. The system stores both generic knowledge of objects and their configurations as well as building and maintaining models of specific environments which it has previously explored. In this paper we demonstrate how the system can be controlled to use these procedures and knowledge bases to direct its attention to find specified objects in a purposive fashion.

In a similar way to any other computer system, the "life-cycle" of a machine vision system can be conveniently divided into two phases: a "bootstrap" phase and a "run" phase. The bootstrap phase consists of executing a pre-determined sequence of special procedures to initialise and orient the system while the run phase involves autonomous dynamic selection and scheduling of procedures to achieve specified visual goals. In this paper a simplifying assumption will be made that the bootstrapping phase involves establishing orientation of the system in an environment where at least some of the objects are known. This is realistic for task domains such as when a vision system is moving in man-made buildings which have previously been partially explored. In these cases much of the framework of the world is fixed i.e. walls, doorways, windows. These permanent landmarks are the objects which can be used for bootstrapping.

Following bootstrapping, object recognition can be simplified by using knowledge from the evolving environmental description and prestored prototypical object configuration models to direct the system attention (at an appropriate scale) to only the most likely image areas for object location. This is an example of purposive use of context within a vision system and requires development of appropriate representations for context description as well as algorithms for using this information to predict the next-look direction of the active camera. In this paper it is shown how this may be achieved using coarse object descriptions, based on bounding box approximations, and descriptions of contexts in local coordinate frames.

The work of this paper is a contribution to the solution of two basic problems in an active object recognition system. The first question relates to the bootstrap phase and is: how can the location and orientation of the sensor be determined given that the system is in a partially known environment?. The second question is relevent to both bootstrapping and the run phase of the system and is: how can the next-look direction be effectively determined given a partial world model, context information and the task of finding a specified object. In particular, how well can these questions be answered given both uncertain geometric information inherent in the coarse models adopted and geometric uncertainties in primitive features extracted from real image data. Not all the tools that are used to address these problems are entirely new. Many of the techniques have been developed and used in both photogrammetry and computer vision for some years. However, the particular applications where they have been previously exploited have influenced their development and one significant contribution of the current work is to show they perform in a more general active vision system and to explore their effectiveness as part of the overall control strategy of such a purposive system.

The next section gives a brief review of the literature relating to estimation of sensor and object pose as well as some of the basic ideas and motivations for active sensing and active vision. Section 3 gives the mathematical ideas used in the implemented methods while Section 4 shows experimental work on the effectiveness of the methods. Tests have been made with both synthetic and real image data and limitations of the methods have been clearly identified. Section 5 summarises the paper and concludes with some suggestions for future work.

23.2   Previous Work

Determination of the transformation between a sensor and objects in the external world is a problem that has received much attention in the photogrammetry and computer vision literature. Several methods have been devised and are distinguished by the image primitives which they use, by the assumed models of optical projection that they adopt or by whether the solution is expressed as a closed form analytic expression or must be found by expensive numerical calculation. In computer vision two problems need to be solved:

– the determination of correspondances between image and object model points,

– the calculation of the transformation given the previously determined image to model correspondences.

Most papers discuss the latter problem and give solutions to the problem in terms of the minimum number of correspondences required to fully constrain the transformation. A prominent early contribution to the literature was that of Fischler and Bolles[Fischler and Bolles, 1980] who showed that in many situations three or four known coplanar point correspondences were sufficient. Other authors such as Barnard[Barnard, 1983] and Kanade[Kanade, 1981] solved the problem for correspondence of configurations of lines meeting at a vertex. Several authors (for example [Dhome *et al.*, 1989; Chen, 1991; Lui *et al.*, 1990]), have further extended and refined these analyses. In relation to establishing stereo camera calibration Tsai[Tsai, 1987] has made a significant contribution, developing regression analysis techniques and distinguishing between the intrinsic (focal length) and extrinsic (distance between cameras, positions, orientations) parameters which have to be found. Of particular interest to this paper is the work of Haralick[Haralick, 1989] and Penna[Penna, 1991] who have considered establishing camera-object pose via matching generic shapes such as rectangles and quadrilaterals of unknown dimensions.

One criticism of many of the analyses cited above is that few of them explicitly include analysis of the likely effect of uncertainties or errors on transformation estimation. There are many sources of such uncertainty including uncertainty in knowledge of camera motion or model location in the world, inaccuracy in feature parameter estimation and gross errors in segmentation of image data. One aspect of the work in this paper is to develop methods which can cope with these real-world factors. Error propagation is a topic which has been more commonly addressed in path planning and mobile robotics applications. Brooks[Brooks, 1982] approaches these problems by the use of a set of local reference frames linked together via uncertain transformations and develops startegies to reduce errors as processing proceeds. Another example of the representation and estimation of spatial uncertainty is the work of Smith and Cheesemann[Smith and Cheeseman, 1986]. A particularly elegant formalisation of the problem is given by Durrant-Whyte[Durrant-Whyte, 1988], [Durrant-Whyte, 1989]. He outlines a theory of uncertain geometry which stresses the importance of transforming information between coordinate systems while maintaining the local topological consistency of geometric features.

Much of the work on transformation between coordinate frames has been given recent relevence for computer vision by the development of the theories and methodologies of the "active vision" paradigm. The basic thesis of the active vision approach is that vision is made feasible by purposively controlling data acquisition and processing [Bajcsy, 1988], [Aloimonos *et al.*, 1987]. Some image analysis problems are ill-posed (i.e. admit no unique solution or are computationally unstable) with respect to a static imaging system. However, they can be made well posed by introducing extra data or additional constraints via controlled sensor motion. The exploitation of this idea demands methods of determining what additional data/constraints are needed and what motions must be executed to use them. The work in this paper provides a contribution to this methodology.

The work in this paper makes use of the idea of scene and context representation via sets of local external coordinate reference frames. The adoption of a single global coordinate system has been common in computer vision but has some difficulties. Several authors, for example Brooks[Brooks, 1982] and Ballard[Ballard, 1991], have suggested advantages for local reference frames. Many important qualitative aspects of scene description are related to spatial proximity of objects (i.e. the cup is *on* the table, the chair is *near* the table etc) and therefore a representation which is based on local proximity seems natural. In addition, it is easy to build a set of local descriptions into a more global structural representation that emphasises the important aspects of the scene at any desired spatial resolution. Finally, local descriptions allow predictions which constrain object search yet, via local registration, avoid the growth of uncertainties across large distances.

23.3   Establishing the Look Direction

In this section the calculations and assumptions underlying the establishment of the camera look direction are described. Individual sub-sections treat aspects related to finding the current-look direction (used in the system bootstrap phase), determining the next-look direction and recovery strategies to cope with situations where the desired transformations lie outside the range of the sensor viewing capabilities.

23.3.1   Current-Look Direction Computation

It is assumed that a partial model of the world has already been determined either via active exploration or by provision to the system of a map or some other representation of aspects of the environment. The basic idea is to use simple, immoveable objects in the scene to establish the camera-world transformation. The particular domain considered is an indoor scene. A generic model of such a scene would include walls, doors and windows and it is the doors and windows which provide the generic structure suitable for bootstrapping purposes. Both doors and windows are reasonably constrained objects in terms of their location, oreientation, shape and size. In this paper it will be assumed that both doors and windows are rectangular in shape, although this or similar methods could be adopted for other simple shapes.

In order to extract features corresponding to doors and windows from an image we have used a sequence of procedures consisting of a Canny edge detector[Canny, 1986], Hough based straight line extractor[Hough, 1962], a probabilistic vertex/junction finder[Matas and Kittler, 1992] and a simple polygon finder[Matas, 1992]. These procedures are reasonably robust for locating the large rectangular structures corresponding to doors or windows. Using the assumption that the camera is upright i.e. oriented approximately vertically with respect to gravity, it is then simple to establish the correspondences between the four corner points of a door or window and its model. From these correspondence and having prior knowledge of the camera focal length, the size of a pixel and the size of the door then the methods of Fischler and Bolles can be used to estimate the object-camera transformation.

Although the standard method of backprojection suggested Fischler and Bolles has been used to determine an estimate of the object-camera transformation, a distinguishing feature of the work presented here is that the uncertainty associated with the result has also been estimated. This has been done in a heuristic Monte-Carlo like way by randomly and repeatedly perturbing the input to the calculation by an amount commensurate with the expected uncertainties and then recalculating the transformation. It has been assumed that uncertainty due to image processing results in Gaussianly distributed errors in the position of the corners that define a door or window.

Figure 23.1 illustrates the geometry used to establish the world-camera transformation. The calculation basically depends on building up two corresponding orthogonal sets of vectors, one expressed in world coordinates and the other in the coordinate frame of the camera. Three corner points of a rectangle suffice for solving the problem. Denoting P_i as the ith corner then its position in the world coordinate frame can be denoted as $\underline{p}_i^w$ while its description in the camera frame is expressed as $\underline{p}_i^c$. Now taking three corner points two vectors can be defined by taking differences between pairs of these. A third orthogonal vector can be constructed by taking the cross-product of two difference vectors. We shall denote these edge vectors by $\underline{e}_i$ with the w and c superscripts to distinguish between the world and camera frames respectively i.e.

$$\underline{e}_1^w = norm\left(\underline{p}_0^w - \underline{p}_1^w\right)$$
$$\underline{e}_2^w = norm\left(\underline{p}_2^w - \underline{p}_1^w\right)$$

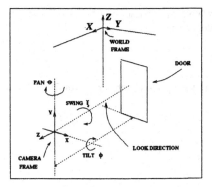

Fig. 23.1. The world and the camera coordinate systems and the camera orientations.

$$\underline{e}_3^w = norm\left(\underline{e}_1^w \times \underline{e}_2^w\right)$$

and

$$\underline{e}_1^c = norm\left(\underline{p}_0^c - \underline{p}_1^c\right)$$
$$\underline{e}_2^c = norm\left(\underline{p}_2^c - \underline{p}_1^c\right)$$
$$\underline{e}_3^c = norm\left(\underline{e}_1^c \times \underline{e}_2^c\right)$$

However, in addition to these vectors the uncertainties associated with them are considered. Denoting the uncertainty in a point coordinate measurement by $\Delta\underline{e}_i$ the relative errors of vectors are given by:

$$\Delta\underline{e}_1^w = \Delta\underline{p}_0^w + \Delta\underline{p}_1^w$$
$$\Delta\underline{e}_2^w = \Delta\underline{p}_2^w + \Delta\underline{p}_1^w$$
$$\Delta\underline{e}_3^w = \Delta\underline{e}_1^w \times \underline{e}_2^w + \underline{e}_1^w \times \Delta\underline{e}_2^w$$

and

$$\Delta\underline{e}_1^c = \Delta\underline{p}_0^c + \Delta\underline{p}_1^c$$
$$\Delta\underline{e}_2^c = \Delta\underline{p}_2^c + \Delta\underline{p}_1^c$$
$$\Delta\underline{e}_3^c = \Delta\underline{e}_1^c \times \underline{e}_2^c + \underline{e}_1^c \times \Delta\underline{e}_2^c$$

where the terms of second order have been neglected. interested in the errors associated with
Using the two bases it is possible to define two matrices:

$$\underline{\underline{E}}^w = \left(\underline{e}_1^w, \underline{e}_2^w, \underline{e}_3^w\right)$$
$$\underline{\underline{E}}^c = \left(\underline{e}_1^c, \underline{e}_2^c, \underline{e}_3^c\right)$$

together with related relative error matrices $\Delta \underline{\underline{E}}^w$ and $\Delta \underline{\underline{E}}^c$. Now the rotation matrix $\underline{\underline{R}}$, which connects the two bases is defined by:

$$\underline{\underline{E}}^w = \underline{\underline{R}} \cdot \underline{\underline{E}}^c$$

Multiplying both sides of the equation by the inverse of $\underline{\underline{E}}^c$, and recalling that for an orthogonal matrix $\underline{\underline{A}}$

$$\left(\underline{\underline{A}}\right)^{-1} = \left(\underline{\underline{A}}\right)^T \tag{1}$$

we can write:

$$\underline{\underline{R}} = \underline{\underline{E}}^w \cdot \left(\underline{\underline{E}}^c\right)^T \tag{2}$$

Once the rotation matrix is calculated, the camera position can be computed using any of the points, P_i, on the rectangle as follows:

$$\underline{t} = \underline{p}_i^w - \underline{\underline{R}} \cdot \underline{p}_i^c \tag{3}$$

Taking into consideration uncertainties, the errors for the rotation matrix and for the camera position can be written, using the equations (2) and (3), as:

$$\Delta \underline{\underline{R}} = \Delta \underline{\underline{E}}^w \cdot \left(\underline{\underline{E}}^c\right)^T + \underline{\underline{E}}^w \cdot \left(\Delta \underline{\underline{E}}^c\right)^T$$
$$\Delta \underline{t} = \Delta \underline{\underline{R}} \cdot \underline{p}_i^c + \Delta \underline{p}_i^c \cdot \underline{\underline{R}} + \Delta \underline{p}_i^w$$

If the look direction is considered as along and opposed to the z-axis of the camera coordinate frame i.e. $(\underline{ld}^c)^T = (0, 0, -1)$ then the look direction vector expressed in the world coordinate system is given by

$$\underline{ld}^w = \underline{\underline{R}} \cdot \underline{ld}^c$$

with a relative error of:

$$\Delta \underline{ld}^w = \Delta \underline{\underline{R}} \cdot \underline{ld}^c$$

23.3.2 Establishing the Next-Look Direction

Once the vision system has completed its bootstrap phase the camera-world transformation is known together with an estimate of its uncertainty. During the run phase of the system the machine may be asked to find specified objects. In accordance with the active vision paradigm it is envisaged that the camera parameters (position, look direction, focal length etc) are adjusted so as to frame only those parts of the scene where the specified object is likely to appear. This purposive calculation of next-look direction uses both prestored context knowledge and the current-look information determined previously.

As mentioned previously a basic idea is to use local reference frames and group objects together into context sets which capture qualitative generic spatial knowledge. For example, it is likely that cups are located on tables and this piece of knowledge can be used to initially constrain the search for a cup to previously identified table areas. Thus the object recognition phase of the system involves a sequence of directed fixations of gaze. Several sub-problems must be solved to achieve transferance of gaze between objects. These include calculation of the field of view required to entirely frame the object (taking appropriate account of known errors and uncertainties), determination of the final look direction, calculation of the camera pan and tilt angles needed to achieve the calculated final-look direction and a recovery strategy for cases where the required motions lie outside the feasible range of camera motions. In subsequent subsections we consider each of these in turn.

Calculating the size of field of view to frame an object In the system discussed in this paper a simple and coarse representation of stored object models has been adopted. Objects are represented for next-look direction calculations via rectangular bounding boxes in the image plane. The problem of determining the maximum field of view necessary to entirely frame an object then corresponds to the finding the maximum angle in the camera frame subtended by two corners of the box. If the points which represent the bounding box of the object are denoted by B_i and the vectors which describe them in the world coordinate system are $\underline{b}_i^w$ then the the maximum field of view required is:

$$\gamma_{max}^{fov} = \max_{i \neq j} \left(\gamma_{ij}^{fov} \right)$$

with γ_{ij}^{fov} defined:

$$\gamma_{ij}^{fov} = \arccos \left(norm \left(\underline{t}_{b_i} \right) \cdot norm \left(\underline{t}_{b_j} \right) \right)$$

where $\underline{t}_{b_i}$ and $\underline{t}_{b_j}$ are the vectors centered in the estimated camera position and pointing respectively at the corners B_i and B_j of the bounding box.

Once the two corners which identify the maximum angle are found, it is necesary to consider how errors in point coordinates and uncertainity in estimated camera position affect this angle. Figure 23.2 and figure 23.3 show how reasonable estimates of this can be calculated assuming that errors in point coordinates can be bounded by spheroids. Figure 23.2 indicates the plane which passes through the two corners and the estimated camera position and shows the projected spheroids of uncertainty. The maximum angle, γ_{max}^{fov}, is increased when errors are taken into account and the angle to frame the object becomes γ_{err}^{fov}. The construction shown in figure 23.3 shows that γ_{err}^{fov} can be calculated as

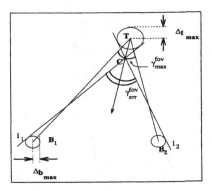

Fig. 23.2. Calculation of maximum angle

$$\gamma_{err}^{fov} = \gamma_{max}^{fov} + \Delta\alpha_1 + \Delta\alpha_2$$

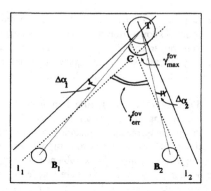

Fig. 23.3. How to compute γ_{err}^{fov}

The two angles $\Delta\alpha_i$ can be computed from

$$\Delta\alpha_i = \arcsin\left(\frac{\Delta t_{max} + \Delta b}{\|\underline{t}_{b_i}\|}\right)$$

where $\underline{t}_{b_i}$ is the vector which is centered at T and points at the corner B_i.

A further source of error which has not been accounted for yet is the uncertainty in the first-look direction. This is given by:

$$\theta_{unc} = |\arccos\left(norm\left(\underline{ld}^w + \Delta\underline{ld}^w\right) \cdot norm\left(\underline{ld}^w - \Delta\underline{ld}^w\right)\right)|$$

Thus the final field of view required to frame an object, γ^{req} is:

$$\gamma^{req} = \gamma_{max}^{fov} + \Delta\alpha_1 + \Delta\alpha_2 + \theta_{unc}$$

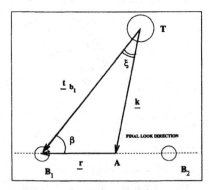

Fig. 23.4. Computing the final look direction.

Calculation of Final-Look direction Calculation of the final look direction vector, $\underline{k}$, is illustrated in figure 23.4 and is given by:

$$\underline{k} = \underline{t}_{b_i} - \underline{r}$$

where the direction of $\underline{r}$ is known from

$$\frac{\underline{r}}{||\underline{r}||} = \frac{\underline{b}_1 - \underline{b}_2}{||\underline{b}_1 - \underline{b}_2||}$$

In order to compute $||\underline{r}||$ the angle β must first be found. It is determined by using:

$$\beta = |\arccos\left(norm\left(\underline{b}_2 - \underline{b}_1\right) \cdot norm\left(\underline{t} - \underline{b}_1\right)\right)|$$

and then applying the sine rule to the triangle $\Delta\left(B_1 T A\right)$ as follows:

$$||\underline{r}|| = \frac{\sin\left(\xi\right)}{\sin\left(\pi - \left(\xi + \beta\right)\right)} \cdot ||\underline{t}_{bi}||$$

where

$$\xi = \left(\gamma^{fov}_{max} + \Delta\alpha_1 + \Delta\alpha_2\right)/2 - \Delta\alpha_1$$

Calculation of Pan and Tilt angles The pan and tilt angles required to direct the optical axis of the camera towards the new look-direction $\underline{k}$ can be computed as shown in figure 23.5. The vectors $\underline{ld}$ and $\underline{k}$ can be projected onto the camera height plane:

$$\underline{ld}^{proj} = \underline{ld} - \left(\underline{ld} \cdot \underline{up}\right)\underline{up}$$
$$\underline{k}^{proj} = \underline{k} - \left(\underline{k} \cdot \underline{up}\right)\underline{up}$$

where $\left(\underline{up}\right)^T = (0,0,1)$ represents the up direction of the camera in the world coordinate system. Then the pan angle is given by:

$$||\theta^{PAN}|| = \arccos\left(norm\left(\underline{ld}^{proj}\right) \cdot norm\left(\underline{k}^{proj}\right)\right)$$

and the direction or sign of the angle can be computed in the camera coordinate system as :

$$sign\left(\theta^{PAN}\right) = \left(norm\left(\underline{ld}^{proj}\right) \times norm\left(\underline{k}^{proj}\right)\right) \cdot \hat{y}_{cam}$$

The tilt angle can be computed in the following three steps:

- First compute the angle ϕ^{ld} between $\underline{ld}$ and its projection $\underline{ld}^{proj}$:

$$\phi^{ld} = sign\left(\left(\underline{ld} - \underline{ld}^{proj}\right) \cdot \hat{z}_w\right)\arccos\left(norm\left(\underline{ld}^{proj}\right) \cdot norm\left(ld\right)\right)$$

- Second compute the angle ϕ^k between $\underline{k}$ and its projection $\underline{k}^{proj}$ as
$$\phi^k = sign\left(\left(\underline{k} - \underline{k}^{proj}\right) \cdot \hat{z}_w\right)\arccos\left(norm\left(\underline{k}^{proj}\right) \cdot norm\left(k\right)\right)$$

- Finally compute the tilt angle ϕ^{TILT} as
$$\phi^{TILT} = \phi^k - \phi^{ld}$$

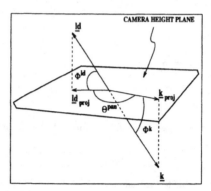

Fig. 23.5. Computing the camera motion.

23.3.3 Recovery Strategies

It is necessary to verify that the pan and tilt angles and the field of view required to frame an object are within the feasible range of the camera. If they are not then additional motions, such as gross translation of the camera, must be made in order to bring the values into allowable ranges.

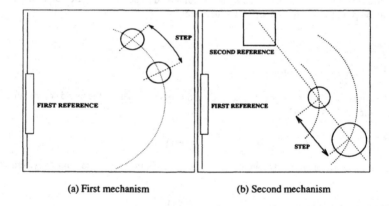

(a) First mechanism   (b) Second mechanism

Fig. 23.6. The step-back mechanisms.

A common example occurs if the camera is too close to the chosen reference object. In this case the required field of view can be outside the available range. Analysis of all possible situations and determination of an optimal movement away from the object could be attempted but would be a very involved calculation and therefore not entirely desirable. In our work a much simpler heuristic recovery strategy has been adopted whereby the camera can move away from the object along one of two predefined paths. The movement is by a fixed step at a time, and after each step the field of view calculation is reverified against the range of available camera parameters. The two defined paths are shown in the figures below. The first step-mechanism of (figure 23.6[a]) is made along the circle centered on the first reference object (the one used to determine the very first estimate of the camera position). In this case it can be assumed that the uncertainty does not change along this path. The second step-mechanism (figure 23.6[b]) is back along the final look direction. In this case the uncertainty can increase if the camera moves away from the first reference object at the same time.

23.4   Experimental Results

In this section results are illustrated and quantitative data is given on the effectiveness of the look direction computations for the active vision system considered. Experiments are shown for both artificial graphics based simulations of environments as well as real world imagery of a laboratory. Typical objects used, or modelled, are doors, cabinets, tables, wall posters and windows. One of the aims of the experiments is to characterise the performance, reliability and limitations of the camera location method used. A further aim is to illustrate the importance of estimating and correctly propagating errors and uncertainties. Finally, it is shown that the use of many local reference frames confers advantages over use of a single global reference frame and that the simple mechanisms used to frame target objects work effectively.

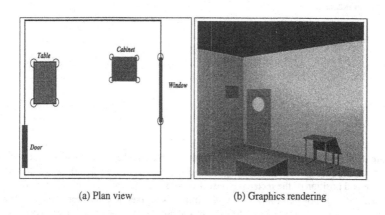

(a) Plan view   (b) Graphics rendering

Fig. 23.7. The modelled environment

A typical indoor environment that is considered in the current study is shown in figure 23.4. This shows a plan view of an indoor laboratory scene in our department and a graphics rendering of one view of the model (using the Rayshade public domain graphics package). The room has a door and contains a small table on top of which are several everday objects (specifically a cup and plate). The position and size of the door are fixed and well known and this can therefore be used to establish the location of the camera during the system bootstrap phase. The location of the table is not so well established as it can be moved around the room to any large enough area of free floor space. However, the location of the cup and plate are almost certainly restricted to the table top and thus if this is established the camera look-direction can be restricted to that small region of the room. The model representation of the scene binds the table, cup and plate together in a local reference frame or context set.

The bootstrap phase of the system involves finding the room door and then using it to infer the camera position via the backprojection method of Fischler and Bolles. The success of this depends both on finding the door in the field of view and the accuracy of the back projection. At system start-up the camera look point could be anywhere in the scene. However it is relatively easy to devise a system which will automatically drive the camera so that it is looking horizontally (this may involve some information from an inertial sensor to define the vertical direction) and the camera has its maximum field of view. In this case, bootstrapping consists of panning the camera around the room in small angular increments until the door is identified.

The accuracy of attainable camera position estimation has been tested by applying the backprojection method to model rectangular objects for a range of object sizes, positions and orientations. The performance of the method has been tested as a function of errors in image point accuracy as well as uncertainty in known model position.

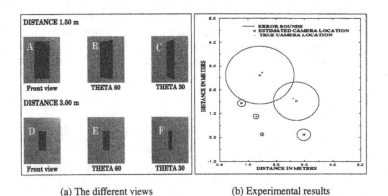

(a) The different views   (b) Experimental results

Fig. 23.8. Some experiments

Figure 23.8[a] shows typical images and results for a rectangular object of size 0.5×0.25 metres located in the lower corner of the room and viewed at distances of 1.5 and 3.0 metres. It is assumed that the size and position of the rectangle in world coordinates is known to an accuracy of 1 cm and the standard deviation of point locations in the image is $\frac{1}{3}$ pixel. The different images correspond to views as the camera is rotated at a fixed radius about the vertical axis. Image (A) represents a fronto-parallel view of the object at 1.5 metres while (B) and (C) correspond to rotations by 30 and 60 degrees respectively. Images (D), (E) and (F) show the same things for the camera at a radius of 3.0 metres. Figure 23.8[b] shows a plan view of the scene with the true camera position, its estimated position and the maximum error bounds (approximated as spheroids) on the estimated position. Several interesting observations can be made. In the case of viewing at 1.5 metres distance the position of the camera is fairly well estimated with reasonably small sizes for the error bounds. The uncertainties grow as the object is viewed at more oblique angles but are still manageable at 60 degrees. However, as the distance between object and camera increases the estimate of the camera position does not necessarily coincide with the true position and the error bounds grow non-linearly. The errors for viewing at 3 metres distance are 3 to 7 times those for 1.5 metres, depending on the angle chosen for comparison. These results clearly demonstrate that care is needed in choosing an object for bootstrapping purposes. Where possible the object-camera distnace should be kept small and the object should not be viewed obliquely.

In addition, to the above experiments with synthetic data, we have performed experiments with real data to use the backprojection technique to calulate camera-object distance as the camera is moved around the scene. Typical results are shown in Figure 23.9 which shows a fronto-parallel approach towards a the face of a small laboratory cabinet. The graph drawn shows the ground truth distance and the bounded estimates of distance for several points along the approach trajectory.

The use of several local reference frames for representing scenes and calculating next look

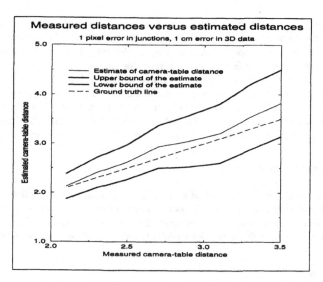

Fig. 23.9. Real data experiment

direction has been tested on real world data. The modelled environment consists of a room with a door and a window on opposite walls. The position of a small cabinet is known and represented with respect to the window. The active vision system initially establishes its orientation in the world by finding the door. In principle, the system could be directed to view the cabinet by composing the transformations from door-window and window-cabinet. However, each of these is somewhat uncertain and therefore the net transformation can have a large error associated with it. A more effective strategy for an active vision system to find the cabinet is first to verify the position of the window and only then perform the motion to direct the camera to look at the cabinet. The introduction of the intermediate local verification step effectively zeros the accumulating errors.

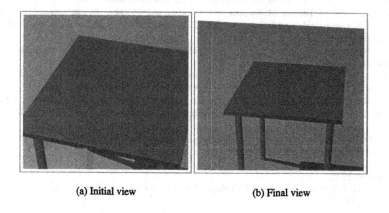

(a) Initial view   (b) Final view

Fig. 23.10. The step back mechanisms

The effectiveness of the recovery strategies for framing a target object is demonstrated in the results summarised in Figure 23.10. This shows the start and end images of a short sequence as the camera is moved backwards along a recovery trajectory. Following bootstrapping the camera has been directed to look in the direction of a table. However, from the known table size and camera parameters it can be inferred that the entire table cannot be seen from the start position. Figure 23.10[a] shows the camera image from the start position and is a clear confirmation that the entire table is not framed from this position. The active system therefore decides to move away from the target object so as to increase the field of view. Each movement is a step of 10 cms and is followed by recalculation of the available field of view. Finally after several moves the system calculates that the entire table can be seen. This is confirmed by the image shown in Figure 23.10[b].

23.5   Conclusions

This paper has considered how the gaze of a simple active camera can be established and how the camera can controlled and directed to look towards objects of interest. Particular attention has been paid to trying to understand sources of uncertainty in these operations and to developing conservative strategies which will ensure that these sources do not result in inappropriate system behaviour.[1]

References

[Aloimonos et al., 1987] J. Aloimonos, A. Bandyopadhyay, and I. Weiss. Active Vision. In Proc. First International Conference on Computer Vision, London, UK, pages 35–54, 1987.

[Bajcsy, 1988] R. Bajcsy. Perception with feedback. In Proc. of DARPA Image Understanding Workshop, pages 279–288, 1988. April, Cambridge Massachussetts.

[Ballard, 1991] D.H. Ballard. Animate vision. Journal of Artificial Intelligence, 48:57 – 86, 1991.

[Barnard, 1983] S.T. Barnard. Iterpreting perspective images. Journal of Artificial Intelligence, 21:435 – 462, 1983.

[Brooks, 1982] R.A. Brooks. Symbolic Error Analysis and Robot Planning. International Journal of Robotics Research, 1:29 – 67, 1982.

[Canny, 1986] J. Canny. A computational approach to edge detection. IEEE Trans. on Pattern Analysis and Machine Intelligence, 8:679 – 698, 1986.

[Chen, 1991] H.H. Chen. Pose determination from line-to-plane correspondences: Existence conditions and closed-form solutions. IEEE Trans. on Pattern Analysis and Machine Intelligence, 13(6):530 – 541, 1991.

[Dhome et al., 1989] M. Dhome, M. Richetin, J-T. Lapreste, and G. Rives. Determination of the attitude of 3d objects from a single pespective view. IEEE Trans. on Pattern Analysis and Machine Intelligence, 11(12):1265 – 1278, 1989.

[Durrant-Whyte, 1988] H.F. Durrant-Whyte. Uncertain geometry in robotics. IEEE Journal of Robotics and Automation, 4:23 – 31, 1988.

[Durrant-Whyte, 1989] H.F. Durrant-Whyte. Uncertain geometry. In D. Kapur and J.L. Mundy, editors, Geometric Reasoning, pages 445 – 481. The MIT Press, Cambridge, Massachusetts, 1989.

[Fischler and Bolles, 1980] M. Fischler and R. Bolles. Random sample consensus: A paradigm for model fitting with applications to image analysis and automated cartography. In Proc. Image Understanding Workshop, pages 71–88, 1980.

[Haralick, 1989] R.M. Haralick. Determining camera parameters from the perspective projection of a rectangle. Pattern Recognition, 22(3):225 – 230, 1989.

[Hough, 1962] P.V.C. Hough. Method and means for recognizing complex patterns. U.S. Patent 3,069,654, 18 December 1962.

[Kanade, 1981] T. Kanade. Recovery of 3d shape of an object from a single view. Journal of Artificial Intelligence, 17, 1981.

[1] This work has been carried out with the financial support from the European Community under ESPRIT project 3038 Vision as Process. Acknowledgment is also due to Craig Kolb, the creator of the software package Rayshade used in our experiments with synthetic data.

[Lui et al., 1990] Y. Lui, T.S. Huang, and O.D. Faugeras. Determination of camera location from 2d to 3d line and point correspondences. *IEEE Trans. on Pattern Analysis and Machine Intelligence*, 12(1):28 – 37, 1990.

[Matas and Kittler, 1992] G. Matas and J. Kittler. Contextual Junction Finder. In *Proc. of British Machine Vision Conference 1992*, 1992. September, Leeds.

[Matas, 1992] J. Matas. Mphil/phd transfer report, to appear. Transfer report, Department of Electronic and Electrical Engineering, University of Surrey, October 1992.

[Penna, 1991] M.A. Penna. Determining camera parameters from the perspective projection of a quadrilateral. *Pattern Recognition*, 24(6):533 – 541, 1991.

[Smith and Cheeseman, 1986] R.C. Smith and P. Cheeseman. On the representation and estimation of spatial unceratinty. *International Journal of Robotics Research*, 5:56 – 68, 1986.

[Tsai, 1987] Roger Y. Tsai. A versatile camera calibration technique for high-accuracy 3d machine vision metrology using off-the-shelf tv cameras and lenses. *IEEE Journal of Robotics and Automation*, RA-3:323 – 344, 1987.

Part VI

Lessons Learned

In this last chapter we will discuss some of the conclusions from the first three years of our project. These lessons fall into three broad classes:

1) Conclusions about vision systems and about the science of machine vision
2) Conclusions concerning specific technical problems, and
3) Lessons about performing basic research in the context of a consortium of research groups distributed over a continent and tied together by an electronic network.

The first section places the VAP project in the context of the evolution of the science of machine vision. After a brief sketch of the history of machine vision, "Vision as Process" is presented as an effort to apply the principals from active vision to the larger problem of control of perception at all levels of a machine vision system. We then review some technical issues that were encountered in the construction of our experimental test-bed system. At the beginning of the project, we deliberately adopted a conservative approach to many of these issues. Using classic techniques in such a system has made the limitations of some of these techniques obvious. We conclude with a brief discussion of the nature of basic research in a consortium.

1. The science of machine vision

Science is a communal activity. Scientific communities are defined as groups of people who share a set of problems and problem solutions (paradigms) [Kuhn 63]. The paradigms of a scientific community evolve as the knowledge and understanding of the community develop. During the early years of a science, the evolution of paradigms and the shift in definition of problems and solutions can be quite rapid. This shift occurs as the problems become better understood and as new problem solutions appear.

The lessons learned from our project are best seen in the perspective of the evolution of the science of matching vision. The science of machine vision has passed through a number of "paradigm shifts" over the last 30 years. While these shifts are subjective, they can be discerned from the titles of the conference, workshops and publications that characterize the field. In the following paragraphs we will broadly outline the evolution of the science of vision in terms of such paradigm shifts. We will then use this evolution to present the problem which was addressed by the project Vision as Process and to examine the results of the project.

Naissance

The science of machine vision has its infancy in the 1950's, with early attempts to use the new computing machines to process and manipulate images. The earliest widely referenced work is the thesis of Roberts [Roberts 63]. Roberts thesis anticipates many of the later phases of machine vision, and in particular, the reconstruction phase which we describe below. During the same period, the community of pattern recognition was in its flower, and vision was often seen as a problem of pattern recognition.

Vision as Recognition

The dominant problem for the period 1965-1975 could be said to be pattern recognition. The problem was seen as one of recognizing patterns in an image. A major problem was segmenting the image into meaningful "chunks" which could then be classified. Associated problems were the invention of features which could be used to define pattern classes and techniques for automatically learning the classes of patterns that existed in the image. Perhaps the best reference for this period is the book of Duda and Hart published in 1972. This book contains chapters which are still relevant today, 20 years after its publication.

Vision as recognition exposed several fundamental problems.The problem of segmenting an image into parts which could be classified, in particular, proved to be unsolvable in a general sense. With time it has become recognized that a meaningful segmentation requires more than measurements made in the image; the proper segmentation can only be defined with respect to the use to which the segments are to be applied. Eventually it was recognized that machine vision required an understanding of the world which was represented in the image. This led to a shift in viewpoint toward vision as an application domain for artificial intelligence.

Vision as Image Understanding

"Vision as recognition" gradually became "vision as image analysis". The recognition that world knowledge was needed for segmentation led to a view of "vision as image understanding". This viewpoint occurred in the context of a period when Artificial Intelligence (AI) changed from an obscure domain with controversial scientific methods, to a revolutionary discipline which promised fundamental changes in computer science. Of course, both of these reputations were exaggerated. After the AI explosion of the early 80's has come the AI winter of the early 90's, or at the least a cold spell.

The technological break-through that triggered the great expansion of AI in the 80's was the development of techniques for programming "expert systems", and in particular a new understanding of techniques for knowledge representation and inference. It was believed that these techniques would furnish the "world knowledge" necessary to analyse and understand images. Some scientists even went so far as to propose that a reasoning system use explicit knowledge of geometry and physics to understand the image. Examples of classic references from this period are collected in a book edited by Hanson and Riseman [Hanson-Riseman 78].

Vision as Image understanding encountered several problems which limited its success. First of all, the task of generating and encoding the necessary world knowledge has proved tractable for only very limited domains. An incredible amount of small unrelated facts are needed for even the simple common sense reasoning that humans use everyday. More difficult still, when a human programmer encodes knowledge for a vision system, he imagines that it perceives with the same low level operations used by his own visual cortex. This is not the case, and programmers are always surprised and disappointed to find that their hypotheses about what a vision system can see are not valid.

Vision as image understanding did not solve the segmentation problem. Indeed, most AI techniques are rather sensitive to the quality of the image segmentation that they interpret. The initial image segmentation remains the great problem that causes many promising algorithms to fail. Overcoming the noise in the image encoding with AI techniques leads to approaches with intractable computational complexity.

Vision as Reconstruction

To understand an image you must see beyond the 2D patterns to the form of the objects which generated these patterns. After all, if the patterns vary with movement and change in lighting, the form itself does not. This observation motivated a scientific movement, still strong today, that argues that general perception requires 3D reconstruction. This viewpoint was elegantly stated by David Marr of MIT, and ardently defended by his colleagues and graduate students. The book *Vision* by Marr [Marr 82] became the reference to which all vision research was compared.

This approach led to a great multiplicity of research into different techniques for measuring shape. "Shape from Shading", "Shape from Motion", "Shape from Texture", "Shape from Stereo" were soon generalised into "Shape from X". Numerous techniques reported in the literature turned out to be computationally unstable and very difficult to duplicate. Eventually it was pointed out that for the case of a single static image, many such processes are ill-posed and computationally expensive [Aloimonos et al. 88]. To make such processes stable and computable in real time, it is necessary to adapt the camera configuration to the scene.

Active Vision

"Vision as recognition" saw vision as recognizing known objects and patterns. "Vision as reconstruction" saw vision as building an internal model of the geometry of a scene. Yet many tasks require neither recognition nor geometry, but simply a measurement for control. In the late 1980s it was demonstrated that a number of difficult and computationally expensive processes could be made robust and computationally simple by the use of controlled camera motion. The term, "active Vision", first used by Bajcsy in 1980 [Bajcsy 80], was promoted by Aloimonos [Aloimonos et al. 88] and others. Similar ideas were demonstrated by Ballard and Brown under the name of "animate vision" [Ballard 89].

Active vision exploits the use of controlled camera motion and controlled processing to provide simple and robust measurements of the scene. Such algorithms use constant or linear complexity algorithms, and limited processing to achieve measurements in real time. Rather than emphasize recognition or reconstruction, active vision emphasizes the measurements needed to accomplish a task.

Vision as Process

Active vision emphasized the use of vision task-oriented processing to perform low level observations in real time. But if some tasks do not require recognition, others do. Some tasks, equally, require the construction of a 3D description. In all cases, it is necessary to make these measurements within a delay determined by the needs of the task. The project "Vision as Process" was formed to apply the lessons of Active Vision to all levels of a machine vision system.

Vision is a process of making observations about the external world. In general this requires the ability to configure the system to make the observation that is required for the task without trying to exhaustively describe everything that can be seen. This leads directly to the need to control the image acquisition and description process. When the vision system must respond with a fixed delay, it becomes essential to limit the data to a small enough "chunk". It is not possible to see everything at the same time. Thus, on a macroscopic level, vision is a sequential process. Our understanding of this problem has been reinforced by the experience of our project to date. What we argued from theoretical considerations in 1989 we have found to be true from experiments in 1992.

Vision is neither exclusively reconstruction nor exclusively recognition. Both recognition and reconstruction are needed for many vision tasks and each can aid the other. However, a third function is more fundamental: control. Vision can provide the measurements needed to control movements and detect events. In many cases it is neither necessary to recognize objects nor to reconstruct their form in order to make the measurements that are needed by a task. On the other hand, in some cases, the ability to recognize an object or to reconstruct a 3D geometry can provide measurements which are very useful for specific tasks. With this viewpoint, recognition and reconstruction become part of the "toolbox" of functions that a vision system can provide.

Understanding vision as a controlled sequential process of observations frees us from an overly strong bias to the "image". For example , it suggests that it is useful to consider using such things as non-Cartesian retinas (for example, log-polar and log-Cartesian retinas), and it suggests the use of descriptions of the visual signal based on spatio-temporal filters. One way to summarize this viewpoint is that the vision system, including the cameras, is a sequence of filters which may be used to reduce the overwhelming mass of possible perceptual information, to select the small subset which is relevant for the current task.

Vision scientists produce theories and techniques. The exploitation of these techniques depends on the requirements of some external task. The design of a vision system must take into account the purpose that the vision system is to serve. The vision system must be designed to "serve" other processes and its design must starts with the constraints imposed by those processes. This leads to a view in which a vision system provides a set of "services" that are available on request to a client "system". A vision system architecture concerns the scheduling and coordination of these services as much as their individual design. Designing a vision system requires a perspective of software systems design. This perspective has been reinforced by the experiences of our project.

2 System architecture and integration

The construction and maintenance of an experimental software system with contributions generated as software experiments by six laboratories has not been an easy task. Integrating even a well specified software system is a messy process. When the components are software experiments, subject to revision whenever the author is inspired, integration can become nearly impossible. Our only hope was to provide a standard software environment in which the components could be implemented. This idea lead the consortium to propose a system composed of modules connected by well defined communication channels.

2.1 The SAVA standard module

In the design of the VAP system, much effort was devoted to the development of a standard module architecture based on the maintenance of a description of the scene. Model maintenance is based on a cyclic process composed of the phases predict, match and update. This approach has been used at all levels of the system with success, although at the symbolic level we found it more convenient to implement the cycle with rules. At the lower levels of the system, a Kalman filter was adopted for the maintenance of parametric descriptions. At the higher levels rules were used for updating a model composed of object hypotheses. As we discovered the importance of demons for control of perception, the use of a standard module architecture made it possible to integrate demon processes at any level of the system. The integration of a rule interpreter as the scheduler for each module proved to be an extremely useful innovation.

Using a rule based system as the control components for each module gave:

1) The possibility of a declarative representation of control
2) The possibility to experiment with control without re-compiling the entire system.
3) The ability to write demons using declarative symbolic reasoning to control processing based on both the data and on external commands.
4) The ability for one module to send an ascii definition of a message handling function to another module, thereby letting the modules define their protocol dynamically.

This has given us an implementation tool for experiments in designing the control component of perception using a formalism based on state transition graphs. It allows each module to have its own goals, and to emit commands to the other module based on both goals and events. This also raises the problem of how to control a heterarchical system.

2.2 Hierarchical vs heterarchical architecture

An important lesson learned in the VAP project concerns the "heterarchical" nature of a vision system. At the time in which the VAP proposal was written in 1989, we believed that we could construct a demonstration system as a sequence of "layers". Each layer was to take perceptual information from below and expectations from above. Layers would then produce interpretations for the layer above and control signals or hypotheses for the layer below. Layers were defined for camera control, image signals, 3D description, object recognition, and a system supervisor. All communication passed through adjacent layers.

From the very early stages, we were led to violate the strict hierarchical nature of these layers. Object recognition often depended more on 2D measurements than on 3D form. Eventually our view of the vision process evolved into an architecture which is closer to the shape of a star. At the center is the process for image acquisition and processing. This process serves the image processing needs of other processes in order to miminize the amount of information transmitted.

Concentrating image processing on special purpose hardware provided a great speed-up and allowed us to avoid an important bottleneck caused by transmitting image data between processes. Interestingly, with image processing consolidated on special purpose hardware, communication appeared as a bigger bottleneck then image processing (albeit an order of magnitude faster). A star shaped architecture avoids the unnecessary passing of data and requests between modules that would have been produced by a strictly layered approach.

Continuous operation also argues against a strict hierarchy. In a strictly hierarchical system, maintenance over time must be based on information available at the level just below. This implies that the dynamic models at all levels must be able to accommodate the needs of all higher level modules. This requires a substantial amount of computational resources. Recognition and maintenance of model composed of a set of known objects can be performed on the basis of low level features such as texture, colour, and groupings of features such as segments and ellipses. These groupings can be performed on demand based on expectations. Verification based image description is both faster and more robust.

A hierarchical architecture also poses problems with time delays. In the strict hierarchical processing model, the availability of data at a high level is delayed for a number of cycles due to sequential processing. This delay poses a problem of stability for control. If control decisions are based on information available at the symbolic level, it must be ensured that the changes that have occurred since the data was generated are taken into account. To overcome the problem of delays, two different approaches may be adopted:

a) Labyrinthic Approach: visual perception for each control task can be built up from a separate chain of processing, resulting, in the limit, in a specialised perceptual process for each action.

b) Modular Approach: Functions may be grouped together according to the level of abstraction of the data, and executed by request from the actions. In this approach, estimation techniques are necessary to allow anticipation to provide a fast response.

The labyrinthic approach looks extremely costly to implement and imposes a time delay equal to the processing time of the entire chain. The modular approach allows a system to be built up incrementally with components reused as new actions are developed. The use of continuous operation and prediction permits such a system to provide much faster response time. The Kalman filter provided a mathematical framework for prediction and estimation.

2.3 System construction

In the VAP project the aim has been to design and implement a vision system that may be used in a variety of domains. The approach has been to construct a system composed of individual modules that can be reconfigured dynamically to suit a specific task. There has been intense debate among the consortium and with our associates over the suitability of such an approach and alternatives. Aloimonos has argued that the "grain size" of our modules is too large. He would have had us build the system up as needed for each task from smaller components. Some of the partners in the consortium have made systems building experiments using such tools as Vipwob (chapter 4), MNT (chapter 3) and AVS. At one point such an approach was proposed to a system composed of modules such as SAVA.

At the completion of our month 33 milestone, the consortium developed a work plan for the next three years. At this occasion the issue was raised concerning what kind of architectures should be constructed. The consensus was to continue to develop an architecture composed of modules maintaining descriptions of the scene at different levels of abstraction. The experience in systems building with different architectures permitted the partners to divorce the design from individual dogmas and to base it on practical experience.

3. Control of perception

Control of perception refers to the selection, execution and evaluation of the sequential observation process. Specifically, the problem of control can be summarized by where, what and how:

> *Where* to look next
> *What* procedures to uses for the observation,
> *What* parameters to use for the procedures (*how*).

Diverse approaches have been tried in the VAP project, even for description at the same representational level.

3.1 Active vision

In the initial VAP proposal, control of the camera system is briefly mentioned and only a small amount of resources were allocated to the topic of active camera control. During the last three years there has been an almost explosive development in the area of active vision. This is also the case for the VAP effort, where three different heads have been designed and implemented.

The starting approach for the VAP effort was traditional static techniques. The emphasis was on integration, continuous operation and the control of the perceptual process. Our experience has confirmed that hypothesis that significant reductions in computational cost may be achieved with an active vision approach. Tough problems such as figure-ground segmentation and description of objects in an object centered representation are simplified when this approach is adopted. At the out set active vision was considered an issue which may be considered independently of the problem of continuous operation and control. Experience has demonstrated that a number of both signal-level and symbolic problems are simplified when they are studied in the context of a controllable sensory system.

3.2 Demons

In the SAVA demonstration system, processing modes are treated as states. Control is organised as a finite state machine, with state transitions driven by two kinds of "events": "control events", and "observation events". Control events are top-down signals which select the procedures and parameters for making the required observation. Perceptual events are data driven measurements which can provoke changes in the procedures and parameters of the system. The detection of perceptual events is performed by demon processes that operate continually within each module. Perceptual events can be anticipated based on the context, and the demons which detect them can be enabled and disabled.

The role of demons for detecting perceptual events remains somewhat controversial within the consortium. Some graduate students who work regularly with the code have remarked that these processes waste computing cycles. Disabling demons permits the system to reduce its computational load (making a more impressive demo!). However, the price for this reduction is the inability to react to unanticipated events. In general, the more we can restrict the environment, the more we can "optimise" the system by eliminating unnecessary event detection. However, our goal is the opposite: To render the system as robust as possible in face of unexpected events. If we could accurately predict the events, we can fine tune the system to only do the computation necessary to event detection when we expect the events to be possible. However, the there is no theory which tells us how to build such knowledge either automatically, or by hand!

In our project we have encountered the difficulty of hand coding world knowledge. We initially imagined that we could adapt some existing "AI" tool for this need. We have been unable to find an appropriate tool. More importantly, we were unable to even specify the requirements for such a tool until we had constructed several working examples of demonstration systems. Eventually C based rule interpreter CLIPS 5.1 provided the implementation tool for control.

3.3 Control of modules

A variety of different methods have been applied for control. To appreciate the different approaches they are briefly summaries below. This presentation demonstrates that many different approaches may be adopted, and that researchers working in different areas have different notion of the concept of control. This does not imply that any of these efforts are irrelevant or incorrect, but it merely demonstrates the scope of the problem.

The VAP project has explored two different approaches for image description:

a) the traditional approach based on multi-scale image representation, edge detection using first and second order derivatives, and sequential line linking, and

b) spatio-temporal filtering with matched (an adaptive) filters for computation of phenomena of interest.

In the first approach, algorithmic control is related to dynamic selection of suitable thresholds (an almost impossible problem), and selection and evaluation of grouping procedures which act as filters. Such techniques may be implemented in hardware to achieve real-time performance, and through use of the control measures mentioned above it is possible to control *where* and *how many* primitives that are extracted.

For the temporal maintenance a Kalman filter is employed. The maintenance of limited sized models is achieved through use of a fixed recency window (i.e., if an item has not been detected the last N frames it is discarded). The search region used in matching may be controlled based on the covariance of the estimated parameters, which results in a controllable performance [Christensen et al. 92]. This control is fairly traditional and based mainly on well known techniques which also are employed by others [Burt 88]. The control is limited to definition of spatial regions of interest (attention regions), the size of the recency window, and parameters such as thresholds for edge detection and line linking.

In the competing approach, the spatio-temporal filtering, a signal processing approach to control is adopted. This implies that adaptive and matched filters are used for the extraction of information of interest. Control information is thus used for tuning of filter parameters which provides a mean for improved robustness. A problem with such techniques is often that full images are processes, which results in a prohibitive algorithmic complexity. These methods may, however, also be applied in regions of interest, which facilitates a reduction in the computational requirements. In addition it has been shown that data driven and relatively inexpensive attention mechanisms may be defined (as described in chapter 12). The use of such attention mechanisms allow integrated multi-scale control of the image description process, which allow for a more dynamic control of the computational complexity. The signal processing approach to control does appear promising for the feature-extraction and integration process (especially when combined with confidence information as described in chapters 10 through 14).

Active vision approaches for the extraction of depth information have been explored. It is evident that control of the sensory parameters allow definition of a 3-D attention regions (defined by the field of view and the depth of the field of view). The introduction of multiple cues in depth estimation (i.e., accommodation, vergence and disparity) allow for a more robust estimation.The parameters are independent, but control of the process might still dynamically shift the contribution of different cues in response to the setting. The control approach adopted in depth estimation has been motivated by algorithmic considerations.

Several approaches have been explored for recognition of objects present techniques which are driven by explicit object model and exploitation of contextual information defined by the domain. In this example, control is the use of verification of models to enable pruning of the set of hypotheses. This is achieved through cuing of models based on data-driven computation of cuing features. In this process the system goal drives the extraction of objects through use of a-priori defined context sets. The context sets are organised hierarchically to enable recognition of objects in the environment. Recognised objects then may drive the recognition of other objects. For specific objects, a-priori models have been defined and a set of verification procedures are applied to verify the presence of the object. In the control environmental constraints are exploited to define attention regions for the image description process (i.e. the table top is assumed planar and with a horizontal orientation). This type of control is strongly tied to the definition of context information that may drive the recognition process.

Another approach which has been explored in the recognition work is based on description in terms of qualitative primitives (see chapter 17). In this work the recognition of volumetric primitives is based on identification of faces. Cuing features are computed at the beginning of the process. The set of cuing features of interest is defined by the present set of goals. Once a set of hypotheses has been formed (based on the availability of closed contour segments), a verification phase is initiated. The verification is based on evaluation of the set of component faces for a particular volumetric primitive. Selection of the faces to be verified is based on a description in an aspect hierarchy, where the likelihood of detecting a particular face or aspect is recorded. The control is thus based on models of the volumetric primitives that are to be recognised in combination with probabilistic information about the detectability of such features. In addition to this search for addition information the module exploits attentions regions as faces to be recognised must be adjacent to the faces that were used in the model invocation process. This type of control is thus based on decision theory and use of attention regions to limit the complexity of the image description process.

In recognition another decision theoretical approach has also been explored. In this approach the scene and object of interest are modelled in a belief network [Jensen et al. 90]. Based on goal information from a supervisory process, features to be detected are then selected based on their discriminatory power. A set of features needed for recognition of an object are then forwarded to the image description modules which compared the utility of these features to the cost of computing them. The features which has the best cost/utility ratio is then computed in an attention regions that is defined by the scene context. This kind of control provides a hierarchical strategy for processing of scenes and the techniques has also been used by others [Rimey-Brown 91]. The basis for this kind of control is standard decision theory as for example described by [Lindley 71].

At the supervisor level the control employed in the skeleton system is based almost exclusively on planning in the artificial intelligence sense. In this approach a set of planning and plan execution rules was defined. Use of this set of rules in combination with contextual information about the scenario allow selection of set of actions that may result in completion of user defined goal commands. The use of contextual and layout information in the computed data provides a basis for subsequent goals.

Despite these individual efforts, a major problem remains to be solved. It is necessary to formulate the control strategies in a common framework that allow evaluation in the context of the full system. Bajcsy has termed this effort the definition of a "theory for the active observer".

4. Specific technical lessons

Building an integrated vision system forced us to implement techniques for camera control, 2D image tracking and description, 3D description and object recognition. In a number of cases these implementations have made it possible to make comments about specific vision techniques.

4.1 Image Description

In the first year of the project it was argued the development of new image description techniques was beyond the scope of the project. As a result, the experimental test-bed system has been constructed using the "standard" technique of edge segments, ellipses and edge-chains. We have found edge information alone to be insufficient as a general image descriptor.

Our original demonstration scenario consisted of a breakfast scene. This scene consisted of a white or checkered table cloth on which we placed a place mat and various typical breakfast objects. A plate surrounded by a knife, fork and spoon in a classical European continental style table setting. A dark bowl was filled with white sugar and placed near the plate. A cylindrical cup was placed near the place setting. Several boxes (sugar, milk) were placed on the table.

In our initial demos, we attempted to recognize and watch cylindrical highly specular objects using groupings of straight line edge segments. As described in chapter 7, edge segment boundaries were determined by a composition of changes in gradient orientation and the area under an edge chain. Even this very stable segmentation rule produces an edge segmentation of an arc which is depends on where the edge chain began, and hence changes with the viewpoint. We attempted to compensate by grouping edge segments into arcs, as described in chapter 8. Even with grouping, both tracking and initial recognition of objects has proven to be only partially reliable. Under such circumstances, the points at which circular arcs are broken to form segments is highly variable, which can change a junction composed of two equal length segments into a very different junction composed of a s and a long segment.

Neither groupings of edge segments nor chains and elliptic arc of edge points have proven sufficient. A much richer set of of pre-attentive image descriptors is needed for robust image description. Such image descriptors should robustly detect reference points on objects despite changes in illumination or viewing position. Also needed are spatio-temporal filters to detect events and measure motion. We are currently experimenting with the design of detectors inspired by models of the complex and hyper-complex cells in the human visual cortex. At the same time Knuttson at Linkoping has developed a new theory which combines perceptual information with an explicit measure of its reliability. While such measures are computationally expensive, they do have linear computational complexities, and can easily be implemented as real time processes using special purpose hardware. We believe that such approaches will eventually provide the basis for robust real time vision.

4.2 On the role of 3D description

At the time that we wrote our project proposal in 1988, it was widely believed that a 3D model of an object provided the view invariant description needed for recognition. As a result we included a 3D modeling process as input to recognition. However, the classic methods for such modeling require precise calibration of the camera intrinsic parameters. Unfortunately continuous control of the aperture, focus, zoom and vergence makes such calibration impossible. The object recognition component in our system was constrained to operate with information measured from the 2D images (chapters 19 through 23). At the same time, the 3D model component lead us to develop new techniques for auto-calibration to object centered coordinates, and the separation of estimation of the position of objects in the scene from kinematics of fixation from the estimation of intrinsic 3D form from active stereo.

From this we have concluded that the major role of 3D is not recognition but perceptual organisation. The 3D position of an object is a kind of spatial indexing for organising the model of the environment and for returning to previously observed objects. At the same time, we have seen that it is possible to recognise without 3D reconstruction. We have yet to conclude whether a qualitative 3D description in a reference frame intrinsic to the object provides a more robust or less costly recognition technique. Current experiments within the consortium should lead us to a conclusion of this issue in the near future.

5. The Nature of Basic Research in a Consortium

The process of performing basic research as a multi-year project by a consortium distributed over a large geographic is a somewhat different experience than the more traditional style of individual work. One of the dominant characteristics of a multi-year, multi-partner project is stability. This can be both a benefit and a handicap. Before a consortium can begin constructing a system it must have a well thought out work program so that the partners can know what they can obtain from other partners. Designing the work-plan forces the each scientist to carefully work out the activities for his research and to justify each part of it among his partners. Criticism and suggests from collaborators can be introduced at a very early stage, avoiding unnecessary expenditures of resources. Within the consortium partners critic each other and help each other to remain objective. The disadvantage is a loss in flexibility. It is difficult to foresee and plan for changes of insight or unexpected results. To counter-act this, the consortium must expect to revise their work-plan at regular intervals to take into account changes in the viewpoint, as well as new results from both within and without the consortium.

The VAP consortium developed the process of "cooperative competition" in constructing its work plan. In such a process, partners work together to develop a common definition of a research problem, possible approaches to a solution, and evaluation criteria for success. Partners then explore individual solutions. The results are then brought together and compared. An objective comparison of results often leads to a synthesis of the possible solutions, bringing together the most successful aspects of each.

The spirit of cooperative competition is evident in the chapters of this book. For a number of the problems that have been addressed, several solutions have often been explored. In part I, we present two individual experiments in software integration environments, MNT (chapter 3) and Vipwob (Chapter 4). Techniques from these two systems were integrated into the common testbed system SAVA II. In a similar manner, competing approaches to image description are presented in parts II and III. The approach in part II is a more classical approach permitted a quick implementation of a real time system, while part III presents a longer term approach which offers the potential to provide description which is both more robust and more appropriate to the problems of continuously operating vision. Three different physical binocular camera heads and one simulated head were developed by the consortium. Part IV presents the most successful of these heads, as well as the description of a control architecture designed to map onto any of the heads. Parts IV and V present alternative approaches to interpretation and control based on a common definition of the problem. In each of these cases, competition led to a more robust systems design and a more objective understanding of the results.

Bibliography

[Aloimonos et al 88] J.Y. Aloimonos, I. Weiss, and A. Bandyopadhyay, "Active Vision", *International Journal of Computer Vision*, Vol. 1, No. 4, Jan. 1988.

[Ballard 89] D.H. Ballard, R.C. Nelson, and B. Yamauchi, "Animate vision". *Optics News*, 15(5):17-25, 1989.

[Bajcsy 80] R. Bajcsy and D Rosenthal, "Visual and Conceptual Focus of Attention", in *Structured Computer Vision*, S. Tanimoto & A. Klinger (Eds.), Academic Press, New York, N.Y., 1980.

[Bajcsy 88] R. Bajcsy, "Active perception" , *IEEE Proceedings*, Vol. 76, No 8, pp. 996-1006, August 1988.

[Burt 88] P. J. Burt, *Smart Sensing in Machine Vision*, Academic Press, 1988.

[Christensen et al. 92] H.I. Christensen, C.S. Andersen, & E. Granum, Control of Perception in Dynamic Computer Vision, First Danish Conference on pattern Recognition and Image Analysis, Copenhagen, June 1992.

[Duda-Hart 73] R. Duda and P. Hart, *Pattern Recognition and Scene Analysis*, Wiley, 1973.

[Hanson-Riseman 78] Hanson, A. and E. Riseman, *"Computer Vision Systems*, Academic Press, 1978.

[Jensen et al 90] F.V. Jensen, J. Nielsen, H.I. Christensen, Use of Causal Probabilistic Networks as High Level Models in Computer Vision, Tech Report R-90-39, Aalborg University, Institute of Electronic Systems, November 1990.

[Kuhn 62] T.S. Kuhn, *The Structure of Scientific Revolutions*, The Univ. of Chicago Press, Chicago, 1962.

[Lindley 71] D.V. Lindley, *Making Decisions*, John Wiley & Sons, London, 1985 (first edition 1971).

[Marr 82] D. Marr, *Vision*, W. H. Freeman Co, San Francisco, 1982.

[Rimey-Brown 91] R.D. Rimey and C. Brown, "Controlling Eye Movements with Hidden Markov Models", *International Journal on Computer Vision*, April 1991.

[Roberts 63] R.G. Roberts, *Machine Perception of Three Dimensional Solids*, PhD Thesis, Stanford, 1963.

ESPRIT Basic Research Series

Springer-Verlag
and the Environment

We at Springer-Verlag firmly believe that an international science publisher has a special obligation to the environment, and our corporate policies consistently reflect this conviction.

We also expect our business partners – paper mills, printers, packaging manufacturers, etc. – to commit themselves to using environmentally friendly materials and production processes.

The paper in this book is made from low- or no-chlorine pulp and is acid free, in conformance with international standards for paper permanency.